Springer Complexity

Springer Complexity is an interdisciplinary program publishing the best research and academic-level teaching on both fundamental and applied aspects of complex systems — cutting across all traditional disciplines of the natural and life sciences, engineering, economics, medicine, neuroscience, social and computer science.

Complex Systems are systems that comprise many interacting parts with the ability to generate a new quality of macroscopic collective behavior the manifestations of which are the spontaneous formation of distinctive temporal, spatial or functional structures. Models of such systems can be successfully mapped onto quite diverse "real-life" situations like the climate, the coherent emission of light from lasers, chemical reaction-diffusion systems, biological cellular networks, the dynamics of stock markets and of the internet, earthquake statistics and prediction, freeway traffic, the human brain, or the formation of opinions in social systems, to name just some of the popular applications.

Although their scope and methodologies overlap somewhat, one can distinguish the following main concepts and tools: self-organization, nonlinear dynamics, synergetics, turbulence, dynamical systems, catastrophes, instabilities, stochastic processes, chaos, graphs and networks, cellular automata, adaptive systems, genetic algorithms and computational intelligence.

The three major book publication platforms of the Springer Complexity program are the monograph series "Understanding Complex Systems" focusing on the various applications of complexity, the "Springer Series in Synergetics", which is devoted to the quantitative theoretical and methodological foundations, and the "SpringerBriefs in Complexity" which are concise and topical working reports, case-studies, surveys, essays and lecture notes of relevance to the field.

In addition to the books in these two core series, the program also incorporates individual titles ranging from textbooks to major reference works.

Springer Series in Synergetics

Founding Editor: H. Haken

The Springer Series in Synergetics was founded by Herman Haken in 1977. Since then, the series has evolved into a substantial reference library for the quantitative, theoretical and methodological foundations of the science of complex systems.

Through many enduring classic texts, such as Haken's *Synergetics and Information and Self-Organization*, Gardiner's *Handbook of Stochastic Methods*, Risken's *The Fokker Planck-Equation* or Haake's *Quantum Signatures of Chaos*, the series has made, and continues to make, important contributions to shaping the foundations of the field.

The series publishes monographs and graduate-level textbooks of broad and general interest, with a pronounced emphasis on the physico-mathematical approach.

For further volumes:
http://www.springer.com/series/712

Tassos Bountis • Haris Skokos

Complex Hamiltonian Dynamics

With a Foreword by Sergej Flach

 Springer

Tassos Bountis
University of Patras
Department of Mathematics
Patras
Greece

Haris Skokos
MPI for the Physics of Complex Systems
Dresden
Germany
and
Physics Department
Aristotle University of Thessaloniki
Thessaloniki
Greece

ISSN 0172-7389
ISBN 978-3-642-27304-9   e-ISBN 978-3-642-27305-6
DOI 10.1007/978-3-642-27305-6
Springer Heidelberg Dordrecht London New York

Library of Congress Control Number: 2012935685

Printed on acid-free paper

Springer is part of Springer Science+Business Media (www.springer.com)

To our wives Angela and Irini, our children Athena and Dimitris and our parents Athena and Christos, Niki and Dimitris.
Their love, dedication and support are a constant source of inspiration and encouragement in our lives.

Foreword

What makes a science book interesting, valuable, useful, and perhaps also worth spending the money to own it? Well, in the case of this book I guess it starts with its title which immediately attracts one's attention. Next, the intrigued reader is invited to inspect a Contents List which combines many familiar-sounding topics and chapters along with completely new and unknown ones, raising our curiosity to find out how these different topics are interrelated. Ultimately, of course, we realize that it is the reading itself which will tell us whether we have found enough new and interesting insights into this domain of the endless world of wonders in science, which the authors call *complex Hamiltonian dynamics*. This brief foreword is thus devoted to tell the reader that all the above conditions are satisfied for the book you hold in your hands (or have downloaded to your electronic device).

To begin with, the short title tells those of you who do not know it already that Hamiltonian dynamics is endlessly complex. Indeed, Hamiltonian models can be formally used for almost any problem in nature, which includes hopelessly complex systems as well. But complexity in Hamiltonian dynamics starts on much lower and seemingly simpler levels. Just get into Chap. 2, where the authors demonstrate in a pedagogical and very readable way some basic and well-known facts of chaos theory for systems with a few degrees of freedom, illustrated by a number of illuminating examples.

While reading the chapters that follow, the concept of the monograph, its title, and the intentions of the authors quickly become clear. In a nutshell, what is addressed here is the border between strongly chaotic and fully regular dynamical systems. The authors use recent progress in the study of relaxation of nonequilibrium states as a playground for applying novel tools. The models are carefully chosen from a set of well-known half-century-old paradigms, which were invented to address basic questions of statistical physics. The fact that a number of these questions still remain unanswered hints at a kind of complexity that is present at seemingly simple levels. It is indeed worth noting that the experienced authors take special care to formulate a number of exercises, which makes this monograph a combination of an introduction into the nonlinear dynamics of many degrees of freedom, a report on recent progress at the forefront of nonlinear science, and an ideal textbook for students and teachers of advanced physics courses.

An interesting attempt to describe and distinguish order from chaos in Hamiltonian systems with many degrees of freedom is given in Chap. 3. The authors discuss various types of fixed and dynamical equilibria and their local stability properties, and smoothly connect them to the issue of global stability of dynamical states. Besides discussing standard ways of characterizing chaos via Lyapunov spectra, the authors also introduce novel methods (called SALI and GALI) which connect tangent dynamics with stability of motion and the nature of the dynamical state under investigation. Both aspects of characterizing equilibria and exploring techniques to distinguish order from chaos are discussed in detail in the two subsequent chapters.

Then comes Chap. 6, where we encounter many applications of the methods introduced earlier to the classical problem of the FPU paradox, first posed by Enrico Fermi, John Pasta, and Stanislav Ulam in their pioneering studies of the early 1950s. The next chapter addresses the fascinating phenomenon of localization and reports on another recent puzzle of great interest at the border between order and chaos, namely, the spreading of nonlinear waves in ordered and disordered media.

The monograph continues in Chap. 8 with a critical discussion and comparison of many systems exhibiting "weak chaos" from the viewpoint of nonextensive statistical mechanics. Finally, the book ends in Chap. 9 with a number of open problems, which should prove quite inspiring to graduate student readers, and concludes with a brief review of additional fascinating topics of Hamiltonian dynamics, which are of great current interest and outstanding potential for practical applications.

In summary, this book constitutes in my opinion a very solid piece of work, which serves several purposes: It is very useful as an introductory textbook that familiarizes the reader with modern methods of analysis applied to Hamiltonian systems of many degrees of freedom and reviews a set of modern research areas at the forefront of nonlinear science. Last but not least, thanks to its pedagogical structure, it should prove easily exploitable as an exercise source for advanced university courses.

Dresden, Germany   *Sergej Flach*

Preface

The main purpose of this book is to present and discuss, in an introductory and pedagogical way, a number of important recent developments in the dynamics of Hamiltonian systems of N degrees of freedom. This is a subject with a long and glorious history, which continues to be actively studied due to its many applications in a wide variety of scientific fields, the most important of them being classical mechanics, astronomy, optics, electromagnetism, solid state physics, quantum mechanics, and statistical mechanics.

One could, of course, immediately point out the absence of biology, chemistry, or engineering from this list. And yet, even in such diverse areas, when the oscillations of mutually interacting elements arise, a Hamiltonian formulation can prove especially useful, as long as dissipation phenomena can be considered negligible. This situation occurs, for example, in weakly oscillating mechanical structures, low-resistance electrical circuits, energy transport processes in macromolecular models of motor proteins, or vibrating DNA double helical structures.

Let us briefly review some basic facts about Hamiltonian dynamics, before proceeding to describe the contents of this book.

The fundamental property of Hamiltonian systems is that they are derived from Hamilton's Principle of Least Action and are intimately related to the conservation of volume, under time evolution in the phase space of their position and momentum variables q_k, p_k, $k = 1, 2, \ldots, N$, defined in the Euclidean phase space $\mathbb{R}^{2N}$. Their associated system of (first-order) differential equations of motion is obtained from a Hamiltonian function H, which depends on the phase space variables and perhaps also time. If H is explicitly time-independent, it represents a first integral of the motion expressing the conservation of total energy of the Hamiltonian system. The dynamics of this system is completely described by the solutions (trajectories or orbits) of Hamilton's equations, which lie on a $(2N - 1)$-dimensional manifold, the so-called energy surface, $H(q_1, \ldots, q_N, p_1, \ldots, p_N) = E$.

This constant energy manifold can be compact or not. If it is not, some orbits may escape to infinity, thus providing a suitable framework for studying many problems of interest to the dynamics of scattering phenomena. In the present book, however, we shall be exclusively concerned with the case where the constant energy manifold is compact. In this situation, the well-known theorems of Liouville-Arnol'd (LA) and Kolmogorov-Arnol'd-Moser (KAM) rigorously establish the following two important facts [19].

The LA theorem: If $N - 1$ global, analytic, single-valued integrals exist (besides the Hamiltonian) that are functionally independent and in involution (the Poisson bracket of any two of them vanishes), the system is called *completely integrable*, as its equations can in principle be integrated by quadratures to a single integral equation expressing the solution curves. Moreover, these curves generally lie on N-dimensional tori and are either periodic or quasiperiodic functions of N incommensurate frequencies.

The KAM theorem: If H can be written in the form $H = H_0 + \varepsilon H_1$ of an ε perturbation of a completely integrable Hamiltonian system H_0, *most* (in the sense of positive measure) quasiperiodic tori persist for sufficiently small ε. This establishes the fact that many near-integrable Hamiltonian systems (like the solar system for example!) are "globally stable" in the sense that most of their solutions around an isolated stable-elliptic equilibrium point or periodic orbit are "regular" or "predictable."

And what about Hamiltonian systems which are far from integrable? As has been rigorously established and numerically amply documented, they possess near their *unstable* equilibria and periodic orbits dense sets of solutions which are called *chaotic*, as they are characterized by an extremely sensitive dependence on initial conditions known as *chaos*. These chaotic solutions also exist in generic near-integrable Hamiltonian systems down to arbitrarily small values of $\varepsilon \to 0$ and form a network of regions on the energy surface, whose size generally grows with increasing $|\varepsilon|$.

In the last four decades, since KAM theory and its implications became widely known, Hamiltonian systems have been studied exhaustively, especially in the cases of $N = 2$ and $N = 3$ degrees of freedom. A wide variety of powerful analytical and numerical tools have been developed to (1) verify whether a given Hamiltonian system is integrable; (2) examine whether a specific initial state leads to a periodic, quasiperiodic, or chaotic orbit; (3) estimate the "size" of regular domains of predominantly quasiperiodic motion; and (4) analyze mathematically the "boundary" of these regular domains, beyond which large-scale chaotic regions dominate the dynamics and most solutions exhibit in the course of time statistical properties that prevail over their deterministic character.

As it often happens, however, physicists are more daring than mathematicians. Impatient with the slow progress of rigorous analysis and inspired by the pioneering numerical experiments of Fermi, Pasta, and Ulam (FPU) in the 1950s, a number of statistical mechanics experts embarked on a wonderful journey in the field of $N \gg 1$ coupled nonlinear oscillator chains and lattices and discovered a goldmine. Much to the surprise of their more traditional colleagues, they discovered a wealth of extremely interesting results and opened up a path that is most vigorously pursued to this very day. They concentrated especially on one-dimensional FPU lattices (or chains) of N classical particles and sought to uncover their transport properties, especially in the $N \to \infty$ and $t \to \infty$ limits.

They were joined in their efforts by a new generation of mathematical physicists aiming ultimately to establish the validity of Fourier's law of heat conduction, unravel the mysteries of localized oscillations, understand energy transport, and explore the statistical properties of these Hamiltonian systems at far from equilibrium situations. They often set all parameters equal, but also seriously pondered the effect of disorder and its connections with nonlinearity. Although most results obtained to date concern ($d = 1$)-dimensional chains, a number of findings have been extended to the case of higher ($d > 1$)-dimensional lattices.

Throughout these studies, regular motion has been associated with quasiperiodic orbits on N-dimensional tori and chaos has been connected to Lyapunov exponents, the maximal of which is expected to converge to a finite positive value in the long time limit $t \rightarrow \infty$. Recently, however, this "duality" has been challenged by a number of results regarding longtime Hamiltonian dynamics, which reveal (a) the role of tori with a dimension as low as $d = 2, 3, \ldots$ on the $2N - 1$ energy surface and (b) the significance of regimes of "weak chaos," near the boundaries of regular regions. These phenomena lead to the emergence of a hierarchy of structures, which form what we call *quasi-stationary states*, and give rise to particularly long-lived regular or chaotic phenomena that manifest a deeper level of complexity with far-reaching physical consequences.

It is the purpose of this book to discuss these phenomena within the context of what we call *complex Hamiltonian dynamics*. In the chapters that follow, we intend to summarize many years of research and discuss a number of recent results within the framework of what is already known about N degrees of freedom Hamiltonian systems. We intend to make the presentation self-contained and introductory enough to be accessible to a wide range of scientists, young and old, who possess some basic knowledge of mathematical physics.

We do not intend to focus on traditional topics of Hamiltonian dynamics, such as their symplectic formalism, bifurcation properties, renormalization theory, or chaotic transport in homoclinic tangles, which have already been expertly reviewed in many other textbooks. Rather, we plan to focus on the progress of the last decade on one-dimensional Hamiltonian lattices, which has yielded, in our opinion, a multitude of inspiring discoveries and new insights, begging to be investigated further in the years to come.

More specifically, we propose to present in Chap. 1 some fundamental background material on Hamiltonian systems that would help the uninitiated reader build some basic knowledge on what the rest of the book is all about. As part of this introductory material, we mention the pioneering results of A. Lyapunov and H. Poincaré regarding local and global stability of the solutions of Hamiltonian systems. We then consider in Chap. 2 some illustrative examples of Hamiltonian systems of $N = 1$ and 2 degrees of freedom and discuss the concept of integrability and the departure from it using singularity analysis in *complex time* and perturbation theory. In particular, the occurrence of chaos in such systems as a result of intersections of invariant manifolds of saddle points will be examined in some detail.

In Chap. 3, we present in an elementary way the mathematical concepts and basic ingredients of equilibrium points, periodic orbits, and their local stability analysis for arbitrary N. We describe the method of Lyapunov exponents and examine their usefulness in estimating the Kolmogorov entropy of certain physically important Hamiltonian systems in the thermodynamic limit, that is, taking the total energy E and the number of particles N very large with $E/N = $ constant. Moreover, we introduce some alternative methods for distinguishing order from chaos based on the more recently developed approach of Generalized Alignment Indices (GALIs) described in detail in Chap. 5.

Chapter 4 introduces the fundamental notions of *nonlinear normal modes* (NNMs), resonances, and their implications for global stability of motion in Hamiltonian systems with a finite number of degrees of freedom N. In particular, we examine the importance of discrete symmetries and the usefulness of group theory in analyzing periodic and quasiperiodic motion in Hamiltonian systems with periodic boundary conditions. Next, we discuss in Chap. 5 a number of analytical and numerical results concerning the GALI method (and its ancestor the Smaller Alignment Index—SALI—method), which uses properties of wedge products of deviation vectors and exploits the tangent dynamics to provide indicators of stable and chaotic motion that are more accurate and efficient than those proposed by other approaches. All this is then applied in Chap. 6 to explain the paradox of *FPU recurrences* and the associated transition from "weak" to "strong" chaos. We introduce the notion of *energy localization* in normal mode space and discuss the existence and stability of low-dimensional "q-tori," aiming to provide a more complete interpretation of FPU recurrences and their connection to energy equipartition in FPU models of particle chains.

In Chap. 7 we proceed to discuss the phenomenon of *localized oscillations* in the configuration space of nonlinear one-dimensional lattices with $N \rightarrow \infty$, concentrating first on the so-called periodic (or translationally invariant) case where all parameters in the on-site and interaction potentials are identical. We also mention in this chapter recent results regarding the effects of *delocalization* and diffusion due to *disorder* introduced by choosing some of the parameters (masses or spring constants) randomly at the initialization of the system.

Next, in Chap. 8 we examine the statistical properties of chaotic regions in cases where the orbits exhibit "weak chaos," for example, near the boundaries of islands of regular motion where the positive Lyapunov exponents are relatively small. We demonstrate that "stickiness" phenomena are particularly important in these regimes, while probability density functions (pdfs) of sums of orbital components (treated as random variables in the sense of the Central Limit Theorem) are well approximated by functions that are far from Gaussian! In fact, these pdfs closely resemble q-Gaussian distributions resulting from minimizing Tsallis' q-entropy (subject to certain constraints) rather than the classical Boltzmann Gibbs (BG) entropy and are related to what has been called *nonextensive statistical mechanics* of strongly correlated dynamical processes.

In this context, we discuss chaotic orbits close to unstable NNMs of multidimensional Hamiltonian systems and show that they give rise to certain very interesting quasi-stationary states, which last for very long times and whose pdfs (of the above type) are well fitted by functions of the q-Gaussian type. Of course, in most cases, as

t continues to grow, these pdfs are expected to converge to a Gaussian distribution ($q \rightarrow 1$), as chaotic orbits exit from weakly chaotic regimes into domains of strong chaos, where the positive Lyapunov exponents are large and BG statistics prevail. Still, we suggest that the complex statistics of these states need to be explored further, particularly with regard to the onset of energy equipartition, as their occurrence is far from exceptional and their long-lived nature implies that they may be physically important in unveiling some of the mysteries of Hamiltonian systems in many dimensions.

The book ends with Chap. 9 containing our conclusions, a list of open research problems, and a discussion of future prospects in a number of areas of Hamiltonian dynamics. Moreover, at the end of every chapter we have included a number of exercises and problems aimed at training the uninitiated reader to learn how to use some of the fundamental concepts and techniques described in this book. Some of the problems are intended as projects for ambitious postgraduate students and offer suggestions that may lead to new discoveries in the field of complex Hamiltonian dynamics in the years ahead.

In the Acknowledgments that follow this Preface, we express our gratitude to a number of junior and senior scientists, who have contributed to the present book in many ways: Some have provided useful comments and suggestions on many topics treated in the book, while others have actively collaborated with us in obtaining many of the results presented here.

Whether we have done justice to all those whose work is mentioned in the text and listed in our References is not for us to judge. The fact remains that, beyond the help we have received from all acknowledged scientists and referenced sources, the responsibility for the accurate presentation and discussion of the scientific field of complex Hamiltonian dynamics lies entirely with the authors.

Patras, Greece   *Tassos Bountis*
Dresden, Germany   *Haris Skokos*

Acknowledgments

Throughout our career and, in particular, in the course of our work described in the present book, we have been privileged to collaborate with many scientists, junior and senior to us in age and scientific experience. It is impossible to do justice to all of them in the limited space available here. Among the senior scientists, Tassos Bountis (T. B.) first wishes to mention Professors Elliott Montroll and Robert Helleman of the University of Rochester, New York, and Professor Joseph Ford of Georgia Institute of Technology, who first introduced him to the wonderful world of Hamiltonian Systems and Nonlinear Dynamics in the USA. Most influential to his particular scientific viewpoint have also been Professors Grégoire Nicolis, Université Libre de Bruxelles, Giulio Casati, University of Insubria, and Hans Capel, University of Amsterdam, and, in Greece, Professors George Contopoulos, University of Athens, and John Nicolis, University of Patras. More recently, T. B. has worked on lattice dynamics based on a number of ideas and results put forth by Dr. Sergej Flach and his group at the Max Planck Institute of the Physics of Complex Systems in Dresden, Professor George Chechin of the Southern Federal University of Russia at Rostov-on-Don, and Professor Constantino Tsallis and his coworkers at the Centro Brasileiro de Pesquisas Fisicas in Rio de Janeiro.

T. B. has benefited significantly from the collaboration of several of his Ph.D. students, whose ingenuity, enthusiasm, and hard work produced many of the results discussed in this book. Their dynamic personalities, congeniality, and pleasant character created a most enjoyable and productive environment at the Department of Mathematics and the Center of Research and Applications of Nonlinear Systems of the University of Patras. T. B., therefore, happily acknowledges many enlightening discussions with Vassilis Rothos, Jeroen Bergamin, Christos Antonopoulos, Helen Christodoulidi, and Thanos Manos, whose work has contributed significantly to the contents of this book. We want to thank them all especially for allowing us to use many figures and results of our common publications in this book.

Haris Skokos (H. S.) also wishes to express his gratitude to a number of scientists who have played a significant role in helping him shape his career as a researcher. More specifically, following a chronological order of his collaborations, he wishes to mention the late Dr. Chronis Polymilis of the University of Athens, who introduced him to the fascinating world of Nonlinear Dynamics during his undergraduate years. Later on, he had the opportunity to closely work with Professor Antonio Giorgilli of the University of Milan, Dr. Panos Patsis of the Academy of Athens, and Dr. Lia Athanassoula of the Observatory of Marseilles, on applications of Hamiltonian dynamics to problems of astronomical interest. H. S. wishes to express his deepest thanks to all of them, both for their friendship and productive collaboration which helped him deepen his knowledge in several theoretical aspects of Chaos Theory and enhanced his ability to work on applications to real physical problems.

The next stop of H. S.'s journey in the tangled world of Chaotic Dynamics was Paris, France, where he had the privilege to work together with Professor Jacques Laskar at the Observatory of Paris. H. S. is grateful to Jacques for broadening his knowledge on chaos detection methods (which is one of his most loved subjects) and symplectic integration techniques as well as for giving him the opportunity to work on the dynamics of real accelerators.

H. S. also considers himself most fortunate for having been able to work in the multidisciplinary environment of the Max Planck Institute of the Physics of Complex Systems in Dresden, where this book was written. He expresses his gratitude to Dr. Sergej Flach not only for giving him the opportunity to extend his activity to the intriguing field of disordered nonlinear systems, but also for his true friendship which made H.S.'s stay in Dresden one of the most pleasant periods of his life.

We also wish to thank Professor Ko van der Weele for critically reading the manuscript and providing many suggestions, Dr. Enrico Gerlach for preparing the MAPLE algorithms for the computation of the SALI and GALI presented in Chap. 5, and Dr. Yannis Kominis for helping us summarize applications of Hamiltonian dynamics to nonlinear optics and plasma physics, in the last part of Chap. 9.

Contents

Acronyms

BEC   Bose Einstein Condensation: Important physical phenomenon related to the behavior of bosons at very low temperatures.

BG   Boltzmann Gibbs: Ensembles of classical equilibrium statistical mechanics proposed by Boltzmann and Gibbs in the late nineteenth and early twentieth centuries.

CLT   Central Limit Theorem: The classical theorem, according to which sums of N identically and independently distributed random variables tend, in the limit $N \to \infty$, to a Gaussian distribution with the same mean and variance as the original variables.

CPU   Central Processing Unit: The part of a computer system that performs the basic arithmetical, logical, and input/output operations required by a computer program. The time required by the CPU for the completion of the tasks of a particular computer program (CPU time) is used to characterize the efficiency of the program.

DNLS   Discrete Nonlinear Schrödinger Equation: The Hamiltonian system of equations emanating from the NLS by the discretization of its second derivative with respect to the space variable.

DDNLS   Disordered Discrete Nonlinear Schrödinger Equation: The DNLS when some of its parameters are chosen randomly and uniformly from within a specified interval.

dof   Degree(s) of freedom: The number of canonical conjugate pairs of position and momentum variables characterizing a Hamiltonian system.

FPU   Fermi, Pasta and Ulam: Names of the researchers who first integrated numerically a chain (one-dimensional lattice) of identical oscillators coupled by quadratic as well as cubic and/or quartic nearest neighbor interaction terms in their Hamiltonian.

FPU$-\alpha$ FPU lattice whose interaction terms in the Hamiltonian, beyond the harmonic ones, are only of the cubic type.

FPU$-\beta$ FPU lattice whose interaction terms in the Hamiltonian, beyond the harmonic ones, are only of the quartic type.

GALI$_k$	Generalized Alignment Index of order $k \geq 2$: Chaos indicator related to the volume of the parallelepiped formed by k deviation vectors in the tangent space of an orbit of a dynamical system.
IPM	In Phase Mode: A particular NNM of one-dimensional lattices, where all particles oscillate identically and in phase.
KAM	Kolmogorov Arnol'd Moser: Name of a theorem that establishes an important rigorous result regarding the behavior of weakly perturbed integrable Hamiltonian systems.
KdV	Korteweg de Vries: Name of an equation exhibiting solitary wave solutions first obtained by Korteweg and de Vries in the 1890s. In the 1960s these waves were named solitons and formed an integral part of the theory of completely integrable evolution equations.
KG	Klein Gordon: It refers to the so-called Klein Gordon potential of classical and quantum physics, which consists of a quadratic and a quartic part.
LA	Liouville Arnol'd: Name of a theorem that establishes an important rigorous result regarding the solvability of integrable Hamiltonian systems of N dof and the existence of N-dimensional tori on which their bounded solutions lie.
LCE(s)	Lyapunov Characteristic Exponent(s): Exponents characterizing the rates of divergence of nearby trajectories of dynamical systems in phase space.
MLE	Maximum Lyapunov Exponent: The LCE with the maximal value.
NLS	Nonlinear Schrödinger Equation: A completely integrable PDE which consists of the linear Schrödinger equation plus a cubic nonlinearity and is of particular importance in the field of nonlinear optics.
NNM(s)	Nonlinear Normal Mode(s): Extension of the normal modes of a linear system of coupled harmonic oscillators in the nonlinear regime.
NME(s)	Normal Mode Eigenvector(s): Eigenvector(s) of a linear problem, used as a basis for the expansion of the solutions of the associated perturbed nonlinear problem.
ODE(s)	Ordinary Differential Equation(s): Differential equations involving derivatives with respect to only one independent variable.
OPM	Out of Phase Mode: A particular NNM of one-dimensional lattices, where all particles oscillate identically out of phase with respect to each other in all neighboring pairs.
PDE(s)	Partial Differential Equation(s): Differential equations involving derivatives with respect to more than one independent variable.
pdf(s)	Probability distribution function(s): Function(s) representing the probability density of observables related to the long time evolution of variables of a dynamical system in chaotic regimes.
PSS	Poincaré Surface of Section: Cross-section of orbits of a Hamiltonian system with certain lower dimensional subspace of its phase space. For example, in the case of a two dof Hamiltonian system the PSS is a two-dimensional plane defined by one pair of position and momentum

coordinates, say (x, p_x), at times when the second position coordinate has a prescribed value, say $y = 0$, and the second momentum a prescribed sign, say $p_y > 0$.

QSS   Quasi-stationary states: Weakly chaotic dynamical states of Hamiltonian systems that persist for very long times and are generally characterized by pdfs that are different from Gaussian.

SALI   Smaller Alignment Index: Chaos indicator related to the area of a parallelogram formed by two deviation vectors in the tangent space of an orbit of a dynamical system. It is analogous to $GALI_2$.

SPO(s)   Simple Periodic Orbit(s): Periodic orbit(s) of a dynamical system returning to their initial state after a single oscillation of its variables.

Chapter 1
Introduction

Abstract Chapter 1 starts by defining a dynamical system in terms of ordinary differential equations and presents the fundamental framework within which one can study the stability of their equilibrium (or fixed) points, as developed by the great Russian mathematician A. M. Lyapunov. The concept of Lyapunov Characteristic Exponents is introduced and two theorems by Lyapunov are discussed, which establish criteria for the asymptotic stability of a fixed point. The mathematical setting of a Hamiltonian system is presented and a third theorem by Lyapunov is stated concerning the continuation of the linear normal modes of N harmonic oscillators, when the system is perturbed by adding to the Hamiltonian nonlinear terms higher than quadratic. Finally, we discuss the meaning of *complexity* in Hamiltonian dynamics, by referring to certain weakly chaotic orbits, which form complicated quasi-stationary states that are well-approximated by the principles of nonextensive statistical mechanics for very long times.

1.1 Preamble

The title of this book consists of three words and it is important to understand all of them. We shall not start, however, with the first one, as it is the most difficult to define. It may also turn out to be the most fascinating and intriguing one in the end. Rather, we will start with the last word: What is dynamics?

Well, it comes from the Greek word "dynamis" meaning "force", so it will be very natural to think of it in the Newtonian context of mechanics as describing the effect of a vectorial quantity proportional to the mass and acceleration of a given body. But that is a very limited interpretation. Dynamics, in fact, refers to any physical observable that is in motion with respect to a stationary frame of reference. And as every student knows this motion may very well be uniform (i.e. with constant velocity) and hence not involve any force at all.

Thus we will speak of dynamics when we are interested to know how the *state* of an observable changes in time. This may represent, for example, the position or

T. Bountis and H. Skokos, *Complex Hamiltonian Dynamics*,
Springer Series in Synergetics, DOI 10.1007/978-3-642-27305-6_1,
© Springer-Verlag Berlin Heidelberg 2012

velocity of a mass particle in the three-dimensional space, the flow of current in an electrical circuit, the concentration of a chemical substance, or the population of a biological species. When we refer to a collection of individual components, we shall speak of the dynamics of a *system*. Most frequently, these components will interact with each other and act interdependently. And this is when matters start to get complicated.

What is the nature of this interdependence and how does it affect the dynamics? Does it always lead to behavior that we call "unpredictable", or *chaotic* (in the sense of extremely sensitive dependence on initial conditions), or does it also give rise to motions that we may refer to as "regular", "ordered" or "predictable"? And what about the intermediate regime between order and chaos, where complexity frequently lies?

It is clear that if we wish to address such questions, we must first introduce a proper mathematical framework for their study and this is provided by the field of Ordinary Differential Equations (ODEs). In this context, we will assume that our system possesses n real dependent variables $(x_k(t), k = 1, 2, \ldots, n$, which constitute a state $x(t) = (x_1(t), \ldots, x_n(t))$ in the *phase space* of the system $D \subseteq \mathbb{R}^n$ and are functions of the single independent variable of the problem: the time $t \in \mathbb{R}$. Their dynamics is described by the system of first order ODEs

$$\frac{dx_k}{dt} = f_k(x_1, x_2, \ldots, x_n), \quad k = 1, 2, \ldots, n, \tag{1.1}$$

which is thus called a *dynamical system*. Since the f_k do not explicitly depend on t the system is called *autonomous*. The functions f_k are defined everywhere in D and, for simplicity, we shall assume that they are analytic in all their variables, meaning that they can be expressed as convergent series expansions in the x_k (with non-zero radius of convergence) near one of their *equilibrium (or fixed) points*, located at the origin of phase space $0 = (0, \ldots, 0) \in D$, where

$$f_k(0) = 0, \quad k = 1, 2, \ldots, n. \tag{1.2}$$

This implies that the right sides of our (1.1) can be expanded in power series:

$$\frac{dx_k}{dt} = p_{k1}x_1 + p_{k2}x_2 + \ldots + p_{kn}x_n + \sum_{m_1, m_2, \ldots, m_n} P_k^{(m_1, \ldots, m_n)} x_1^{m_1} x_2^{m_2} \ldots x_n^{m_n} \tag{1.3}$$

where $k = 1, 2, \ldots, n$ and the $m_k > 0$ are such that $m_1 + m_2 + \ldots + m_n > 1$. Since the f_k are analytic, the existence and uniqueness of the solutions of (1.1), for any initial condition $x_1(0), \ldots, x_n(0)$ (where the f_k are defined) is guaranteed by the classical theorems of Euler, Cauchy and Picard [46, 97]. Furthermore, the series in (1.3) are convergent for $|x_k| \leq A_k$ and by a well-known theorem of analysis (e.g. see [155, p. 273]) its coefficients are bounded by:

$$| P_k^{(m_1, \ldots, m_n)} | \leq \frac{M_k}{A_1^{m_1} A_2^{m_2} \ldots A_n^{m_n}}, \tag{1.4}$$

where M_k is an upper bound of the modulus of all terms of the form $x_1^{m_1} x_2^{m_2} \ldots x_n^{m_n}$ entering in (1.3).

What can we say about the solutions of these equations in a small neighborhood of the equilibrium point (1.2), where the series expansions of the f_k converge? Is the motion "regular" or "predictable" there? This is the question of *stability of motion* first studied systematically by the great Russian mathematician A. M. Lyapunov, more than 110 years ago [232]. In the section that follows we shall review his famous method that led to the proof of the existence of *periodic solutions*, as continuations of the corresponding oscillations of the linearized system of equations.

This method is *local* in character, as it describes solutions in a very small region around the origin and for finite intervals in time. It generalizes the treatment and results found in Poincaré's thesis [273], of which Lyapunov learned at a later time, as he explains in the Introduction of [232].

1.2 Lyapunov Stability of Dynamical Systems

Let us first define the notions of stability that Lyapunov had in mind: The first and simplest one concerns what may be called *asymptotic stability*, as it refers to the case where all solutions $x_k(t)$ of (1.3), starting within a domain of the origin given by $|x_k(t_0)| \leq A_k$, tend to 0 as $t \to \infty$. A less restrictive situation arises when we can prove that for every $0 < \varepsilon < \varepsilon_0$, no matter how small, all solutions starting at $t = t_0$ within a neighborhood of the origin $K(\varepsilon) \subseteq B(\varepsilon)$, where $B(\varepsilon)$ is a "ball" of radius ε around the origin, remain inside $B(\varepsilon)$ for all $t \geq t_0$.

This weaker condition is often called *neutral or conditional stability* and will be of great importance to us, as it frequently occurs in *conservative dynamical systems* (among which are the Hamiltonian ones). These conserve phase space volume and hence cannot come to a complete rest at any value of t, finite or infinite (except for some special solutions lying on stable invariant manifolds, as we discuss later). Conditional stability, in fact, characterizes precisely the systems for which Lyapunov could prove the existence of families of periodic solutions around the origin by relating them directly to a conserved quantity called *integral of the motion*. For Lyapunov, the existence of integrals was a means to an end, unlike Poincaré, who considered *integrability* as a primary goal in trying to show the *global stability* of the motion of a dynamical system.

Let us illustrate Lyapunov's method for establishing the stability of a fixed point, by considering the simple example of a one-dimensional system, $n = 1$, described by a single ODE of the Riccati type:

$$\frac{\mathrm{d}x}{\mathrm{d}t} = -x + x^2. \tag{1.5}$$

The main idea is to write its solution as a series:

$$x(t) = x^{(1)} + x^{(2)} + \ldots + x^{(n)} + \ldots, \tag{1.6}$$

whose leading term $x^{(1)}$ is the general solution of the linear part of (1.5), considered naturally as the most important part of the solution in a small region around the fixed point $x = 0$. This term is $x^{(1)} = a\mathrm{e}^{-t}$ and incorporates the only arbitrary constant needed for the general solution of the problem. Substituting (1.6) into (1.5) we thus obtain an infinite set of linear inhomogeneous equations for the $x^{(k)}$

$$\frac{\mathrm{d}x^{(k)}}{\mathrm{d}t} = -x^{(k)} + \sum_{j=1}^{k} x^{(j)} x^{(j-k)}, \quad k > 1, \tag{1.7}$$

from which we can easily obtain particular solutions, since the homogeneous part is already represented by $x^{(1)}(t)$. These are: $x^{(k)} = -(-a)^k \mathrm{e}^{-kt}$ and hence the general solution of (1) becomes:

$$x(t) = x^{(1)} + x^{(2)} + \ldots = a\mathrm{e}^{-t} - a^2\mathrm{e}^{-2t} + a^3\mathrm{e}^{-3t} - \ldots = q - q^2 + q^3 - \ldots, \tag{1.8}$$

where we have set $q = a\mathrm{e}^{-t}$. This expression clearly converges for $|q| < 1$, which gives the region of initial conditions, at $t = 0$ with $|a| < 1$, where these solutions exist. Furthermore, in this region, all solutions tend to 0 as $t \to \infty$ and therefore 0 is an asymptotically stable *equilibrium state*.

Let us now see how all this works for arbitrary n: First, we write again the general solution in the form of a series

$$x_k(t) = x_k^{(1)} + x_k^{(2)} + \ldots, \quad k = 1, 2, \ldots, n, \tag{1.9}$$

and substitute in (1.3) separating the linear system of equations for the $x_k^{(1)}$:

$$\frac{\mathrm{d}x_k^{(1)}}{\mathrm{d}t} = p_{k1}x_1^{(1)} + p_{k2}x_2^{(1)} + \ldots + p_{kn}x_n^{(1)}, \tag{1.10}$$

from the remaining ones

$$\frac{\mathrm{d}x_k^{(m)}}{\mathrm{d}t} = p_{k1}x_1^{(m)} + p_{k2}x_2^{(m)} + \ldots + p_{kn}x_n^{(m)} + R_k^{(m)}, \quad m > 1, \tag{1.11}$$

where the $R_k^{(m)}$ contain only terms $x_s^{(j)}$, $s = 1, 2, \ldots, n$ and $j = 1, 2, \ldots, m - 1$, which have already been determined at lower orders. Thus, we first need to study the linear system (1.10) and obtain its general solution as a linear combination of n independent particular solutions

$$x_k^{(1)}(t) = a_1 x_{k1}(t) + a_2 x_{k2}(t) + \ldots + a_n x_{kn}(t), \quad k = 1, 2, \ldots, n, \tag{1.12}$$

in such a way that the initial conditions $x_k^{(1)}(t_0) = a_k$ are satisfied. We then use (1.12) to insert in (1.11) and find particular solutions of the corresponding linear inhomogeneous system for every m, so that $x_k^{(m)}(t_0) = 0$.

Lyapunov then obtains explicit expressions for the $x_k^{(m)}(t)$ and makes the additional assumption that the sets $A_1, A_2, \ldots, A_n$ and $M_1, M_2, \ldots, M_n$ (see the discussion at the beginning of the chapter), considered as functions of t, have non-zero upper bound for each A_k and non-zero lower bound for each M_k, for all $t_0 \leq t \leq T$ and any T. Thus, he was able to show that the solution (1.9) written as power series in the quantities a_k is *absolutely convergent*, as long as the $|a_k|$ do not exceed some limit depending on T.

Observe now that, even though we have found the general solution, it is impossible to discuss the question of stability of the origin, unless we know something about the behavior of the solutions of the linear system (1.10) as functions of t, as we did for the simple Riccati equation. To proceed one would have to compare first every one of the n independent solutions of (1.10) to an exponential function of time, with the purpose of identifying the particular exponent that would enter in such a relationship.

To achieve this, Lyapunov had the ingenious idea of introducing what he called the *characteristic number* of a function $x(t)$, as follows: First form the auxiliary function $z(t) = x(t)e^{\lambda t}$ and define as the characteristic number λ_0 of x(t), that value of λ for which z(t) vanishes for $\lambda < \lambda_0$ and becomes unbounded for $\lambda > \lambda_0$, as $t \to \infty$. Thus, this number represents the "rate" of exponential decay (or growth) of $x(t)$, as time becomes arbitrarily large. In fact, it is closely connected to the *negative* of what we call today the *Lyapunov characteristic exponent* (LCE) [310]. Adopting the convention that functions $x(t)$ whose $z(t)$ vanishes *for all* λ have $\lambda_0 = \infty$ and those for which $z(t)$ is unbounded *for all* λ have $\lambda_0 = -\infty$, one can thus define a unique characteristic number for every function $x(t)$.

Proceeding then to derive characteristic numbers for sums and products of functions arising in the solutions of systems of linear ODEs, Lyapunov defines what he calls a *regular system of linear ODEs* by the condition that the sum of the characteristic numbers of its solutions equals the negative of the characteristic number of the function

$$\exp\left(-\int \sum_{k=1}^{n} p_{kk}(t)dt\right) \tag{1.13}$$

(see [232, p. 43]). Thus, he was led to one of his most important theorems:

Theorem 1.1. *(see [232, p. 57]) If the linear system of differential equations (1.10) is regular and the characteristic numbers of its independent solutions are all positive, the equilibrium point at the origin is asymptotically stable.*

It is important to note that the condition of *regularity* of a linear system, as defined above, can be shown to hold for all systems, whose coefficients p_{jk}, $j, k = 1, 2, \ldots, n$ are constant or periodic functions of t, and is, in fact, also verified for a much larger class of linear systems. Moreover, since we will be exclusively concerned here with the case of constant coefficients, it is necessary to identify the meaning of characteristic numbers for our problem at hand. Indeed, as the reader can easily verify, they are directly related to the eigenvalues of the $n \times n$ matrix $J = (p_{jk})$, $j, k = 1, 2, \ldots, n$, obtained as the roots of the characteristic equation

$$\det(J - \lambda I_n) = 0, \tag{1.14}$$

$\lambda_1, \lambda_2, \ldots, \lambda_n$, I_n being the $n \times n$ identity matrix. In modern terminology, therefore, for a dynamical system (1.1), with an equilibrium point at $(0, 0, \ldots, 0)$ and a constant Jacobian matrix

$$J_{k,j} = p_{kj} = \frac{\partial f_k}{\partial x_j}(0, \ldots, 0), \quad j, k = 1, 2, \ldots, n, \tag{1.15}$$

the above theorem translates to the following well-known result:

Theorem 1.2. *(see [170, p. 181]) If all eigenvalues of the matrix J have negative real part less than $-c$, $c > 0$, there is a compact neighborhood U of the origin, such that, for all $(x_1(0), x_2(0), \ldots, x_n(0)) \in U$, all solutions $x_k(t) \to 0$, as $t \to \infty$. Furthermore, one can show that this approach to the fixed point is exponential: Indeed, if we denote by $|\cdot|$ the Euclidean norm in $\mathbb{R}^n$ and define $x(t) = (x_1(t), x_2(t), \ldots, x_n(t))$, it can be proved, using simple topological arguments, that for all $x(0) \in U$, $|x(t)| \le |x(0)|e^{-ct}$, and $|x(t)|$ is in U for all $t \ge 0$.*

Let us finally observe that, since Lyapunov expressed his solutions as sums of exponentials of the form

$$x_s^{(m)} = \sum C_s^{(m_1, m_2, \ldots, m_n)} e^{(m_1 \lambda_1 + m_2 \lambda_2 + \ldots + m_n \lambda_n)t}, \tag{1.16}$$

with $0 < m_1 + m_2 + \ldots + m_n < m$, when we substitute these expressions in the equations of motion (1.10), (1.11) and try to solve for the coefficients, $C_s^{(m_1, m_2, \ldots, m_n)}$, we need to divide by the expression

$$m_1 \lambda_1 + m_2 \lambda_2 + \ldots + m_n \lambda_n. \tag{1.17}$$

If this quantity ever becomes zero, this will lead to *secular terms* in the above series which will grow linearly in time. This can happen for example if one (or more) of the $\lambda_j = 0$, or if the system possesses one (or more) pairs of imaginary eigenvalues. Note that such so-called *resonances* will never occur, as long as the conditions of Theorem 1.2 above are satisfied and thus, in the case of $Re(\lambda_j) > 0$ all secular terms are avoided and the convergence of the corresponding series is proved.

Assuming that the linear terms of these series have constant coefficients, Lyapunov then paid particular attention to the case where one (or more) of the eigenvalues of the matrix of these coefficients have *zero real part*. This was the beginning of what we now call *bifurcation theory* (see e.g. [165, 170]), as it constitutes the turning point between *stability* of the fixed point (all eigenvalues have negative real part) and *instability* (at least one eigenvalue has positive real part).

Besides the above so-called *first method*, which is local and parallels similar work by Poincaré, Lyapunov also introduced a *second method*, which is entirely his own and is based on the following idea [232]: Instead of focusing on the solutions of the direct problem, one constructs specific functions of these solutions, with well-defined geometric properties, whose evolution in time reveals indirectly the properties of the actual solutions for all $t > t_0$! This is the famous *method of Lyapunov functions*, as we know it today. It is global in the sense that it does not refer to a finite time interval and applies to relatively large regions of phase space around the equilibrium point. However, it does rely strongly on one's resourcefulness to construct the appropriate partial differential equation (PDE) satisfied by these functions and solve them by power series expansions.

We will not dwell at all here on Lyapunov's second method, as it applies almost exclusively to non-conservative (hence non-Hamiltonian) dynamical systems. The interested reader can find excellent accounts of the theory and applications of Lyapunov's functions, not only in Lyapunov's treatise [232] but also in a number of excellent textbooks on the qualitative theory of ODEs and dynamical systems [170, 265, 346]. There is, however, one more result of Lyapunov's theory [232] concerning simple periodic solutions of *Hamiltonian systems*, which will be very useful to us in the chapters that follow and is described separately in the next section.

1.3 Hamiltonian Dynamical Systems

In this book we shall often speak of conditional (or neutral) stability where there is (at least) one pair of purely imaginary eigenvalues (all others being real and negative), implying the possibility of the existence of periodic orbits about the origin. This allows us to apply the above theory to the case of *Hamiltonian dynamical systems* of N *degrees of freedom* (dof), where $n = 2N$ and the equations of motion (1.1) are written in the form

$$\frac{dq_k}{dt} = \frac{\partial H}{\partial p_k}, \quad \frac{dp_k}{dt} = -\frac{\partial H}{\partial q_k}, \quad k = 1, 2, \ldots, N, \tag{1.18}$$

where $q_k(t)$, $p_k(t)$, $k = 1, 2, \ldots, 2N$ are the *position and momentum coordinates* respectively and H is called the *Hamiltonian function*. If H does not explicitly depend on t, it is easy to see from (1.18) that its total time derivative is zero and thus represents a first integral (or *constant of the motion*), whose value equals the *total energy* of the system E.

In most cases treated in this book, we will assume that our Hamiltonian can be expanded in power series as a sum of homogeneous polynomials H_m of degree $m \geq 2$

$$H = H_2(q_1, \ldots, q_N, p_1, \ldots, p_N) + H_3(q_1, \ldots, q_N, p_1, \ldots, p_N) + \ldots = E, \tag{1.19}$$

so that the origin $q_k = p_k = 0$, $k = 1, 2, \ldots, N$ is an equilibrium point of the system. $H(q_k(t), p_k(t)) = E$ thus defines the so-called (constant) *energy surface*, $\Gamma(E) \subset \mathbb{R}^{2N}$, on which our dynamics evolves.

Let us now assume that the linear equations resulting from (1.18) and (1.19), with $H_m = 0$ for all $m > 2$, yield a matrix, whose eigenvalues all occur in conjugate imaginary pairs, $\pm i\omega_k$ and thus provide the *frequencies* of the so-called *normal mode* oscillations of the linearized system. This means that we can change to *normal mode coordinates* and write our Hamiltonian in the form of N *uncoupled* harmonic oscillators

$$H^{(2)} = \frac{\omega_1}{2}(x_1{}^2 + y_1{}^2) + \frac{\omega_2}{2}(x_2^2 + y_2^2) + \ldots + \frac{\omega_N}{2}(x_N^2 + y_N^2) = E, \quad (1.20)$$

where x_k, y_k, $k = 1, 2, \ldots, N$ are the new position and momentum coordinates and ω_k represent the normal mode frequencies of the system.

Theorem 1.3. *[232] If none of the ratios of these eigenvalues, ω_j/ω_k, is an integer, for any $j, k = 1, 2, \ldots, N$, $j \neq k$, the linear normal modes continue to exist as periodic solutions of the nonlinear system (1.18) when higher order terms $H_3, H_4, \ldots$ etc. are taken into account in (1.19).*

These solutions have frequencies close to those of the linear modes and are examples of what we call *simple periodic orbits* (*SPOs*), where all variables oscillate with the same frequency $\omega_k = 2\pi/T_k$, returning to the same values after a single maximum (and minimum) in their time evolution over one period T_k.

What is the importance of these particular SPOs which are called *nonlinear normal modes*, or NNMs? Once we have established that they exist, what can we say about their stability under small perturbations of their initial conditions? How do their stability properties change as we vary the total energy E in (1.19)? Do such changes only affect the motion in the immediate vicinity of the NNMs or can they also influence the dynamics of the system more globally? Are there other simple periodic orbits of comparable importance that may also be useful to study from this point of view? These are the questions we shall try to answer in the chapters that follow.

After discussing in Chap. 2 the fundamental concepts of *order* and *chaos* in Hamiltonian systems of few dof, we shall examine in Chap. 3 the topics of *local and global stability* in the case of arbitrary (but finite) number of dof N. It is here that we will introduce the spectrum of LCEs for Hamiltonian systems, following the discussion of the previous section. In particular, we shall dwell on the relation of the LCEs to the chaotic behavior of individual solutions or *orbits* of the system and use them to evaluate the so-called *Kolmogorov entropy*. More specifically, we will demonstrate on concrete examples that this entropy is *extensive* in regimes of "*strong chaos*", i.e. it grows linearly with N in the thermodynamic limit ($E \rightarrow \infty$, $N \rightarrow \infty$ with E/N fixed) and is hence compatible with Boltzmann Gibbs (BG) statistical mechanics.

Then in Chap. 4 we focus on the importance of NNMs in Hamiltonian systems possessing discrete symmetries and discuss their relevance with regard to physical models oscillating in one, two or three spatial dimensions. Our main example will be the Fermi Pasta Ulam (FPU) chain under periodic boundary conditions, but the results will be quite general and can apply to a wide variety of systems, which are of interest to solid state physics. Next, we will describe in detail in Chap. 5 the method of the Generalized Alignment Indices $GALI_k$, $k = 1, 2, \ldots, 2N$ indicators. The GALIs are ideal for identifying domains of chaos and order not only in N dof Hamiltonian systems but also $2N$-dimensional ($2ND$) symplectic maps and represent a significant improvement over many other similar techniques used in the literature. For example, these indicators can also be employed to locate $s(\leq N)$-dimensional tori on which the motion is governed by s rationally independent frequencies and provide conditions for the "breakdown" of these tori and hence the destabilization of the corresponding quasiperiodic orbits.

In Chap. 6, we return to a very important physical property of Hamiltonian lattices, namely arrays of *nonlinear oscillators* coupled to each other by nearest neighbor interactions in one or more spatial dimensions. In particular, we revisit the famous FPU lattice and investigate the phenomenon of *FPU recurrences* at low energies, using Poincaré-Linstedt perturbation theory and the GALI method. We show that these are connected with *"weakly chaotic"* regimes dominated by *low-dimensional tori*, whose energy profile is *exponentially localized* in Fourier space. We then study, at higher energies, *diffusion* phenomena and the transition to *delocalization* and *equipartition* of the total energy among the N normal modes.

Then, in Chap. 7 we discuss different types of *localization* and diffusion phenomena this time in the configuration space of one-dimensional nonlinear lattices (chains of anharmonic oscillators). In particular, we start with the case where all lattice parameters are equal and show how exponentially localized oscillations called *discrete breathers* arise. These SPOs, which are often linearly stable and provide a barrier to energy transport, can in fact be constructed by methods involving *invariant manifold* intersections that are described in Chap. 2. However, when translational invariance is broken by introducing random *disorder* in the parameters of the system, nonlinearity has a delocalizing effect and very important diffusive phenomena are observed, which persist for extremely long times.

In Chap. 8, we turn to the investigation of the type of *statistics* that characterizes the dynamics in "weakly chaotic" regimes, where slow diffusion processes are observed. Perhaps not surprisingly, we find that, as t grows, the *probability distribution functions* (pdfs) associated with sums of chaotic variables in these domains do *not* quickly tend to a Gaussian at equilibrium, as expected by BG statistical mechanics. Rather, they go through a sequence of *quasi-stationary states* (QSS), which are well-approximated by a family of q-Gaussian functions and share some remarkable properties in many examples of multi-dimensional Hamiltonian systems. In these examples, it appears that "weakly chaotic" dynamics is connected with *"stickiness"* phenomena near the boundary of islands of regular motion called *"edge of chaos"*, where LCEs are very small and orbits get trapped for extremely long times. Finally, Chap. 9 presents our conclusions, a list of open research problems and a number of other promising directions not treated in this book, which are expected to shed new light on the fascinating and important subject of complex Hamiltonian dynamics.

1.4   Complex Hamiltonian Dynamics

In attempting to understand the complexity of chaotic behavior in Hamiltonian systems, it is important to recall some basic facts about equilibrium thermodynamics in the framework of BG theory of statistical ensembles. These topics form the backbone of the results discussed in Chap. 8, but it is instructive to briefly review them here. As is well-known in BG statistics, if a system can be at any one of $i = 1, 2, \ldots, W$ states with probability p_i, its entropy is given by the famous formula

$$S_{BG} = -k \sum_{i=1}^{W} p_i \ln p_i , \tag{1.21}$$

where k is Boltzmann's constant, provided, of course,

$$\sum_{i=1}^{W} p_i = 1. \tag{1.22}$$

The BG entropy satisfies the property of *additivity*, i.e. if A and B are two *independent* systems, the probability to be in their union is $p_{i,j}^{A+B} = p_i^A p_j^B$ and this necessitates that the entropy of the joint state be additive, i.e.

$$S_{BG}(A + B) = S_{BG}(A) + S_{BG}(B). \tag{1.23}$$

At thermal equilibrium, the probabilities that optimize the BG entropy subject to (1.22), given the energy spectrum E_i and temperature T of these states are:

$$p_i = \frac{e^{-\beta E_i}}{Z_{BG}}, \quad Z_{BG} = \sum_{i=1}^{W} e^{-\beta E_i} , \tag{1.24}$$

where $\beta = 1/kT$ and Z_{BG} is the so-called BG partition function. For a continuum set of states depending on one variable, x, the optimal pdf for BG statistics subject to (1.22), zero mean and a given variance V is, of course, the well-known Gaussian $p(x) = e^{-x^2/2V}/\sqrt{2V}$. Another important property of the BG entropy is that it is *extensive*, i.e.

$$lim_{N \to \infty} \frac{S_{BG}}{N} < \infty. \tag{1.25}$$

This means that the BG entropy grows *linearly* as a function of the number of dof N of the system. But then, what about many physically important systems that are *not extensive*?

As we discuss in Chap. 8, it is for these type of situations that a different form of entropy has been proposed [332], the so-called *Tsallis entropy*

$$S_q = k \frac{1 - \sum_{i=1}^{W} p_i^q}{q-1} \text{ with } \sum_{i=1}^{W} p_i = 1 \qquad (1.26)$$

depending on an index q, where $i = 1, \ldots, W$ counts the states of the system occurring with probability p_i and k is the Boltzmann constant. Just as the Gaussian distribution represents an extremal of the BG entropy (1.21) for a continuum set of states $x \in \mathbb{R}$, the *q-Gaussian distribution*

$$P(x) = a e_q^{-\beta x^2} \equiv a \left[1 - (1-q)\beta x^2 \right]^{\frac{1}{1-q}} , \qquad (1.27)$$

is obtained by optimizing the Tsallis entropy (1.26), where q is the entropic index, β is a free parameter and a a normalization constant. Expression (1.27) is a generalization of the Gaussian, since in the limit $q \to 1$ we have $\lim_{q \to 1} e_q^{-\beta x^2} = e^{-\beta x^2}$.

The Tsallis entropy is, in general, *not additive*, since it can be shown that $S_q(A+B) = S_q(A) + S_q(B) + k(1-q)S_q(A)S_q(B)$ and hence is also *not extensive*. It thus offers us the possibility of studying cases where different subsystems like A and B above are never completely uncorrelated, as is the case e.g. with many realistic physical systems where long range forces are involved [332]. Furthermore, as we argue in Chap. 8, the persistence of non-Gaussian QSS in "weakly chaotic" regimes of Hamiltonian systems represents another important case of highly correlated dynamics that must be more deeply understood. It appears, therefore, that a framework formally analogous to BG thermodynamics is needed for nonextensive systems as well, where the exponentials in (1.24) are replaced by q-exponentials, as defined in (1.27).

In the present book, complex Hamiltonian dynamics does not only refer to the non-Gaussian statistics of "weakly chaotic" motion. Indeed, one may argue that the quasi-stationary character of *regular* or *ordered* motion, as described for example in the analysis of FPU recurrences presented in Chap. 6, constitutes another complicated phenomenon. More generally, our thesis is that complexity in Hamiltonian systems lies in the transition from local to global behavior, exactly where one would expect it to be, you might say. The point we are trying to make is that this is not merely a question of mathematical interest. As is by now widely recognized, it is precisely in this type of transition that important phenomena of great physical relevance are found that need to be further explored.

Complexity is, of course, difficult to describe by a single definition. Broadly speaking, it characterizes systems of nonlinearly interacting components that can exhibit collective phenomena like pattern formation, self-organization and the emergence of states that are not observed at the level of any one individual component. Clearly, therefore, Hamiltonian systems are not complex in the sense there exists a solid mathematical framework for their accurate description. We do not need to invent new and inspiring models that will capture the main features of the dynamics, as we do in the flocking of birds, for example, or the spread of an epidemic.

In Hamiltonian dynamics we shall speak of complexity when new phenomena arise, which are not expected by the classical theory of Newtonian mechanics, or statistical physics. Our claim is that such phenomena do occur, for example, in regimes where different *hierarchies* of chaotic behavior are detected, where physical processes like diffusion and energy transport crucially depend on the rate of separation of nearby trajectories as a function of time. Even if this separation grows exponentially, it may do so in ways that are indistinguishable from a power-law due to strong correlations present for very long times. In fact, it appears that this kind of evolution is connected with the intricacies of a *geometric* complexity of phase space dynamics caused by the presence of invariant Cantor sets located at the boundaries of islands of regular motion around stable periodic solutions of the system.

But this does not exhaust the sources of complexity in Hamiltonian dynamics. As we will discover, besides varying degrees of chaos, there also exist hierarchies of order in N dof Hamiltonian systems. Indeed, around discrete breathers and near NNMs responsible for FPU recurrences at low energies, one often finds that invariant tori exist of dimension *much lower* than the number N expected by classical theorems. As we show in this book, one way to approach this issue is to study the stability of tori, using for example the GALI method, or group theoretical techniques in cases where the system possesses discrete symmetries. Much work, however, still needs to be done before these questions are completely understood.

What other complex phenomena are still waiting to be discovered in Hamiltonian dynamics? No one knows. If we judge, however, by the progress that has been achieved to date and the rapidly growing interest of theoretical and experimental researchers, we would not be surprised if new results came to the surface, whose importance may rival analogous discoveries in non-Hamiltonian systems. The challenge is before us and the future of complex Hamiltonian dynamics looks very bright indeed.

Chapter 2
Hamiltonian Systems of Few Degrees of Freedom

Abstract In Chap. 2 we provide first an elementary introduction to some simple examples of Hamiltonian systems of one and two degrees of freedom. We describe the essential features of phase space plots and focus on the concepts of periodic and quasiperiodic motions. We then address the questions of integrability and solvability of the equations, first for linear and then for nonlinear problems. We present the important integrability criteria of Painlevé analysis in *complex time* and show, on the non-integrable Hénon-Heiles model, how chaotic orbits arise on a Poincaré Surface of Section of the dynamics in phase space. Using the example of a periodically driven Duffing oscillator, we explain that chaos is connected with the intersection of invariant manifolds and describe how these intersections can be analytically studied by the perturbation approach of Mel'nikov theory.

2.1 The Case of $N = 1$ Degree of Freedom

One of the first physical systems that an undergraduate student encounters in his (or her) science studies is the harmonic oscillator. This describes the oscillations of a mass m, tied to a spring, which exerts on the mass a force that is proportional to the negative of the displacement q of the mass from its equilibrium position ($q = 0$), as shown in Fig. 2.1. The dynamics is described by Newton's second order differential equation

$$m\frac{\mathrm{d}^2 q}{\mathrm{d}t^2} = -kq, \tag{2.1}$$

where $k > 0$ is a constant representing the hardness (or softness) of the spring. Equation 2.1 can be easily solved by standard techniques of linear ODEs to yield the displacement $q(t)$ as an oscillatory function of time of the form

$$q(t) = A\sin(\omega t + \alpha), \quad \omega = \sqrt{k/m}, \tag{2.2}$$

T. Bountis and H. Skokos, *Complex Hamiltonian Dynamics*,
Springer Series in Synergetics, DOI 10.1007/978-3-642-27305-6_2,
© Springer-Verlag Berlin Heidelberg 2012

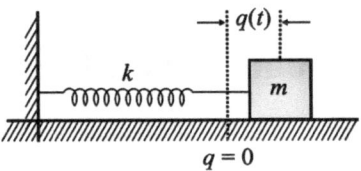

Fig. 2.1 A harmonic oscillator consists of a mass m moving horizontally on a frictionless table under the action of a force which is proportional to the negative of the displacement $q(t)$ of the mass from its equilibrium position at $q = 0$ ($k > 0$ is a proportionality constant)

Fig. 2.2 Plot of the solutions of the harmonic oscillator Hamiltonian (2.4) in the (q, p) phase space, as a one-parameter family of curves, for different values of the total energy E

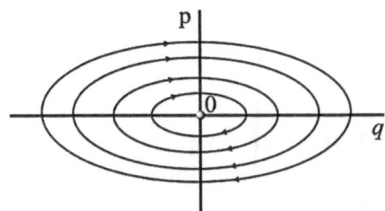

where A and α are free constants corresponding to the amplitude and phase of oscillations and ω is the frequency.

If we now recall from introductory physics that $p = m\mathrm{d}q/\mathrm{d}t$ represents the momentum of the mass m, we can rewrite (2.1) in the form of two first order ODEs

$$\frac{\mathrm{d}q}{\mathrm{d}t} = \frac{p}{m} = \frac{\partial H}{\partial p}, \quad \frac{\mathrm{d}p}{\mathrm{d}t} = -kq = -\frac{\partial H}{\partial q}, \tag{2.3}$$

where we have already anticipated that the dynamics is derived from the Hamiltonian function

$$H(q, p) = \frac{p^2}{2m} + k\frac{q^2}{2} = E, \tag{2.4}$$

which is an integral of the motion (since $\mathrm{d}H/\mathrm{d}t = 0$), whose value E represents the total (kinetic plus potential) energy of the system. Thus, this mass-spring system is our first example of a Hamiltonian system of $N = 1$ dof.

We will not attempt to solve (2.3). Had we done so, of course, we would have recovered (2.2). Instead, we will consider the solutions of the problem *geometrically* as a one-parameter family of trajectories (or orbits) given by (2.3) and plotted in the (q, p) phase space of the system in Fig. 2.2. Thus, as this figure shows, both variables $q(t)$ and $p(t)$ oscillate periodically with the same frequency ω. Their orbits are ellipses whose (semi-) major and minor axes are determined by the parameters k, m and the value of E.

Note now that the energy integral (2.4) can also be used to bring us one step closer to the complete solution of the problem, by solving (2.4) for p and obtaining an equation where the variable q is separated from the time t. Integrating both sides of this equation in terms of the respective variables, we arrive at the expression

$$\int_0^q \frac{dq}{\sqrt{2E/m - \omega^2 q^2}} = t + c, \qquad (2.5)$$

which defines implicitly the solution $q(t)$ as function of t, c being the second free constant of the problem (the first one is E). The reader now recognizes the left side of (2.5) as the inverse sine function, whence we finally obtain

$$q(t) = \sqrt{\frac{2E}{k}} \sin(\omega(t + c)), \qquad (2.6)$$

where the constants E and c can be related to A and α, by direct comparison with (2.2).

The process described above is called *integration by quadratures* and was made possible by the crucial existence of the integral of motion (2.4).

Of course, the harmonic oscillator is a linear system whose dynamics is very simple. It possesses a single *equilibrium or fixed point* at $(0,0)$, where all first derivatives in (2.3) vanish. This point is called *elliptic* and all closed curves about it in Fig. 2.2 describe oscillations whose frequency ω is independent of the choice of initial conditions.

Let us turn, therefore, to a more interesting one dof Hamiltonian system representing the motion of a simple pendulum shown in Fig. 2.3. Its equation of motion according to classical mechanics is

$$ml^2 \frac{d^2\theta}{dt^2} = -mlg \sin \theta, \qquad (2.7)$$

where the right side of (2.7) expresses the restoring torque due to the weight mg of the mass (g being the acceleration due to gravity) and the left side describes the time derivative of the angular momentum of the mass. If we now write this equation as a system of two first order ODEs, as we did for the harmonic oscillator, we find again that they can be cast in Hamiltonian form

$$\frac{dq}{dt} = p = \frac{\partial H}{\partial p}, \quad \frac{dp}{dt} = -\frac{g}{l} \sin q = -\frac{\partial H}{\partial q}, \qquad (2.8)$$

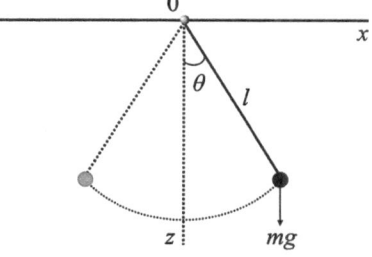

Fig. 2.3 The simple pendulum is moving in the two-dimensional plane (x, z), but its motion is completely described by the angle θ and its derivative $d\theta/dt$

Fig. 2.4 Plot of the curves of the energy integral (2.9) in the (q, p) phase space. Note, the presence of additional equilibrium points at $(\pm\pi, 0)$, which are of the saddle type. Observe also the curve S separating oscillatory motion (L) about the central elliptic point at $(0, 0)$ from rotational motion (R)

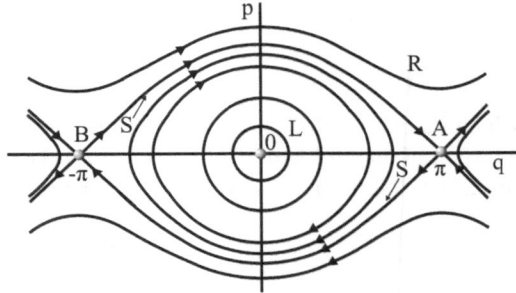

where $q = \theta$ and l is the length of the pendulum. Of course, we cannot solve these equations as easily as we did for the harmonic oscillator. Still, as you will find out by solving Exercise 2.1, the solution of (2.8) can be obtained via the so-called Jacobi elliptic functions [6, 107, 169], which have been thoroughly studied in the literature and are directly related to the solution of an anharmonic oscillator with cubic nonlinearity, about which we will have a lot to say in later sections.

Still, we can make great progress in understanding the dynamics of the simple pendulum through its Hamiltonian function, which provides the energy integral

$$H(q, p) = \frac{p^2}{2} + \frac{g}{l}(1 - \cos q) = E. \tag{2.9}$$

Plotting this family of curves in the (q, p) phase space for different values of E we now obtain a much more interesting picture than Fig. 2.2 depicted in Fig. 2.4. Observe that, besides the elliptic fixed point at the origin, there are two new equilibria located at the points $(\pm\pi, 0)$. These are called *saddle points* and *repel* most orbits in their neighborhood along hyperbola-looking trajectories. For this reason points A and B in Fig. 2.4 are also called *unstable*, in contrast to the $(0, 0)$ fixed point, which is called *stable*. More precisely, in view of the theory presented in Chap. 1, $(0, 0)$ is characterized by what we called conditional (or neutral) stability.

Let us now discuss the *linear stability* of fixed points of a Hamiltonian system within the more general framework outlined in Sect. 1.2. According to this analysis, we first need to linearize Hamilton's system of ODEs (1.18) about a fixed point, which we assume to be located at the origin $\mathbf{0} = (0, 0, \ldots, 0)$ of the $2N$D phase space. The resulting set of linear ODEs is written in terms of the Jacobian matrix $J = \{J_{jk}\}$:

$$\dot{x}_j = \sum_{k=1}^{2N} J_{jk} x_k, \quad j = 1, 2, \ldots, 2N, \tag{2.10}$$

whose elements (show this as a simple exercise starting with (1.18))

$$j = 1, 2, \ldots, N: \quad J_{jk} = \frac{\partial^2 H}{\partial p_j \partial q_k}(\mathbf{0}), \quad k = 1, \ldots, N,$$

$$J_{jk} = \frac{\partial^2 H}{\partial p_j \partial p_{k-N}}(\mathbf{0}), \quad k = N + 1, \ldots, 2N,$$

$$j = N + 1, \ldots, 2N: \quad J_{jk} = -\frac{\partial^2 H}{\partial q_{j-N} \partial q_k}(\mathbf{0}), \quad k = 1, \ldots, N,$$

$$J_{jk} = -\frac{\partial^2 H}{\partial q_{j-N} \partial p_{k-N}}(\mathbf{0}), \quad k = N + 1, \ldots, 2N,$$

(2.11)

where x_j represents small variations away from the origin ($q_j = p_j = 0$, $j = 1, \ldots, N$) of the q_j variables for $j = 1, \ldots, N$ and the p_j variables for $j = N + 1, \ldots, 2N$.

For Hamiltonian systems of $N = 1$ dof, it is clear from (2.11) that the elements of the 2×2 Jacobian matrix $J = \{J_{jk}\}$, $j, k = 1, 2$ are

$$J_{11} = -J_{22} = \frac{\partial^2 H}{\partial q \partial p}(0, 0), \quad J_{12} = \frac{\partial^2 H}{\partial p^2}(0, 0), \quad J_{21} = -\frac{\partial^2 H}{\partial q^2}(0, 0). \quad (2.12)$$

In the case of the harmonic oscillator (2.3) this matrix has two imaginary eigenvalues $\lambda_1 = i\omega$ and $\lambda_2 = -i\omega$, and hence corresponds to the case of conditional (neutral) stability defined in Sect. 1.2. As a local property, it characterizes the fixed point at the origin of Fig. 2.2 as elliptic. However, since the original problem is linear, this type of stability is also *global* and implies that the solutions are oscillatory not only near the origin but in the full phase plane (q, p).

In the case of the pendulum Hamiltonian, however, the situation is very different. First of all, there are more than one fixed points. In fact, there is an infinity of them located at $(\theta_n, 0) = (\pm n\pi, 0)$, $n = 0, 1, 2, \ldots$, where $(\dot{q}, \dot{p}) = (0, 0)$ in the (q, p) phase space. The linearized equations of motion about these fixed points are:

$$\begin{pmatrix} \dot{x}_1 \\ \dot{x}_2 \end{pmatrix} = \begin{pmatrix} J_{11} & J_{12} \\ J_{21} & J_{22} \end{pmatrix} \begin{pmatrix} x_1 \\ x_2 \end{pmatrix}$$

$$= \begin{pmatrix} 0 & 1 \\ -\frac{g}{l}\cos\theta_n & 0 \end{pmatrix} \begin{pmatrix} x_1 \\ x_2 \end{pmatrix} = \begin{pmatrix} 0 & 1 \\ -\frac{g}{l}(-1)^n & 0 \end{pmatrix} \begin{pmatrix} x_1 \\ x_2 \end{pmatrix}. \quad (2.13)$$

Thus, for $n = 2k$ even (k any integer) the Jacobian matrix J has two complex conjugate (imaginary) eigenvalues $\lambda_1 = i\omega_0$, $\lambda_2 = -i\omega_0$ and corresponds to elliptic fixed points at $(2k\pi, 0)$, while for $n = 2k + 1$, the equilibria are of the *saddle type* with real eigenvalues $\lambda_1 = \omega_0$, $\lambda_2 = -\omega_0$, where $\omega_0 = \sqrt{g/l}$ (see points A, B in Fig. 2.4). Near these saddle points, the motion is governed by the two real eigenvectors corresponding to the above two eigenvalues. Along the eigenvector corresponding to $\lambda_1 > 0$, solutions of (2.13) move exponentially away from the fixed point, while along the $\lambda_2 < 0$ eigenvector they exponentially converge to the fixed point.

The important observation here is that, in the case of the simple pendulum, the solutions of the linearized equations of motion (2.13) are valid only *locally* within an infinitesimally small neighborhood of the fixed points. The reason for this is that the full equations of the system (2.8) are nonlinear and hence there is no guarantee that the linear dynamics near the fixed points is *topologically equivalent* to the solutions of the system when the nonlinear terms are included.

By topological equivalence, we are referring to the existence of a homeomorphism (i.e. a continuous and invertible map, with a continuous inverse) that maps orbits of (2.13) in a one-to-one way to orbits of (2.8), see [170, 265, 346]. How do we know that? How can we be sure that if we expand the right hand sides of (2.8) in powers of q, p, using e.g. $\sin q = q - q^3/6 + \dots$ as we did in Sect. 1.2, we will obtain solutions which are topologically equivalent to those of the linearized equations (2.13)?

The answer to this question is provided by the Hartman-Grobman theorem [170, 175, 257, 265, 346], which states that such an equivalence can be established provided all eigenvalues of the Jacobian matrix have real part different from zero. Now, as we learned from Lyapunov's theory, if all eigenvalues satisfy $\text{Re}(\lambda_i) < 0$, the equilibrium is asymptotically stable. However, if there is at least one eigenvalue whose real part is positive, the general solutions of the nonlinear system deviate exponentially away from the fixed point.

Well, this settles the issue about the saddle points at $((2k + 1)\pi, 0)$ of the simple pendulum. It does not help us, however, with the elliptic points at $(2k\pi, 0)$, since the linearized equations about them have $\text{Re}(\lambda_i) = 0$, $i = 1, 2$ and the Hartman-Grobman theorem does not apply. Here is where the integral of the motion represented by the Hamiltonian function (2.9) comes to the rescue. It establishes the existence of a family of closed curves globally, i.e. within a large region around these elliptic points.

In our problem, this region extends all the way to the *separatrix* curve S joining the *unstable manifold* of saddle point A to the *stable manifold* of saddle point B (and the unstable manifold of B with the stable one of A), as shown in Fig. 2.4. These manifolds are analytic curves which are tangent to the corresponding eigenvectors of the linearized equations and are called *invariant*, since all solutions starting on such a manifold remain on it for all $t > 0$ (or $t < 0$).

Invariant manifolds, however, do not need to join smoothly to form separatrices, as in Fig. 2.4. In fact, as we explain in the next sections, invariant manifolds of Hamiltonian systems of $N \geq 2$ dof generically intersect each other transversally (i.e. with non-zero angle) and give rise to chaotic regions in the $2N$D phase space $\mathbb{R}^{2N}$. If, on the other hand, these manifolds do join smoothly, this suggests that the system has N analytic, single-valued integrals and is hence completely integrable in the sense of the LA theorem, as in the case of the simple pendulum.

2.2   The Case of $N = 2$ Degrees of Freedom

Following the above discussion, it would be natural to extend our study to Hamiltonian systems of two dof, joining at first two harmonic oscillators, as shown in Fig. 2.5 and applying the approach of Sect. 2.1. We furthermore assume that our oscillators have equal masses $m_1 = m_2 = m$ and spring constants $k_1 = k_2 = k$ and impose fixed boundary conditions to their endpoints, as shown in Fig. 2.5. Clearly, Newton's equations of motion give in this case:

$$m\frac{d^2 q_1}{dt^2} = -kq_1 - k(q_1 - q_2), \quad m\frac{d^2 q_2}{dt^2} = -kq_2 + k(q_1 - q_2), \qquad (2.14)$$

where $q_i(t)$ are the particles' displacements from their equilibrium positions at $q_i = 0$, $i = 1, 2$. If we also introduce the momenta $p_i(t)$ of the two particles in terms of their velocities, as we did for the single harmonic oscillator, we arrive at the following fourth order system of ODEs

$$\frac{dq_1}{dt} = \frac{p_1}{m}, \quad \frac{dp_1}{dt} -kq_1 -k(q_1 - q_2), \quad \frac{dq_2}{dt} = \frac{p_2}{m}, \quad \frac{dp_2}{dt} -kq_2 +k(q_1 - q_2), \quad (2.15)$$

which is definitely Hamiltonian, since it is derived from the Hamiltonian function

$$H(q_1, p_1, q_2, p_2) = \frac{p_1^2}{2m} + \frac{p_2^2}{2m} + k\frac{q_1^2}{2} + k\frac{q_2^2}{2} + k\frac{(q_1 - q_2)^2}{2} = E \qquad (2.16)$$

(see (1.18)), expressing the integral of total energy E. Note that we may think of the two endpoints of the system as represented by positions and momenta q_0, p_0 and q_3, p_3 that vanish identically for all t.

What can we say about the solutions of this system? Are they all periodic, as in the case of the single harmonic oscillator? Is the dynamics more complicated in the four-dimensional phase space of the $q_i(t)$, $p_i(t)$, $i = 1, 2$ and in what way? Can we still use the integral (2.16) to solve the system by quadratures? Note, first of all, that, once we specify a value for the total energy E, our variables $q_i(t)$, $p_i(t)$, $i = 1, 2$ are no longer independent, as the motion actually evolves on a three-dimensional so-called *energy surface* $\Gamma(E)$. If we choose to solve (2.16) for one of them, say

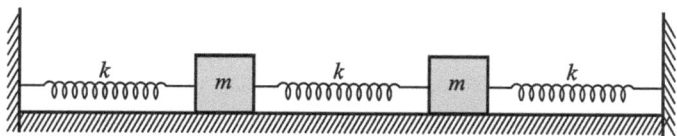

Fig. 2.5 Two coupled harmonic oscillators of equal mass m and force constant k, with displacements $q_1(t)$, $q_2(t)$ from their equilibrium position, moving horizontally on a frictionless table with their endpoints firmly attached to two immovable walls

p_1, in terms of the other three, we do not arrive immediately at an equation whose variables can be separated and thus integration by quadratures is impossible. What would happen, however, if a *second integral* of the motion were available? Could the method of quadratures lead us to the solution in that case?

Your immediate reaction, of course, would be to say no. Two integrals could only serve to express one variable, say p_1 again, in terms of two others, say q_1 and q_2, suggesting again that integration by quadratures is impossible. But, here is where appearances deceive. Note that the difficulty arises because the equations of motion (2.14), as given to us by Newtonian mechanics, are *coupled* in the variables q_1, q_2. Is it perhaps possible to use a suitable coordinate transformation to separate them and write our equations as a system of two uncoupled harmonic oscillators?

2.2.1 Coordinate Transformations and Solution by Quadratures

It is not very difficult to answer this question. Indeed, if we perform the change of variables

$$Q_1 = \frac{q_1 + q_2}{\sqrt{2}}, \quad Q_2 = \frac{q_1 - q_2}{\sqrt{2}}, \quad P_1 = \frac{p_1 + p_2}{\sqrt{2}}, \quad P_2 = \frac{p_1 - p_2}{\sqrt{2}} \qquad (2.17)$$

(the factor $1/\sqrt{2}$ will be explained later), we see that, adding and subtracting by sides the two equations in (2.14) (dividing also by m and introducing $\omega = \sqrt{k/m}$), splits the problem to two harmonic oscillators

$$\frac{dQ_i}{dt} = \frac{P_i}{m}, \quad \frac{dP_i}{dt} = -\omega_i^2 Q_i, \quad i = 1, 2, \quad \omega_1 = \omega, \quad \omega_2 = \sqrt{3}\omega, \qquad (2.18)$$

which have *different* frequencies ω_1, ω_2. Observe that the new Hamiltonian of the system becomes

$$K(Q_1, P_1, Q_2, P_2) = \frac{P_1^2}{2m} + \frac{P_2^2}{2m} + k\frac{Q_1^2}{2} + 3k\frac{Q_2^2}{2} = E. \qquad (2.19)$$

We now see that the factor $1/\sqrt{2}$ was introduced in (2.17) to make the new equations of motion (2.18) have the same *canonical form* as two uncoupled harmonic oscillators, while the new Hamiltonian K is expressed as the sum of the Hamiltonians of these oscillators. In this way, we may say that changing from q_i, p_i to Q_i, P_i variables we have performed a canonical coordinate transformation to our system.

In this framework, we immediately realize that our problem possesses not one but *two* integrals of the motion which may be thought of as the energies of the two oscillators

$$F_i(Q_1, P_1, Q_2, P_2)) = \frac{P_i^2}{2m} + k_i \frac{Q_i^2}{2} = E_i, \quad i = 1, 2, \tag{2.20}$$

with $k_1 = k$, $k_2 = 3k$, while E_i are two free parameters of the system to be fixed by the initial conditions $q_i(0)$, $p_i(0)$, $i = 1, 2$. Thus, we may now proceed to apply the method of quadratures to each of these oscillators, as described in Sect. 2.1 to obtain the general solution of the system in the form

$$Q_i(t) = \sqrt{\frac{2E_i}{k_i}} \sin(\omega_i(t + c_i)), \quad i = 1, 2, \tag{2.21}$$

c_i being the other two free parameters needed for the complete solution of the four first order ODEs (2.18). Naturally, if we wish to write our general solution in terms of the original variables of the problem, we only need to invert (2.17) to find $q_1(t)$, $q_2(t)$. We also remark that the above considerations show that the only fixed point of the system lies at the origin of phase space and is elliptic (why?).

Let us make a crucial observation at this point: Note that the successful application of the above analysis rests on the existence of two integrals of motion (2.20) associated with the free constants E_1, E_2 connected with the *amplitudes* of the corresponding oscillations. This shows that, for a system of two dof, two integrals are necessary and sufficient for its integration by quadratures, as the remaining two arbitrary constants c_1, c_2 merely represent phases of oscillation and are not as important as E_1 and E_2. To understand this better, note that c_1, c_2 are *not* single-valued and hence do not belong to the class of integrals required by the LA theorem for complete integrability.

All this suggests that it would be advisable to introduce one more transformation to the so-called *action-angle* variables I_i, θ_i, $i = 1, 2$ as follows

$$Q_i = \sqrt{\frac{2\omega_i I_i}{k_i}} \sin \theta_i, \quad P_i = \sqrt{2m\omega_i I_i} \cos \theta_i, \quad i = 1, 2 \tag{2.22}$$

and write our integrals (2.20) as $F_i = I_i \omega_i$ so that the new Hamiltonian (2.19) may be expressed in a form

$$G(I_1, \theta_1, I_2, \theta_2) = G(I_1, I_2) = I_1 \omega_1 + I_2 \omega_2 \tag{2.23}$$

that is *independent* of the angles θ_i, which also demonstrates the irrelevance of the phases of the oscillations in (2.21). Evidently, our action-angle variables also satisfy Hamilton's equations of motion

$$\frac{d\theta_i}{dt} = \frac{\partial G}{\partial I_i} = \omega_i, \quad \frac{dI_i}{dt} = -\frac{\partial G}{\partial \theta_i} = 0, \tag{2.24}$$

which imply that the new momenta I_i are constants of the motion, while the θ_i are immediately integrated to give $\theta_i = \omega_i t + \theta_{i0}$, where θ_{i0} are the two initial phases. Thus, the change to action-angle variables defines a canonical transformation and the oscillations (2.22) are called *linear normal modes* of the system.

Let us now discuss the solutions of this coupled system of linear oscillators. Using (2.17) and (2.21) we see that they are expressed, in general, as linear combinations of two trigonometric oscillations with frequencies $\omega_1 = \sqrt{k}$, $\omega_2 = \sqrt{3k}$. If these frequencies were *rationally dependent*, i.e. if their ratio were a rational number $\omega_1/\omega_2 = m_1/m_2$ (m_1, m_2 being positive integers with no common divisor) all orbits would *close* on two-dimensional *invariant tori*, like the one shown in Fig. 2.6 and the motion would be periodic (what are the m_1, m_2 of the orbit shown in Fig. 2.6?). Note that in our example, this could only happen for initial conditions such that E_1 or E_2 is zero, whence the solutions would execute in-phase or out-of-phase oscillations with frequency ω_2, or ω_1 respectively.

In the general case, though, E_1 and E_2 are both non-zero and the oscillations are *quasiperiodic*, in the sense that they result from the superposition of trigonometric terms whose frequencies are *rationally independent*, as the ratio $\omega_2/\omega_1 = \sqrt{3}$ is an irrational number. Hence, the orbits produced by these solutions in the 4-dimensional phase space are *never closed* (periodic). Unlike the orbit shown in Fig. 2.6, they never pass by the same point, covering eventually uniformly the two-dimensional torus of Fig. 2.6 specified by the values of E_1 and E_2.

Thus, in the spirit of the LA theorem we have verified for a Hamiltonian system of two coupled harmonic oscillators that two global, single-valued, analytic integrals are necessary and sufficient to completely integrate the equations of motion by quadratures and moreover that the general solutions lie on two-dimensional tori, since the motion on the four-dimensional phase space of the problem is bounded about an elliptic fixed point at the origin $q_i = p_i = 0, i = 1, 2$.

Let us now turn to the case of a system of two coupled *nonlinear* oscillators. We could choose to do so by introducing nonlinear terms in the spring forces involved in the above two linear oscillators. However, we prefer to postpone such a study for later chapters, where we will deal extensively with one-dimensional chains of $N > 2$ coupled nonlinear oscillators. Thus, we shall discuss here one of the most famous Hamiltonian two dof system, originally due to Hénon and Heiles [164]

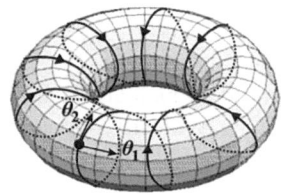

Fig. 2.6 If the ratio of the frequencies of a two dof integrable Hamiltonian system is a *rational* number, the orbits are periodic (and hence close upon themselves) on two-dimensional invariant tori, such as the one shown in the figure. However, if the ratio ω_1/ω_2 is irrational, the orbits never close and eventually cover uniformly two-dimensional tori in the four-dimensional phase space of the coordinates $q_i, p_i, i = 1, 2$

$$H = \frac{1}{2m}(p_1^2 + p_2^2) + \frac{1}{2}(\omega_1^2 q_1^2 + \omega_2^2 q_2^2) + q_1^2 q_2 - \frac{C}{3} q_2^3. \qquad (2.25)$$

This describes the motion of a star of mass m in the axisymmetric potential of a galaxy whose lowest order (quadratic) terms are those of two uncoupled harmonic oscillators and the only higher order terms present are cubic in the q_1, q_2 variables (for a most informative recent review on the applications of nonlinear dynamics and chaos to galaxy models see [99]). Note that the value of m in (2.25) can be easily scaled to unity by appropriately rescaling time, while an additional parameter before the $q_1^2 q_2$ term in (2.25) has already been removed by a similar change of scale of all variables q_i, p_i.

Thus, introducing the more convenient variables $q_1 = x$, $q_2 = y$, $p_1 = p_x$, $p_2 = p_y$, we can rewrite the above Hamiltonian in the form

$$H = \frac{1}{2}(p_x^2 + p_y^2) + V(x, y) = \frac{1}{2}(p_x^2 + p_y^2) + \frac{1}{2}(Ax^2 + By^2) + x^2 y - \frac{C}{3} y^3 = E,$$
$$(2.26)$$

where E is the total energy and we have set $\omega_1^2 = A > 0$ and $\omega_2^2 = B > 0$. It is instructive to write down Newton's equations of motion associated with this system

$$\frac{d^2 x}{dt^2} = -\frac{\partial V}{\partial x} = -Ax - 2xy, \quad \frac{d^2 y}{dt^2} = -\frac{\partial V}{\partial y} = -By - x^2 + Cy^2. \qquad (2.27)$$

Before going further, let us first write down here all possible fixed points of the system, which are easily found by setting the right hand sides in (2.27) equal to zero:

$$\text{(i)} \ (x, y) = (0, 0), \quad \text{(ii)} \ (x, y) = \left(0, \sqrt{\frac{B}{C}}\right),$$
$$(2.28)$$
$$\text{(iii)} \ (x, y) = \left(\pm \frac{\sqrt{2BA + CA^2}}{2}, -\frac{A}{2}\right)$$

(the momentum coordinates corresponding to all these points are, of course, zero). The local (or linear) stability of these equilibria will be discussed later, with reference to some very important special cases of the Hénon-Heiles problem. However, there is one of them whose character can be immediately deduced from the above equations: It is, of course, the origin (i), where all nonlinear terms in (2.27) do not contribute to the analysis. Thus, infinitesimally close to this point, the equations become identical to those of two uncoupled harmonic oscillator (since $A > 0$, $B > 0$) and therefore the equilibrium at the origin is *elliptic*.

Let us remark now that (2.26) represents a first integral of this system. If we could also find a second one, as in the problem of the two harmonic oscillators, with all the nice properties required by the LA theorem, the problem would be completely integrable and we would be able to integrate the (2.27) by quadratures and obtain the general solution. Unfortunately, in nonlinear two dof

Hamiltonian systems, the existence of such an integral is far from guaranteed. In fact, all indications for the Hénon-Heiles system is that a second integral for all A, B, C does not exist.

"How does one know that?", you may rightfully ask. Well, this is a very good question. A first answer may be given by the fact that in many cases that have been integrated numerically by many authors, even close to the equilibrium point (i), one finds, besides periodic and quasiperiodic orbits, a new kind of solution that appears "irregular" and "unpredictable", one may even say random-looking [226]! These are the orbits we have called chaotic. They tend to occupy densely three-dimensional regions in the four-dimensional phase space and depend very sensitively on initial conditions, in the sense that starting at almost any other point in their immediate vicinity produces another trajectory (also chaotic), which deviates exponentially from the original (reference) orbit as time increases.

This would not happen, of course, if the problem was completely integrable, since in that case—at least close to the elliptic equilibrium point at the origin—all phase space would be occupied by two-dimensional tori, which are uniformly covered by periodic or quasiperiodic solutions. These have nothing "irregular" or "unpredictable" about them. In fact, it can be shown that the distance between nearby orbits in their vicinity deviate linearly from each other in phase space [226].

Still, if one works only numerically, it is possible to miss chaotic orbits altogether, since their associated regions can be very thin and may not even appear in the computations. To speak about complete integrability therefore (or the absence thereof) one must have some analytical theory to support rigorously one's statements. One such theory is provided by the so-called *Painlevè analysis*, which investigates a very fundamental property of the solutions related to their *singularities in complex time* [98, 152, 279].

2.2.2 Integrability and Solvability of the Equations of Motion

It is of course very difficult to look at a pair of nonlinear coupled second order ODEs like (2.27) and try to integrate them by some clever change of variables, as we did in the case of two coupled linear oscillators (2.14). Still, the careful reader will notice (by inspection, as we say) that there is one obvious case where (2.27) can be directly integrated: $A = B$, $C = -1$. Under this condition, a transformation to sum and difference variables reduces the Hénon-Heiles equations to the simple *uncoupled* form

$$\frac{d^2 X}{dt^2} = -AX - \sqrt{2}X^2, \quad \frac{d^2 Y}{dt^2} = -AY + \sqrt{2}Y^2, \quad X = \frac{x+y}{\sqrt{2}}, \quad Y = \frac{x-y}{\sqrt{2}} \tag{2.29}$$

(see (2.17))."This is great!", you would exclaim: Now, we can multiply these equations respectively by dX/dt and dY/dt and integrate them to arrive at the following integrals of the motion

$$F_1 = \frac{1}{2}\left(\frac{\mathrm{d}X}{\mathrm{d}t}\right)^2 + \frac{A}{2}X^2 - \frac{\sqrt{2}}{3}X^3, \quad F_2 = \frac{1}{2}\left(\frac{\mathrm{d}Y}{\mathrm{d}t}\right)^2 + \frac{A}{2}Y^2 + \frac{\sqrt{2}}{3}Y^3. \quad (2.30)$$

It is easy to see now that these two integrals are *not* independent from the Hamiltonian (2.26). Indeed, if we also define the momentum coordinates $P_X = \mathrm{d}X/\mathrm{d}t$, $P_Y = \mathrm{d}Y/\mathrm{d}t$ and observe that $p_x = (P_X + P_Y)/\sqrt{2}$, $p_y = (P_X - P_Y)/\sqrt{2}$ we easily verify that $H = F_1 + F_2$ and hence that the change to new variables is a canonical transformation.

Furthermore, it is possible to prove that the two integrals (2.30) are independent *from each other*. This can be done by demonstrating that *the tangent space* spanned by the partial derivatives of F_1 with respect to X, P_X and F_2 with respect to Y, P_Y at almost all points where $F_1 = F_2$ has the full dimensionality of the phase space of the system. Moreover, the Poisson bracket $\{F_1, F_2\}$ vanishes which implies that the two integrals are *in involution* [19, 226]. We conclude, therefore, according to the LA theorem, that the system is completely integrable and hence its general solution can be obtained by quadratures.

Try it: If you go back to (2.30), solve for $\mathrm{d}X/\mathrm{d}t$ and $\mathrm{d}Y/\mathrm{d}t$ and integrate by separation of variables, as we did for the two coupled harmonic oscillators, you will be led to elliptic integrals, which are the inverse of the so-called Jacobi elliptic functions [6, 107, 169]. The most fundamental ones of them are $\mathrm{sn}(\mu t, \kappa^2)$ and $\mathrm{cn}(\mu t, \kappa^2)$ and their parameters μ, κ are related to the parameter A of our problem. When their so called *modulus* $0 \leq \kappa \leq 1$ tends to 1, the sn and cn tend to trigonometric functions (sine and cosine respectively), while $\mu \to \sqrt{A}$, which is the frequency of oscillations of our variables near the origin, see (2.29). On the other hand, when $\kappa \to 1$, the period of sn and cn tends to infinity and they become hyperbolic sinh and cosh functions (see [6, 107, 169] and Exercise 2.1).

So everything is fine with the $A = B$, $C = -1$ case of the Hénon-Heiles equations. What do we do, however, for other choices of these parameters? Can we still hope to come up with some "clever" transformations of variables, such as the ones employed above, which will help us uncouple and eventually solve the nonlinear ODEs of the problem? The answer, unfortunately, is negative. As clever as we may be, resourcefulness has its limits. Inspection alone is not sufficient to make progress here. We need to develop a more systematic approach, such as the one provided by the so called Painlevè analysis of (2.27) [98, 152, 279].

Let us, therefore, think more analytically and try to understand more about the mathematics of the integrable case we discovered above. Our motivation is the following: What if integrability (and hence also solvability) of our equations is connected with the *mathematical simplicity* of the solutions of these equations? In other words, what problems can we solve using the known mathematical functions available in the literature today? And since we already discovered elliptic functions in one example of our problem, why not try to find other integrable cases of the Hénon-Heiles system, by *requiring* that their solutions belong to the same "class" as the Jacobi elliptic functions?

This is not such a crazy question. In fact, it was first asked by the famous Russian mathematician S. Kowalevskaya in the 1880s, who was thus able to discover one more integrable case of the rotating body, in addition to the other three that had already been found by Euler and Jacobi. To this day, these are the *only four* known cases, where the rotating body can be explicitly integrated. So the moral of the story is clear: If you want to become famous, find a new integrable case of a physically important system of differential equations!

To understand what Kowalevskaya did we need to start from a remarkable mathematical property of the Jacobi elliptic functions: When their independent variable $u = \mu t$ is complex, their *only singularities* in the complex t-plane are poles. In other words, these functions are not analytic everywhere in the complex domain, and thus are more general than the ordinary exponentials and trigonometric functions of elementary mathematics. They are, however, *single-valued*, in the sense that when evaluated on a closed loop around their singularities they recover their values after every turn.

The recipe, therefore, appears rather simple: First express the solutions $x(t)$, $y(t)$ of (2.27) about a singularity at $t = t_*$ as series expansions that begin with a possible pole of order r and s respectively. This gives

$$x(t) = \sum_{k=r}^{\infty} a_k \tau^k \quad y(t) = \sum_{k=s}^{\infty} b_k \tau^k, \quad \tau = t - t_*, \tag{2.31}$$

where r, s are integers and at least one of them is *negative*. These series are assumed to converge within a finite radius $0 < |\tau| = |t - t_*| < \rho$ in the complex t-plane and thus provide the complete solution of the problem near this singularity. It follows, therefore, that among the coefficients a_k, b_k three must be arbitrary, as t_* is already one of the free constants of the solution.

Substituting these series in (2.27) and balancing the most divergent terms leads to the equations

$$r(r-1)a_r \tau^{r-2} = -2a_r b_s \tau^{r+s}, \quad s(s-1)b_s \tau^{s-2} = -a_r^2 \tau^{2r} + C b_s^2 \tau^{2s}, \tag{2.32}$$

which imply that there are *two* possible dominant behaviors (and hence two such singularity types) for this problem, namely

(i) $r = s = -2$: $a_{-2} = \pm\sqrt{18 + 9C}$, $b_{-2} = -3$

(ii) $s = -2$, $r > s$: $a_r =$ arbitrary, $b_{-2} = \dfrac{r(1-r)}{2} = \dfrac{6}{C}$. $\tag{2.33}$

Note that, in type (ii), the power r of the leading term of the expansion of $x(t)$ is directly related to the value of the parameter C. Since we have assumed that our series involve only *integer* powers of τ, the only C values that ensure this are $C = -1, -2, -6$. That's great! The first one of them is already the beginning of the thread that leads to the integrable case discovered earlier in this section by pure inspection.

What remains to be done now is entirely algorithmic and can even be programmed on the computer: Take each one of the above cases and compute the coefficients of the Laurent series of the solutions near the corresponding singularity type. As we have already explained, at most three of the a_k, b_k should be arbitrary. And here is where the most important stumbling block of the method lies: When you try to evaluate these coefficients you will discover that they satisfy a degenerate system of linear algebraic equations that is *not* homogeneous and hence is most likely *incompatible* for an arbitrary choice of the parameters A, B, C! So what do we do now?

Well, the first thing you can do is *choose* these parameter values in such a way that these algebraic equations do become compatible. You will thus be rewarded by all the possible integrable cases that this procedure can identify. When all is said and done, you will be pleasantly surprised to find that besides the case you already know about there are two new ones for which you can in principle now integrate the Hénon-Heiles equations completely. Thus all the known integrable cases of the problem can be summarized as follows:

$$\text{Case } 1: \quad A = B, \quad C = -1,$$
$$\text{Case } 2: \quad A, B \text{ free}, \quad C = -6, \qquad (2.34)$$
$$\text{Case } 3: \quad B = 16A, \quad C = -16.$$

Well, to be completely honest, Case 3 does not arise automatically from the above procedure. In fact, when you use $C = -16$ in the condition of singularity type (ii) you find that the value of the exponent r is a half integer. But that does not mean all is lost. It may be that if we switch from x to a variable $u = x^2$, this new variable may have a true Laurent series expansion with the required number of free constants and that is precisely what happens here. Thus all three cases listed in (2.34) above are successfully identified by the Painlevè analysis.

And what about $C = -2$? Well, in this case we are not so lucky. Although everything begins promisingly, we cannot find the required free constants unless we also introduce *logarithmic terms* in our series expansions. This means that the solutions of the Hénon-Heiles system for $C = -2$ contain logarithmic singularities already at leading order and hence the above method does not apply. Indeed, when we integrate numerically the equations in this case we find that they possess chaotic solutions and hence a second global, analytic and single-valued integral most likely does not exist [56].

In conclusion, we can say that if a system of ODEs has the so-called *Painlevè property*, i.e. its solutions have *only poles* as *movable* singularities, it is expected to be completely integrable, even explicitly solvable in terms of known functions. The term "movable" implies that the location of a singularity t_* is one of the free constants to be specified by the initial conditions of the problem. It serves to differentiate movable singularities from the so-called *fixed* ones, which appear explicitly in the equations of motion [107, 169].

Of course, identifying integrable cases by the Painlevè analysis only tells us where to look for N dof Hamiltonian systems whose solutions are globally ordered and predictable. It does not tell us how to find the N integrals that must exist (according to the LA theorem) and which are necessary to solve the equations of motion by quadratures. But do we really want to get into all this trouble?

First of all the integrals are not at all easy to find. Even for the simple-looking Hénon-Heiles system , the best one can do is assume that the second integral F is a polynomial in the momentum and position variables and solve for its coefficients by direct differentiation. What happens, however, if F turns out to be an infinite series? Well, in the Hénon-Heiles we are lucky. The second integral in Cases 2 and 3 of (2.34) truncates to a polynomial of the form [60, 156]

$$
\begin{aligned}
\text{Case 2}: F &= x^4 + 4x^2y^2 - 4p_x^2 y + 4xp_x p_y + 4Ax^2 y \\
&\quad + (4A - B)(p_x^2 + Ax^2), \\
\text{Case 3}: F &= 3p_x^4 + 6(A + 2y)x^2 p_x^2 - 4x^3 p_x p_y - 4x^4(Ay + y^2) \\
&\quad + 3A^2 x^4 - \frac{2}{3}x^6.
\end{aligned} \tag{2.35}
$$

Now what? If we wish to arrive at the complete solution of the problem we must somehow use (2.35) to uncouple the variables of the problem and solve two one dof systems as we did in integrable Case 1. But that is a far less trivial task in Cases 2 and 3. So, again the question relentlessly arises: Why do it?

After all, integrable cases, you may argue, are so rare, that they can hardly be expected to come up in realistic physical situations. Why worry about them? Why not try to find out instead what happens to the Hénon-Heiles for all other A, B, C values apart from the ones of (2.34)? Well, don't be impatient, we can always do that numerically, of course. Before resorting to this alternative, however, it is worth noting that integrable systems are especially useful for two important reasons: First, they frequently arise as close approximations of many physically realistic problems (the solar system being the most famous example) and second, their basic dynamical features do *not* change very much under small perturbations, as the KAM theorem [19] assures us and as we know by now from an abundance of examples.

Let's take for instance the best-studied case of the Hénon-Heiles system, which is both physically interesting and (most likely) non-integrable, $A = B = 1, C = 1$ [164, 226]

$$
H = \frac{1}{2}(p_x^2 + p_y^2) + \frac{1}{2}(x^2 + y^2) + x^2 y - \frac{1}{3}y^3 = E. \tag{2.36}
$$

Let us select a value of the total energy small enough, within the interval $0 < E < 1/6$, for which it is known that all solutions of (2.36) are bounded. To visualize these solutions, let us plot in Fig. 2.7 the intersections of the corresponding orbits with the

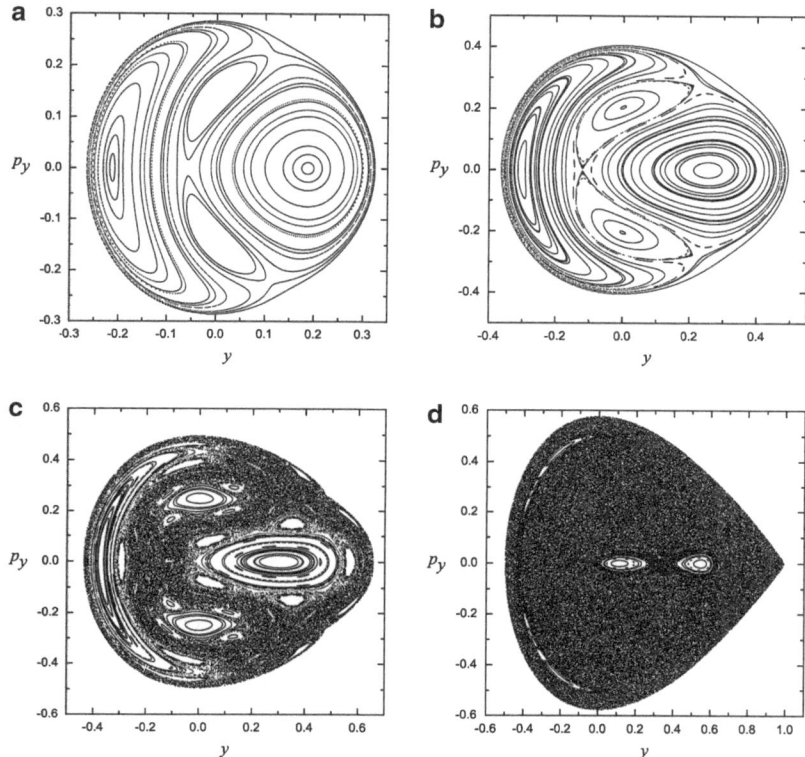

Fig. 2.7 Orbit intersections with the PSS (y, p_y) of the solutions of the Hénon-Heiles system with Hamiltonian (2.36), at times t_k, $k = 1, 2, \ldots$, where $x(t_k) = 0$, $p_x(t_k) > 0$ and for fixed values of the energy E. Note in panels (**a**) and (**b**), where $E = 1/24$ and $E = 1/12$ respectively, no chaotic orbits are visible and a second integral of motion besides (2.36) appears to exist. When we raise the energy however to $E = 1/8$ in (**c**), this is clearly seen to be an illusion. Ordered (i.e. quasiperiodic) motion is restricted within islands surrounded by chaos, whose size diminishes rapidly as the energy increases further. Thus, in (**d**) where E attains the value $E = 1/6$ above which orbits escape to infinity, the domain of chaotic motion extends over most of the available energy surface

plane (y, p_y) every time $t = t_k$, $k = 1, 2, \ldots$ at which $x(t_k) = 0$, $p_x(t_k) > 0$. This is what we call a *Poincaré surface of section* (PSS) for which we shall have a lot more to say in the remainder of this chapter. Let us not forget, after all, that it was Poincaré himself who first pointed out the complexity of the solutions of Hamiltonian systems such as (2.36) in his celebrated study of the non-integrability of the three-body problem of celestial mechanics [274].

Do you see any chaos in panel (a) of this figure, where $E = 1/24$? No. Yet, it is *there*, between the invariant curves that correspond to intersections of two-dimensional tori of quasiperiodic orbits, in the form of thin chaotic layers that are too small to be visible. To see these chaotic solutions, one would have to either increase the resolution of Fig. 2.7a or raise the energy. Let us do the latter. Observe in this way, in Fig. 2.7b, c, that we would have to reach energy values as high as

$E = 1/8$ before we begin to see widespread chaotic regions on the PSS. Note also that these regions grow rapidly as E approaches the escape energy $E = 1/6$, where chaos seems to spread over the full domain of allowed motion, while regular orbits are restricted within small "islands" of quasiperiodic motion that seem to vanish as the energy increases.

How can we better understand these dynamical phenomena? What is the degree of their complexity? Is Hamiltonian dynamics just a battle between order and chaos and all we need to do is find out who is "winning" by estimating their respective domains of "influence"? Well, even that would not be such a small accomplishment! Clearly, however, to make progress in this direction, we would have to know more about the "border" that separates order from chaos in Hamiltonian systems. And, as it often happens in mathematical physics, to understand a problem deeper, we need to analyze first in detail its simplest occurrences.

2.3 Non-autonomous One Degree of Freedom Hamiltonian Systems

Perhaps the simplest non-integrable Hamiltonian system in existence is a single periodically driven anharmonic oscillator, studied by Georg Duffing [116], a German engineer who worked on nonlinear vibrations. This oscillator is described by what is called nowadays Duffing's equation of motion [160, 205, 346]

$$\ddot{q}(t) + \omega_0^2 q(t) + \alpha q^2(t) + \beta q^3(t) + \gamma \sin \Omega t = 0, \qquad (2.37)$$

where we denote from now on time differentiation by a "dot" over the differentiated variable. In (2.37) $q(t)$ again represents the displacement of this one-dimensional oscillator from its zero position. Furthermore, we can also denote its momentum by $p = \dot{q}$ and write the above equation as a driven one dof Hamiltonian system of the form

$$\dot{q} = p = \frac{\partial H}{\partial p}, \quad \dot{p} = -\omega_0^2 q - \alpha q^2 - \beta q^3 - \gamma \sin \Omega t = -\frac{\partial H}{\partial q}, \qquad (2.38)$$

where the Hamiltonian function

$$H = H(q, p, t) = \frac{1}{2}(p^2 + \omega_0^2 q^2) + \frac{\alpha}{3} q^3 + \frac{\beta}{4} q^4 + \gamma q \sin \Omega t, \qquad (2.39)$$

for $\gamma \neq 0$ depends explicitly on t. Thus, H is no longer an integral of the motion.

Indeed, when $\gamma \neq 0$ things quickly start to get complicated. First of all, $q = p = 0$ is no longer an equilibrium position since the sinusoidal term in (2.37) will immediately drive the oscillator away from that position. In fact, *no* equilibria exist at all, since it is clearly not possible to find *any* points q_0, p_0 that remain fixed when we set $\dot{q} = \dot{p} = 0$ in these equations. More importantly, of course, the phase space of the system is *not* two-dimensional. Observe that it is no longer sufficient to specify a point $(q(t_0), p(t_0))$ in the (q, p) plane and expect to get a unique solution lying in that plane. You also need to specify exactly what time t_0 you are talking about, since the dynamics is no longer invariant under time translation, as in the cases of the autonomous Hamiltonian systems considered so far.

The motion, therefore, evolves in the three-dimensional space, q, p, t and (2.39) is sometimes referred to as a system of *one and a half* dof. In fact it is no different than a two dof Hamiltonian system, since, in the terminology of Sect. 2.1, we can represent the nonlinear oscillator in (2.37) by one action-angle pair I_1, θ_1 and think of q, p as coupled to a second (linear) oscillator whose action I_2 is fixed, while its angle coordinate is $\theta_2 = \Omega t$. Mathematically speaking, we say that the phase space is a *cylinder*, $\mathbb{R}^2 \times S^1$, where every point in the (q, p) plane is associated with an angle on the unit circle S^1.

To understand how this periodic driving in (2.37) affects the dynamics we will consider the periodic diving term as a small perturbation in the two special cases $\beta = 0$ and $\alpha = 0$ separately.

2.3.1 The Duffing Oscillator with Quadratic Nonlinearity

Let us set $\beta = 0, \alpha = -1, \omega_0 = 1$ in (2.38) and examine the solutions of the system

$$\dot{q} = p, \quad \dot{p} = -q + q^2 - \gamma \sin \Omega t \qquad (2.40)$$

for $0 \leq \gamma \ll 1$. When $\gamma = 0$ it is easy to see that there are two equilibria: (1) $q = p = 0$, (2) $q = 1, p = 0$. Linearizing the equations of motion about them, as described in Sect. 2.1, we easily find that (1) is a (stable) elliptic point and (2) is an (unstable) saddle. On the other hand, the dynamics of this one dof system in the full phase plane (q, p) is described by the family of curves $H = H(q, p) = (1/2)(p^2 + q^2) - (1/3)q^3 = E$ parameterized by the value of the energy E, as shown in Fig. 2.8a.

Observe that, as in the case of the simple pendulum, here also a region of oscillations exists around the origin, which extends all the way to the saddle at $(1, 0)$ and is separated from the outside part of the phase plane (where all solutions escape to infinity) by an invariant curve S called the separatrix. The only difference is that in the pendulum problem S joins two different saddles and represents a so-called *heteroclinic* solution of the equations, while in the quadratic oscillator of (2.40), S joins $(1, 0)$ to itself and is called *homoclinic* solution (these terms come from the Greek words "cline" meaning "bed" and "hetero" and "homo" which mean, of course, "different" and "same" respectively). These invariant curves, and the homoclinic (or heteroclinic) orbits that lie on them, constitute perhaps the single most important object of study in nonlinear Hamiltonian dynamics.

Let us try to understand why: As Poincaré first pointed out, when $\gamma \neq 0$, the elliptic point at $(0, 0)$ and the saddle point at $(1, 0)$ in Fig. 2.8a shift to slightly different locations P and Q on the PSS

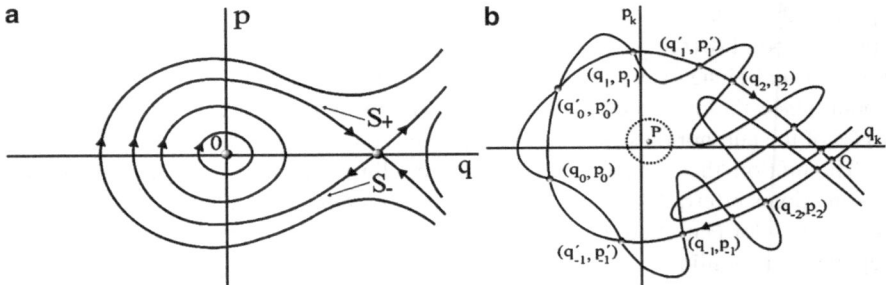

Fig. 2.8 (a) The (q, p) phase plane of the quadratic oscillator (2.40) in the $\gamma = 0$ case. Note the elliptic point at the origin and the saddle point at $(1, 0)$. The separatrix S here is a single invariant curve described by a homoclinic orbit joining the unstable and stable manifolds of the saddle at $(1, 0)$. (b) When $\gamma \neq 0$, however, these manifolds no longer join smoothly but intersect each other infinitely often on a PSS $\Sigma_{t_0} = (q(t_k), p(t_k))$, $t_k = kT + t_0, k = 1, 2, \ldots$, where $T = 2\pi/\Omega$ is the period of the forcing term

$$\Sigma_{t_0} = \{q(t_k), p(t_k)), \quad t_k = kT + t_0, \ k = \pm 1, \pm 2, \ldots\}, \quad T = 2\pi/\Omega \quad (2.41)$$

(see Fig. 2.8b). These are no longer fixed points of the differential equations, but have become intersections of $T(= 2\pi/\Omega)$-periodic orbits of the system (2.40), whose solutions evolve now in the three-dimensional space $\mathbb{R}^2 \times S^1$, as explained above [160, 226].

In addition, the stable and unstable manifolds S_+ and S_- of the point Q intersect transversally at a point (q_0, p_0) in Fig. 2.8b, lying furthest away from Q. This implies the simultaneous occurrence of a double infinity of such points, denoted by (q_k, p_k), $k = \pm 1, \pm 2, \ldots$, with $(q_k, p_k) \to P$ as $k \to \pm\infty$, which constitute in fact *two distinct* homoclinic orbits, one with $k = 0, \pm 2, \ldots, \pm 2m, \ldots$ and one with $k = \pm 1, \ldots, \pm(2m + 1), \ldots$, m being a positive integer.

The reason for this is that the time evolution of the system on the PSS (2.41) is more conveniently described by the Poincaré map [170, 265, 346]

$$P_{t_0} : \Sigma_{t_0} \to \Sigma_{t_0}, \quad t_0 \in (0, 2\pi/\Omega), \quad (2.42)$$

defined by following the trajectories of the systems and monitoring their intersections with Σ_{t_0}. P and Q are respectively an elliptic and a saddle fixed point of this map, since they satisfy $P_{t_0}(P) = P$ and $P_{t_0}(Q) = Q$. Most importantly, the map (2.42) is a *a diffeomorphism*, i.e. it is continuous and at (least once) continuously differentiable with respect to the phase space variables q, p. It is also *invertible*, i.e. $P_{t_0}^{-1}$ exists and is also a diffeomorphism. This means that it maps, forward and backward in time, neighborhoods of the PSS that exhibit diffeomorphically equivalent dynamics. Thus, an intersection point of the invariant manifolds is mapped by P_{t_0} (or $P_{t_0}^{-1}$) to the next (or previous) such intersection point, provided the manifolds in their neighborhoods have the *same orientation*.

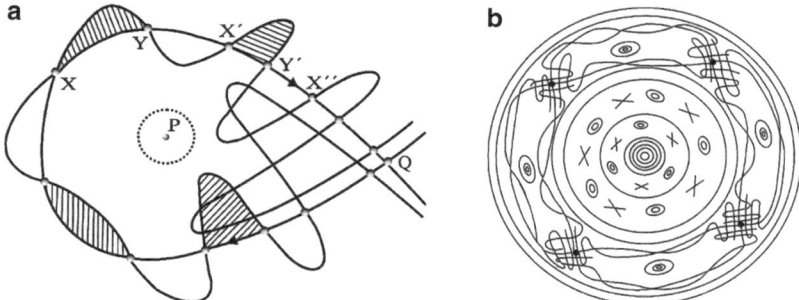

Fig. 2.9 (**a**) The successive action of (2.42) maps each shaded lobe of the figure to the next one, closer to the point Q, thus taking points from the region "inside" the invariant manifolds to the region "outside" and eventually to infinity. (**b**) Schematic picture of the dynamics around the elliptic fixed point of the Poincaré map (2.42), on the surface of section Σ_{t_0}, when $\gamma \neq 0$

This explains why, for example, as the homoclinic point X is mapped to X', X'', etc. (and its companion Y is mapped to Y', Y'', etc.) in Fig. 2.9a, a shaded lobe is mapped to the next one closer to the saddle point Q, by successive applications of the map P_{t_0}. This results in the *transport* of points which were once "inside" the manifolds S_+, S_- to the region "outside". On the other hand, as Fig. 2.9a also shows, there are corresponding lobes which were originally "outside" and are now mapped "inside" the domain of the elliptic point at P. Furthermore, the Hamiltonian nature of the problem implies, according to Liouville's theorem [226, 265, 346], that the system is conservative in the sense that it preserves phase space volume. The consequence of this is that the areas of the lobes shown in Fig. 2.9 are equal to each other.

Meanwhile, in the *homoclinic tangle* formed by the intersecting manifolds near the saddle point Q there is a wealth of dynamical phenomena about which a lot is known. Besides the homoclinic orbits, there is a countable infinity of unstable periodic solutions, while there also exist invariant Cantor sets, for which it can be rigorously proved by the Smale horseshoe theory [160, 265, 346] that any two nearby points belonging to this set lead to orbits that separate from each other exponentially fast in phase space.

Of course, if instead of one saddle point our Poincaré map had two (see next subsection) the intersecting manifolds would give rise to *heteroclinic* orbits, in the same way as described above, with lobes of equal areas transporting points from "inside" the invariant manifolds "outside" and vice versa. Heteroclinic tangles would thus be formed, where invariant sets of chaotic orbits can be similarly proved to exist. The important observation is that this does not happen only near saddle fixed points of the Poincaré map (2.42). It also happens, at smaller scales, near every unstable m-periodic orbit of the system of period $T_m = mT$, $m = 2, 3, \ldots$, intersecting the PSS (2.41) at m points (see Fig. 2.9b).

As we know from the KAM theorem, "most" (in the sense of positive measure) invariant curves around the origin of (2.40) at $\gamma = 0$ are preserved for sufficiently small $\gamma \neq 0$. Furthermore, a very important theorem by Birkhoff [226, 346] also assures us that the invariant curves of the $\gamma = 0$ case corresponding to periodic orbits break up into an even number of stable and unstable m-periodic orbits, which appear on the surface of section as *elliptic* and *saddle fixed points* respectively of the mth power of the Poincaré map $P_{t_0}^m$.

This is schematically shown in Fig. 2.9b. Are you impressed? You should be. This is indeed a wonderful sign of a complex phenomenon, as it rigorously demonstrates that there are chaotic regions around saddle points present at all scales! Around every elliptic fixed point of *any power* of the Poincaré map (hence around every small island shown in Fig. 2.9b), there are smaller islands and around them even smaller ones etc., with saddles and tangles of heteroclinic chaos in between down to infinitesimally small scales!

Did you expect all this to happen, as soon as we allowed $\gamma \neq 0$? Clearly, turning on the driving force in (2.40) was not such an innocent move. The Hamiltonian ceased to be an integral and the dimensionality of the motion was increased to three, where an amazing variety of new phenomena are possible, the most important of them being the occurrence of chaotic orbits.

2.3.2   The Duffing Oscillator with Cubic Nonlinearity

Let us now repeat the above exercise for the Duffing oscillator (2.37) with only cubic nonlinearity. Setting $\alpha = 0$, the equation of motion becomes

$$\ddot{q}(t) = -\delta q(t) - \beta q^3(t) + \gamma \sin \Omega t, \tag{2.43}$$

where we have replaced the parameter before the linear term by δ, so as to be able to consider also the case $\delta < 0$. Although the phenomena we encounter here are qualitatively the same as in the example of the quadratic Duffing oscillator, we will have the chance to exploit the additional symmetry of the dynamics under reflection about the axis $q = 0$ (when $\gamma = 0$). This is relevant in many physical systems (like vibrating structures fixed on a horizontal uniform floor), where motions to the right or left are equally favorable. With (2.43), we also have the opportunity to discuss a case that has many common features with the pendulum problem and examine the interesting dynamics of the so-called double-well potential that arises frequently in many problems of physics.

Let us begin by considering the case $\delta = -\beta = 1$, where the above system is written in q, p phase space coordinates as

$$\dot{q} = p, \quad \dot{p} = -q + q^3 + \gamma \sin \Omega t. \tag{2.44}$$

Evidently, in the unforced case $\gamma = 0$, this anharmonic oscillator possesses three fixed points: (1) An elliptic one, $P = (0,0)$ and two saddle points (2) $Q_1 = (-1,0)$ and (3) $Q_2 = (1,0)$. The phase space plot is thus very similar with that of the pendulum (see Fig. 2.4), only in the case of the pendulum $Q_1 = (-\pi,0)$, $Q_2 =$

$(\pi, 0)$ and the picture is repeated about every elliptic point at $(2m\pi, 0)$ and its two saddles at $((2m \pm 1)\pi, 0)$.

Note that our separatrices here represent heteroclinic orbits joining the saddle points at $(\pm 1, 0)$. It is not difficult to show (see Exercise 2.3) that, in the unforced case, both cubic and quadratic oscillators (2.40), (2.44), possess the Painlevè property and are explicitly solvable in terms of Jacobi elliptic functions. This means that their solutions along these separatrices are given in terms of simple hyperbolic functions. We thus have the opportunity to apply to these solutions the so-called *Mel'nikov theory* [160,346], which allows us to study exactly what happens to these heteroclinic orbits when $\gamma \neq 0$. In Problem 2.3 we ask you to carry out such a calculation, following the steps outlined below.

First of all, recall that the saddle points of our cubic oscillator (just as in the case of the quadratic oscillator) are unstable periodic orbits of (2.44) with period $T = 2\pi/\Omega$, which persist on the PSS (2.41) as saddle fixed points of the Poincaré map (2.42). These points continue to possess stable and unstable manifolds, which no longer join smoothly, as in the unforced case. Mel'nikov's approach works within the framework of perturbation theory for $|\gamma| \ll 1$ and yields expressions for the heteroclinic solutions of the perturbed problem, both for the stable as well as unstable manifolds of these saddle fixed points, as series expansions in powers of γ.

We thus obtain *different* expansions which are uniformly valid in time intervals $(-\infty, t_1)$, for the unstable manifold of the Q_1 point and (t_2, ∞) for the stable manifold of the point Q_2, where $t_2 > t_1$. Examining these solutions at times $t_0 \in (t_1, t_2)$ where both series are valid, Mel'nikov's theory shows, *to first order in* γ, that the points at which the corresponding manifolds intersect correspond to the roots of the function

$$M(t_0) = \int_{-\infty}^{\infty} (f_1 g_2 - f_2 g_1)(\hat{x}(t - t_0), t)dt, \qquad (2.45)$$

which is proportional to the *distance* between the two manifolds on the PSS (2.41), as a function of t_0 (see Problem 2.3). To derive (2.45) we have written our equations of motion in the form

$$\dot{q} = f_1(q, p) + \gamma g_1(q, p, t), \quad \dot{q} = f_2(q, p) + \gamma g_2(q, p, t),$$
$$g_i(q, p, t) = g_i(q, p, t + T), \qquad (2.46)$$

$i = 1, 2$, while $\hat{x}(t - t_0) = (\hat{q}(t - t_0), \hat{p}(t - t_0))$ represents the exact heteroclinic (or homoclinic) solution of the unforced $\gamma = 0$ case.

This is fantastic! It means that we can now study analytically what happens to these stable and unstable manifolds as one turns on the perturbation of the periodic forcing in the above systems. Indeed, based on our knowledge of the exact solution on these manifolds for $\gamma = 0$ we can verify, as Poincaré predicted, that the Melnikov integral (2.45) is an *oscillatory* function of t_0 with infinitely many roots, $M(t_{0i}) = 0$, $i = 0, \pm 1, \pm 2, \ldots$. These correspond to the intersection points of the manifolds at which the heteroclinic (or homoclinic) orbits of the perturbed system are located.

Let us see how all this works in the case of the cubic oscillator (2.44). As the reader can easily verify, for $\gamma = 0$, the separatrices (or heteroclinic solutions) of this problem, $S_{\pm}$, are given by the hyperbolic functions

$$\hat{\mathbf{x}}(t - t_0) = (\hat{q}(t - t_0), \hat{p}(t - t_0)) = \pm \left(\tanh \left(\frac{1}{\sqrt{2}}(t - t_0) \right), \frac{1}{\sqrt{2}} \operatorname{sech}^2(t - t_0) \right). \tag{2.47}$$

If we use the above expressions to substitute in (2.45) we obtain an integral

$$M(t_0) = \int_{-\infty}^{\infty} \hat{p}(t - t_0) \sin \Omega t \, dt = \frac{1}{\sqrt{2}} \int_{-\infty}^{\infty} \operatorname{sech}^2 t \sin \Omega(t + t_0) dt, \tag{2.48}$$

that is not elementary. Its evaluation requires that we extend the integration to a strip in the complex t-plane and use Cauchy's residue theorem (see Problem 2.3). The result is a simple formula

$$M(t_0) = \frac{\pi \Omega / 2}{\sinh(\pi \Omega / 2)} \sin \Omega t_0, \tag{2.49}$$

which, when multiplied by γ and divided by the magnitude of the unperturbed vector field $\sqrt{f_1^2(\tau) + f_2^2(\tau)}$ provides an estimate of the *distance* between the stable and unstable manifolds at $t = t_0 + \tau$ (compare with the numerical results of Exercise 2.4). The above formulas demonstrate that $M(t_0)$ has infinitely many zeros (due to the sine function in (2.49)), while the amplitude of its oscillations *grows exponentially* as τ increases and we move closer to the saddle points Q_1, Q_2.

It is important to note that Mel'nikov's theory does *not* require that the vector fields $\hat{\mathbf{f}} = (f_1, f_2)$ or $\hat{\mathbf{g}} = (g_1, g_2)$ be Hamiltonian! All it demands is that the unperturbed system possesses a solution $\hat{\mathbf{x}}(t)$ that is either a homoclinic or heteroclinic orbit. In particular, if our Duffing oscillators involve a small *dissipative* term of the form $-\eta p$, say, with $\eta > 0$ of the same order as γ, we can include it in the periodic perturbation $\hat{\mathbf{g}} = (g_1, g_2)$ and study its crucial role in *inhibiting* the intersection of the invariant manifolds. Indeed, it can be shown that each η provides a threshold that γ must exceed for Smale horseshoe chaos to occur in the system! To understand exactly how this important effect can be studied using Mel'nikov's theory, the reader is encouraged to solve Problem 2.4.

In fact, Mel'nikov's theory does not only apply to perturbations of planar Hamiltonian systems. It can be extended to dynamical systems of arbitrary dimension, $n \geq 2$, which can be written in the form

$$\dot{\mathbf{x}}(t) = \mathbf{f}(\mathbf{x}(t)) + \gamma \mathbf{g}(\mathbf{x}, t), \quad \mathbf{g}(\mathbf{x}, t) = \mathbf{g}(\mathbf{x}, t + T), \tag{2.50}$$

for $\mathbf{x}(t) = (x_1(t), \ldots, x_n(t))$ and γ a sufficiently small real parameter. In the particular case where the unperturbed equations are derived from an integrable

Hamiltonian system, possessing n independent integrals in involution F_i, $i =$ $1, 2, \ldots, n$, we obtain a Mel'nikov vector function, $\mathbf{M}(t_0, \nu) = (M_1, \ldots, M_n)$, whose components are given by [90, 287, 288]

$$M_i(t_0, \nu) = \int_{-\infty}^{\infty} (\nabla F_i(\hat{\mathbf{x}}(t, \nu)), \mathbf{g}(\hat{\mathbf{x}}(t, \nu), t - t_0)) dt, \tag{2.51}$$

where $(,)$ denotes the inner product and ν is a (generally) n-dimensional parameter entering in the homoclinic (or heteroclinic) solution $\hat{\mathbf{x}}(t, \nu)$). Now, however, matters are more complicated, as we no longer have the geometric visualization of the invariant manifolds and the distance between them that was available in the $n = 2$ case. Thus, we shall not proceed any further with the discussion of Mel'nikov's analysis and refer the interested reader to the literature for more details [90, 287, 288, 346].

Exercises

Exercise 2.1. (a) Perform the variable transformation $x = \sin(\theta/2)$ to the equation of motion of the simple pendulum (2.7) and show that this ODE becomes

$$\ddot{x} = -\left(\omega^2 + \frac{E}{2}\right) x + 2\omega^2 x^3, \tag{2.52}$$

where $\omega = \sqrt{g/l}$. Using the theory of Jacobi elliptic functions [6, 107, 169], prove that the solution of this ODE can be expressed as the Jacobi elliptic sine function

$$x(t) = C \operatorname{sn}(\mu(t - t_0), \kappa), \quad \mu^2(1 + \kappa^2) = \omega^2 + \frac{E}{2}, \quad \kappa^2 \mu^2 = \omega^2 C^2, \tag{2.53}$$

which corresponds to the initial conditions $x(t_0) = 0$, $\dot{x}(t_0) = \mu C$, while the energy equation gives $2\mu^2 C^2 = E$. Eliminating C, μ from these equations, obtain an expression relating the modulus κ of the elliptic sine function to the total energy E and show that in the limit $E \to 0$, $\kappa \to 0$, and $x(t)$ becomes the usual trigonometric sine function. Prove also that in the limit $\kappa \to 1$ the solution on the separatrix of Fig. 2.4 reduces to the hyperbolic tangent function.

(b) Show that in the oscillatory domain the period T of the pendulum is given in terms of the elliptic integral of the first kind

$$K(\kappa^2) = \int_0^{\pi/2} \frac{d\phi}{\sqrt{1 - \kappa^2 \sin^2 \phi}}, \tag{2.54}$$

and use this result to evaluate the first 2–3 terms in an expansion of T in powers of κ^2. Finally, use the Fourier representation of the elliptic sine function in terms of trigonometric sines to plot $\operatorname{sn}(u, \kappa^2)$ as a function of its argument u for several values of the modulus κ^2 (see [6, 107, 169]).

Exercise 2.2. Consider a system of two coupled harmonic oscillators of the type considered in Sect. 2.2, Fig. 2.5. Call $k = k_1$ the constant of the springs by which the two masses are tied to the walls and $k = k_2$ the constant of the spring that connects them to each other. Find values k_1, k_2 such that the general solution of the system is periodic and determine the period in every case.

Exercise 2.3. Show that the unforced quadratic and cubic Duffing oscillators (2.40), (2.44) (for $\delta = \pm 1$) possess the Painlevè property and solve the two problems completely in terms of Jacobi elliptic functions. Find explicit expressions for the solutions in all parts of the phase plane corresponding to regions of bounded and unbounded motion, as well as on the separatrices between these regions.

Exercise 2.4. Consider a Duffing cubic oscillator on which periodic forcing is applied parametrically as follows

$$\dot{q} = p, \quad \dot{p} = -q(1 + \gamma \sin(\Omega t)) + q^3. \tag{2.55}$$

This is a case where the spring constant (or the length of an associated pendulum model) is taken to vary sinusoidally about its constant value. Note that the elliptic point at the origin and the two saddle points at $(\pm 1, 0)$ remain at their place when $\gamma \neq 0$.

(a) Linearize (2.55) about the points $(\pm 1, 0)$ and obtain the corresponding linear stable and unstable manifolds $E_s^\pm$, $E_u^\pm$. Keeping $\gamma = 0$ locate points on these manifolds very close to the fixed points, which if propagated forward (or backward) in time by (2.55) will accurately trace the two separatrices $S_\pm$ representing the upper and lower heteroclinic orbits joining the two saddles.

(b) When $\gamma \neq 0$, construct the associated 2×2 linearized Poincaré map L (and its inverse L^{-1}) by solving numerically (2.55) forward and backward in time and plotting points very close to $(\pm 1, 0)$ on the PSS Σ_{t_0} at $t = \pm 2\pi/\Omega$. Evaluate the linear stable and unstable manifolds $E_s^\pm$, $E_u^\pm$ of the 2×2 matrices L and L^{-1}.

(c) Place many points on these linear manifolds very close to $(\pm 1, 0)$ and propagate them forward (and backward) in time by solving numerically (2.55) and plotting their iterates at $t = 2m\pi/\Omega$, $m = \pm 1, \pm 2, \ldots$. Thus show that the corresponding nonlinear manifolds no longer join smoothly but intersect each other repeatedly forming heteroclinic tangles, within which dense sets of chaotic orbits can be found according to Smale horseshoe dynamics.

Problems

Problem 2.1. Perform all the calculations described in Sect. 2.2 to derive and study the general solution of the Hénon-Heiles problem in the integrable case $A = B$, $C = -1$. Show that the two integrals in (2.30) are independent and in involution. Plot the invariant curves in the (X, P_X) and (Y, P_Y) surfaces of section and relate them to the exact solutions of the problem given in terms of Jacobi elliptic functions (see Exercise 2.1). Discuss the dynamics of the system for all values of the energy $E > 0$.

Problem 2.2. Complete all steps of the singularity analysis in the complex t-plane and show that the only cases of the Hénon-Heiles Hamiltonian system possessing the Painlevè property are the ones listed in (2.34).

Problem 2.3. Use Mel'nikov's theory as described in [160, 346] to derive an expression for the distance between stable and unstable manifolds as a function of the Mel'nikov integral and the magnitude of the unperturbed vector field along the invariant manifolds. Apply this theory to the case of the cubic Duffing oscillator (2.44) as outlined in Sect. 2.3.2, evaluate the corresponding Melnikov integral and prove the result shown in (2.49). Repeat this analysis for the parametrically driven Duffing oscillator (2.55) and compare the results.

Problem 2.4. Study the cubic Duffing oscillator (2.43) in the case $\delta = -\beta = -1$, which describes a physical problem with a double well potential. Assume furthermore that the periodic perturbation is of the form

$$\dot{q} = p, \quad \dot{p} = q - q^3 + \varepsilon(-\eta p + \gamma \sin \Omega t), \qquad (2.56)$$

where $-\eta p$ (with $\eta > 0$) is a term introducing dissipation. Draw the phase plane curves of the $\varepsilon = 0$ case and show that the origin is a saddle fixed point whose stable and unstable manifolds are joined by two symmetric homoclinic solutions "embracing" two elliptic points at $(0, \pm 1)$. Apply Mel'nikov's theory to show that for $\varepsilon \neq 0$ these manifolds split into homoclinic orbits when η and γ satisfy a certain inequality. Now solve (2.56) numerically, using the approach described in Exercise 2.4, to examine the validity of the inequality you derived for $0 < \varepsilon \ll 1$. How accurate are its predictions for the appearance of horseshoe chaos in the problem?

Chapter 3
Local and Global Stability of Motion

Abstract In this chapter, we discuss in a unified way equilibrium points, periodic orbits and their stability, which constitute *local* concepts of Hamiltonian dynamics together with ordered and chaotic motion, which are the concern of a more global type of analysis. Using the Fermi Pasta Ulam β (FPU$-\beta$) model as an example, we study the destabilization properties of some of its simple periodic orbits and connect them to wider aspects of the motion in phase space. We show numerically that the spectrum of positive Lyapunov exponents, characterizing a region of strong chaos, becomes invariant in the thermodynamic limit, verifying that the Kolmogorov-Sinai entropy, defined as the sum of these exponents, is an extensive property of the system. The chapter ends with an introduction of the smaller alignment index (SALI) and the generalized alignment index (GALI) criteria for distinguishing ordered from chaotic motion.

3.1 Equilibrium Points, Periodic Orbits and Local Stability

3.1.1 Equilibrium Points

Let us note first that Hamilton's equations of motion (1.18) can be written more compactly in the form

$$\dot{\mathbf{x}} = \frac{d\mathbf{x}}{dt} = \begin{pmatrix} 0_N & I_N \\ -I_N & 0_N \end{pmatrix} \nabla H(\mathbf{x}) = \Omega \nabla H(\mathbf{x}), \quad \mathbf{x} = (\mathbf{q}, \mathbf{p}), \qquad (3.1)$$

where I_N and 0_N denote the $N \times N$ identity and zero matrices respectively. This is not just a convenient notation. It introduces, in fact, the important matrix Ω, which is fundamental in establishing the *symplectic structure* of Hamiltonian dynamics. First of all, it enjoys a number of important properties:

(i) $\Omega^{\mathrm{T}} = -\Omega$ (antisymmetry), (ii) $\Omega^{\mathrm{T}} = \Omega^{-1}$ (orthogonality), (iii) $\Omega^{-1} = -\Omega$,
$$(3.2)$$

T. Bountis and H. Skokos, *Complex Hamiltonian Dynamics*,
Springer Series in Synergetics, DOI 10.1007/978-3-642-27305-6_3,
© Springer-Verlag Berlin Heidelberg 2012

based on which one defines the *group of symplectic $n \times n$ matrices M* as those that satisfy the condition

$$M = M\Omega M^{\mathrm{T}} \quad \Rightarrow \quad M^{\mathrm{T}} = -\Omega M^{-1}\Omega, \quad M^{-1} = \Omega M^{\mathrm{T}}\Omega^{\mathrm{T}}, \qquad (3.3)$$

with superscripts T and -1 denoting the transpose and inverse of a matrix respectively. Using the definition of Ω in (3.1), and its properties listed in (3.2) and (3.3) it is easy to show that $\det \Omega = 1$ and $\det M = \pm 1$. In fact, it can be proved that the determinant of a symplectic matrix is exactly 1, i.e. $\det M = 1$ (see Exercise 3.1).

As we discussed in Chap. 1, the simplest solutions of a Hamiltonian system are its equilibrium (or fixed points) $(\bar{\mathbf{q}}, \bar{\mathbf{p}})$, at which the right hand sides of Hamilton's equations vanish,

$$\frac{\partial H(\bar{\mathbf{q}}, \bar{\mathbf{p}})}{\partial \mathbf{p}_k} = 0, \quad \frac{\partial H(\bar{\mathbf{q}}, \bar{\mathbf{p}})}{\partial q_k} = 0, \ k = 1, 2, \ldots N. \qquad (3.4)$$

Thus, given a Hamiltonian system our first task is to find all its fixed points solving the nonlinear equations (3.4). Next, we need to examine the dynamics near each one of these points.

To do this we write the solutions of (3.1) as small deviations about one of the fixed points as

$$\mathbf{x}(t) = (\bar{\mathbf{q}}, \bar{\mathbf{p}}) + \boldsymbol{\xi}(t), \quad ||\boldsymbol{\xi}(t)|| \ll \varepsilon, \qquad (3.5)$$

(where ε is the maximum of the Euclidean norms $||\bar{\mathbf{q}}||, ||\bar{\mathbf{p}}||$), substitute in (3.1) and linearize Hamilton's equations about this point to obtain

$$\dot{\boldsymbol{\xi}}(t) = A(\bar{\mathbf{q}}, \bar{\mathbf{p}})\boldsymbol{\xi}(t), \qquad (3.6)$$

where higher order terms in $\boldsymbol{\xi}$ have been omitted due to the smallness of the norm of $\boldsymbol{\xi}(t)$ noted in (3.5). The constant matrix A represents the Jacobian of $\Omega \nabla H(\mathbf{x})$ evaluated at the fixed point, as explained in (2.10) and (2.11). The important observation here is that A can be written as the product $A = \Omega S$, where $S = S^{\mathrm{T}}$ is a symmetric matrix. Thus, it can be easily shown that $M = exp(A)$ is symplectic and A is called a *Hamiltonian matrix*.

Clearly, the solutions of the linear system (3.6) will determine the local stability character of $(\bar{\mathbf{q}}, \bar{\mathbf{p}})$ by telling us what kind of dynamics occurs in the vicinity of this equilibrium point. As explained in Chap. 1, we shall call this fixed point *linearly stable* if all the solutions of (3.6) are bounded for all t. To find out under what conditions this is true, let us write the general solution of this system as

$$\boldsymbol{\xi}(t) = \mathrm{e}^{At}\boldsymbol{\xi}(0) = X(t)\boldsymbol{\xi}(0), \qquad (3.7)$$

where $X(t)$ is called the *fundamental matrix* of solutions of (3.6), with $X(0) = I_{2N}$, the identity matrix. Clearly, the boundedness properties of these solutions depend on the eigenvalues of the Hamiltonian matrix A. What do we know about these eigenvalues? Many things, it turns out.

Observe that starting from the characteristic equation they satisfy $\det(A - \mu I) = 0$ and using the properties of symmetric matrices and Ω

$$\det(A - \mu I) = \det(\Omega S - \mu I) = \det(\Omega S - \mu I)^{\mathrm{T}} = \det(S^{\mathrm{T}}\Omega^{\mathrm{T}} - \mu I) =$$

$$\det(-S\Omega - \mu I) = (-1)^{2N}\det(S\Omega + \mu I) = \det[\Omega(S\Omega + \mu I)\Omega^{-1}] =$$

$$\det(\Omega S + \mu I) = \det(A + \mu I) = 0, \quad (3.8)$$

we discover that if μ is an eigenvalue of A so is $-\mu$. This implies that Hamiltonian systems can never have asymptotically stable (or unstable) fixed points. We, therefore, conclude that a necessary and sufficient condition for such a equilibrium point to be stable is that all the eigenvalues of its corresponding matrix A have zero real part! And since A is a real matrix, if it possesses an eigenvalue $\mu = \alpha + i\beta$ (with α, β real), it will also have among its eigenvalues: $\mu = \alpha - i\beta$, $\mu = -\alpha + i\beta$ $\mu = -\alpha - i\beta$.

This is great! Now we understand why in all the $N = 1$, $N = 2$ dof Hamiltonian systems studied in Chap. 2, all stable fixed points have A matrices with purely imaginary eigenvalues and the solutions in their neighborhood execute simple harmonic motion. Furthermore, we also realize that a stable fixed point of a Hamiltonian system can become unstable by two kinds of *bifurcations*:

1. A pair of imaginary ($\pm i\beta$) eigenvalues splitting *on the real axis* into an (α, $-\alpha$) pair, or
2. An eigenvalue pair ($\pm i\beta$) splitting into four eigenvalues ($\pm\alpha \pm i\beta$) in the complex plane, in a type of *complex instability*.

Bifurcation (1) leads to an equilibrium of the *saddle* type, since (α, $-\alpha$) correspond to two real eigenvectors, along which the solutions of (3.6) converge or diverge exponentially from the fixed point. These eigenvectors identify the so-called stable and unstable *euclidean* manifolds, respectively, E^s, E^u, of the fixed point. When continued under the action of the full nonlinear equations of motion (3.1) these become the stable and unstable invariant manifolds W^s, W^u, which may intersect each other (or invariant manifolds of other saddle fixed points) and cause the horseshoe type of homoclinic (or heteroclinic) chaos we already encountered in the examples of Chap. 2.

By contrast, bifurcation (2) occurs more rarely because it requires that the four imaginary eigenvalues of a stable equilibrium point, ($\pm i\beta_1, \pm i\beta_2$), be *degenerate*, i.e. $\beta_1 = \beta_2$. It also does not arise in Hamiltonian systems of $N = 1$ dof (why?). As we will see in all the examples of Hamiltonian lattices analyzed in this book, bifurcation (1) is a lot more common and will thus quite appear frequently in the pages that follow.

At this point it is of crucial importance to make the following remark: Regarding the stable and unstable manifolds of a saddle point, there are important theorems, which establish that the E^s and E^u are *tangent* to W^s and W^u at the fixed point and have respectively the same number of dimensions [175, 257]. This fact is of great practical importance! It allows us to 'trace out' numerically with great accuracy the W^s and W^u, starting with many initial conditions distributed on the E^s and E^u very close to the fixed point and integrating (3.1) forward in time to obtain W^u and backward to obtain W^s. As we will find out in Chap. 7, this strategy will prove extremely useful when we need to construct discrete breather solutions using the (homoclinic or heteroclinic) intersections of W^s and W^u invariant manifolds of invertible maps.

3.1.2 Periodic Orbits

Enough said of equilibrium points. It is time now to discuss the next most important type of solution of Hamiltonian systems, which is their periodic orbits. You might expect, of course, the mathematics here to become more involved and you would be right. However, as we will soon find out, the wonderful instrument of the Poincaré map and its associated surfaces of section will come to the rescue and make the analysis a lot easier. Let us begin by giving a more general definition of the Poincaré map than the one we used in Chap. 2.

In particular, we will assume that our n-dimensional dynamical system, cast in the general form $\dot{\mathbf{x}} = \mathbf{f}(\mathbf{x})$ (see (1.1)) has a periodic solution $\hat{\mathbf{x}}(t) = \hat{\mathbf{x}}(t + T)$ of period T. Let us choose an arbitrary point along this orbit $\hat{\mathbf{x}}(t_0)$ and define a PSS at that point as follows

$$\Sigma_{t_0} = \{\mathbf{x}(t) \ / \ (\mathbf{x}(t) - \hat{\mathbf{x}}(t_0)) \cdot \mathbf{f}(\hat{\mathbf{x}}(t_0)) = 0\} . \tag{3.9}$$

Thus, Σ_{t_0} is a $(n - 1)$-dimensional plane which intersects the given periodic orbit at $\hat{\mathbf{x}}(t_0)$ and is vertical to the direction of the flow at that point. Clearly now a Poincaré map can be defined on that plane as before, by

$$P : \Sigma_{t_0} \to \Sigma_{t_0}, \quad \mathbf{x}_{k+1} = P\mathbf{x}_k, \quad k = 0, 1, 2, \ldots \tag{3.10}$$

for which $\mathbf{x}_0 = \hat{\mathbf{x}}(t_0)$ is a fixed point, since $\mathbf{x}_0 = P\mathbf{x}_0$. We now examine small deviations about this point,

$$\mathbf{x}_k = \hat{\mathbf{x}}_0 + \boldsymbol{\eta}_k, \quad ||\boldsymbol{\eta}_k|| \ll \varepsilon, \tag{3.11}$$

(where ε is of the same magnitude as $||\hat{\mathbf{x}}_0||$), substitute (3.11) in (3.10) and linearize the Poincaré map to obtain

$$\boldsymbol{\eta}_{k+1} = DP(\hat{\mathbf{x}}_0)\boldsymbol{\eta}_k, \tag{3.12}$$

where we have neglected higher order terms in η and $DP(\hat{\mathbf{x}}_0)$ denotes the Jacobian of P evaluated at $\hat{\mathbf{x}}_0$.

"But we don't know anything about P!", you will rightfully argue. Don't worry! We always have the *variational equations* of the original differential equations derived by writing $\mathbf{x}(t) = \hat{\mathbf{x}}(t) + \boldsymbol{\xi}(t)$, whence linearizing (1.1) about this periodic orbit leads to the system

$$\dot{\boldsymbol{\xi}}(t) = A(t)\boldsymbol{\xi}(t), \quad A(t) = A(t+T), \tag{3.13}$$

where $A(t)$ is the Jacobian matrix of $\mathbf{f}(\mathbf{x})$ evaluated at the periodic orbit $\mathbf{x}(t) = \hat{\mathbf{x}}$. The crucial question, of course, we must face now is: How are the two linear systems (3.12) and (3.13) related to each other?

The careful reader will have certainly observed that we have used different notations for the small deviations about the periodic orbit: $\boldsymbol{\xi}(t)$ in the continuous time setting of differential equations and η_k in the discrete time setting of the Poincaré map. This is not just because they represent different quantities, it is also to emphasize that their dimensionality as vectors in the n-dimensional phase space $\mathbb{R}^n$ ($n = 2N$ for a Hamiltonian system) is different: $\boldsymbol{\xi}(t)$ is n-dimensional, while η_k is $(n-1)$-dimensional! Now you may be feeling again somewhat pessimistic. How are we ever to match these two small deviation variables?

The answer will come from what is called *Floquet theory* [107, 265, 346]. First we realize that since (3.13) is a linear system of ODEs it must possess, in general, n linearly independent solutions, forming the columns of the $n \times n$ *fundamental solution* matrix $M(t, t_0)$ in

$$\boldsymbol{\xi}(t) = M(t, t_0)\boldsymbol{\xi}(0), \quad M(t, t_0) = M(t + T, t_0) \tag{3.14}$$

(see (3.7)). Now, we don't know what these solutions are. If we were, however, to change our basis at the point $\hat{\mathbf{x}}(t_0)$ so that one of the directions of motion is along the direction *vertical* to the PSS (3.9), we would observe that the nth column of the matrix $M(T, t_0)$ has zero elements except at the last entry which is 1. Thus, if we eliminate from this matrix its nth row and nth column, it turns out that its $(n-1) \times (n-1)$ submatrix is none other than our beloved Poincaré map (3.10)! Surprised? That is the relation between the two approaches we were seeking.

"So what?" you may ask. How does that help us? Well, it means that if we could compute the so-called monodromy matrix $M(T, t_0)$ numerically we could evaluate its eigenvalues, $\mu_1, \ldots, \mu_{n-1}$ (the last one being $\mu_n = 1$), which are those of the Poincaré map and determine the stability of our periodic orbit as follows: If they are all on the unit circle, i.e. $|\mu_i| = 1$, $i = 1, \ldots, n-1$, the periodic orbit is (linearly) *stable*, while if (at least) one of them satisfies $|\mu_j| > 1$ the periodic solution is *unstable* (see Exercise 3.2).

But how do we compute the monodromy matrix $M(T, t_0)$? It is not so difficult. Let us first set $t_0 = 0$ for convenience and observe from (3.14) that $M(0, 0) = I_n$. All we have to do now is integrate numerically the variational equations (3.14) from $t = 0$ to $t = T$, n times, each time for a *different* initial vector $(0, \ldots, 0, 1, 0, \ldots, 0)$

with 1 placed in the ith position, $i = 1, 2, \ldots, n$. Note that since these equations are linear numerical integration can be performed to *arbitrary accuracy* and is also not too-time consuming for reasonable values of the period T.

Once we have calculated $M(T, 0)$, we may proceed to compute its eigenvalues and determine the stability of the periodic orbit according to whether at least one of these eigenvalues has magnitude greater than 1. In fact, for Hamiltonian systems, it can be shown that $M(T, 0)$ is a $2N \times 2N$ symplectic matrix and hence the product of its eigenvalues is equal to 1 (see Exercise 3.1). Furthermore, symplectic matrices possess the very important property that half of their eigenvalues are the *inverse* of the remaining half, i.e. their full eigenvalue spectrum has the form $\mu_1, \ldots, \mu_n, 1/\mu_1, \ldots, 1/\mu_n$ (see Exercise 3.1).

These are very powerful results. First they provide us with a precise and easily implementable strategy for determining the stability of any periodic orbit of a Hamiltonian system of N dof. They also demonstrate that for such a periodic orbit to be (linearly) stable, all eigenvalues of its monodromy matrix must in general be *complex* and lie on the unit circle. For, if one of them is real (and different from 1), there will be an eigenvalue greater than 1 and nearby displacements from the orbit will grow exponentially at least in one direction in the $2N$D phase space $\mathbb{R}^{2N}$.

In Problems 3.1 and 3.2 at the end of the chapter we urge the reader to apply the above theory to a periodically driven cubic Duffing oscillator and a special case of the $N = 2$ dof Hénon-Heiles Hamiltonian. Only if you solve these problems you will be able to appreciate the more elaborate applications of Floquet theory in the sections that follow. More importantly, however, these problems will also teach you that stability of periodic orbits once relinquished is not lost forever! It may indeed be recovered (even infinitely) many times as we vary an important parameter like the total energy of the system E.

3.2   Linear Stability Analysis

Now that we have learned how to study the linear stability properties of periodic solutions of Hamiltonian systems, it is time to wonder about the implications of this analysis regarding the more "global" dynamics, which is really what we are interested in. Let us turn, therefore, immediately to the class of one-dimensional lattices (or chains) of coupled oscillators, to which this book is primarily devoted.

Our first example is the famous Fermi Pasta Ulam (FPU)$-\beta$ model described by the N dof Hamiltonian [43]

$$H = \frac{1}{2} \sum_{j=1}^{N} p_j^2 + \sum_{j=0}^{N} \frac{1}{2}(x_{j+1} - x_j)^2 + \frac{1}{4}\beta(x_{j+1} - x_j)^4 = E, \qquad (3.15)$$

where x_j are the displacements of the particles from their equilibrium positions, and $p_j = \dot{x}_j$ are the corresponding canonically conjugate momenta, β is a positive real constant and E is the total energy of the system. Note that by not including any cubic nearest neighbor interactions in (3.15), we have kept an important symmetry of the system under the interchange $x_j \rightarrow -x_j$, which will make our analysis simpler. However, most of what we shall be discussing can also be studied when cubic interactions are included with an α coefficient before them in what is called the FPU-α model (see Chaps. 6 and 7).

The second example we shall use as an illustration is called the BEC Hamiltonian and is obtained by a discretization of a PDE of the nonlinear Schrödinger type called the Gross-Pitaevskii equation [106],

$$ih\frac{\partial \Psi(x,t)}{\partial t} = -\frac{h^2}{2}\frac{\partial^2 \Psi(x,t)}{\partial x^2} + V(x)\Psi(x,t) + g|\Psi(x,t)|^2\Psi(x,t), \ i^2 = -1 \tag{3.16}$$

where h is Planck's constant, g is a positive constant (repulsive interactions between atoms in the condensate) and $V(x)$ is an external potential. This equation is related to the phenomenon of Bose-Einstein Condensation (BEC) [187]. Considering the simple case $V(x) = 0$, $h = 1$ and discretizing the complex variable $\Psi(x,t) \equiv \Psi_j(t)$, we may approximate the second order derivative by

$$\Psi_{xx} \simeq \frac{\Psi_{j+1} + \Psi_{j-1} - 2\Psi_j}{\delta x^2}, \ \Psi_j(t) = q_j(t) + ip_j(t), \ j = 1,2,\ldots,N, \tag{3.17}$$

and by defining

$$|\Psi(x,t)|^2 = q_j^2(t) + p_j^2(t), \tag{3.18}$$

obtain a set of ODEs for the canonically conjugate variables, p_j and q_j, described by the BEC Hamiltonian [318, 331]

$$H = \frac{1}{2}\sum_{j=1}^{N}(p_j^2 + q_j^2) + \frac{\gamma}{8}\sum_{j=1}^{N}(p_j^2 + q_j^2)^2 - \frac{\varepsilon}{2}\sum_{j=1}^{N}(p_j p_{j+1} + q_j q_{j+1}) = E, \tag{3.19}$$

where $\gamma > 0$ and $\varepsilon = 1$ are constant parameters, $g = \gamma/2 > 0$, with $\delta x = 1$ and E the total energy. In fact, besides the energy integral the BEC Hamiltonian (3.19) possesses one more constant of the motion given by the norm quantity

$$F = \sum_{j=1}^{N}(p_j^2 + q_j^2) \tag{3.20}$$

and therefore chaotic behavior can only occur for $N \geq 3$.

Let us focus on a special class of periodic solutions we have called Simple Periodic Orbits (SPOs), which have well-defined symmetries and are known in closed form. By SPOs, we mean periodic solutions where all variables return to their initial state after only one maximum and one minimum in their oscillation. In particular the SPOs we shall be concerned with are the following:

I. For the FPU and BEC with *periodic boundary conditions*:

$$x_{N+k}(t) = x_k(t), \quad \forall t, k. \tag{3.21}$$

(a) The Out-of-Phase Mode (OPM) for FPU, with N even (often called the π-mode)

$$\hat{x}_j(t) = -\hat{x}_{j+1}(t) \equiv \hat{x}(t), \quad j = 1, \ldots, N. \tag{3.22}$$

(b) The OPM for BEC, with N even

$$q_j(t) = -q_{j+1}(t) = \hat{q}(t), \quad j = 1, \ldots, N. \tag{3.23}$$

(c) The In-Phase Mode (IPM) for BEC

$$q_j(t) = \hat{q}(t), \quad p_j(t) = \hat{p}(t), \quad \forall j = 1, \ldots, N, \ N \in \mathbb{N} \text{ and } N \geq 2. \tag{3.24}$$

II. For the FPU model and *fixed boundary conditions*:

$$x_0(t) = x_{N+1}(t) = 0, \quad \forall t. \tag{3.25}$$

(a) The SPO1 mode, with N odd,

$$\hat{x}_{2j}(t) = 0, \quad \hat{x}_{2j-1}(t) = -\hat{x}_{2j+1}(t) \equiv \hat{x}(t), \quad j = 1, \ldots, \frac{N-1}{2}. \tag{3.26}$$

(b) The SPO2 mode, with $N = 5 + 3m$, $m = 0, 1, 2, \ldots$ particles,

$$x_{3j}(t) = 0, \quad j = 1, 2, 3 \ldots, \frac{N-2}{3},$$
$$x_j(t) = -x_{j+1}(t) = \hat{x}(t), \quad j = 1, 4, 7, \ldots, N-1. \tag{3.27}$$

In later chapters, we will see that the above SPOs are examples of what we call Nonlinear Normal Modes (NNMs), whose theory is discussed in detail in Chap. 4.

To appreciate the importance of these periodic orbits, let us consider the $N = 2$ case of the BEC Hamiltonian, which is completely integrable. Plotting in Fig. 3.1 its PSS (x, p_x) (for $y = 0$, $p_y > 0$), we observe that the IPM and OPM orbits (intersecting the vertical axis at the points P_1, P_2 respectively) play a major role in the global dynamics: In Fig. 3.1a they are both stable with large size islands of periodic and quasiperiodic tori around them. On the other hand, the local picture changes dramatically when one of them becomes unstable, e.g. when we increase the value of the norm integral (3.20), as in Fig. 3.1b. Of course, the $N = 2$ dof BEC Hamiltonian is integrable and no chaos arises due to the destabilization of the OPM orbit. One may very well wonder, however, what would happen after a similar bifurcation of the OPM orbit in the BEC system with $N \geq 3$ dof? How widespread would chaos be around the unstable orbit in that case?

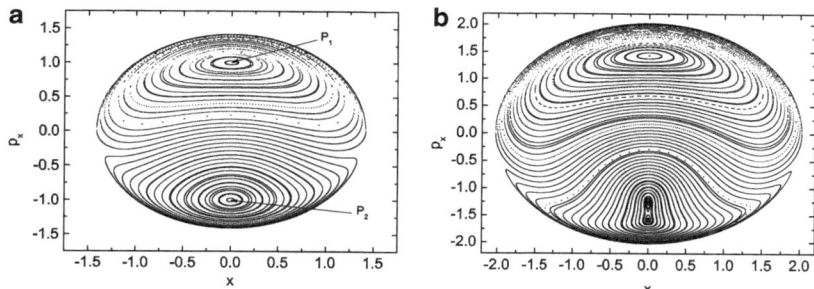

Fig. 3.1 PSS of the BEC Hamiltonian, showing the IPM and OPM orbits intersecting the vertical axis at the points P_1 and P_2 respectively. (**a**) The value of the norm integral is $F = 2$, while the SPOs correspond to different values of the Hamiltonian. (**b**) The value of the norm integral is $F = 4.1$, where the OPM on the negative p_x axis has become unstable yielding two stable SPOs one above and one below it (after [52])

Fortunately, the FPU system with fixed boundary conditions is one of those examples where we can directly apply Lyapunov's Theorem 1.3 of Chap. 1: More specifically, we can use it to *prove the existence* of SPOs as continuations of the linear normal modes of the system, whose frequencies have the well-known form [43, 88, 128, 133, 226]

$$\omega_q = 2 \sin\left(\frac{\pi q}{2(N+1)}\right), \quad q = 1, 2, \ldots, N. \tag{3.28}$$

This is so because the linear mode frequencies (3.28) are seen to satisfy Lyapunov's non-resonance condition for the ω_qs, stated in Theorem 1.3 for all q and general values of N. Thus, our SPO1 and SPO2 orbits, as NNMs of the FPU Hamiltonian, are identified by the indices $q = (N+1)/2$ and $q = 2(N+1)/3$ respectively.

As we discussed in the previous section, linear stability analysis of periodic solutions can be performed by studying the eigenvalues of the monodromy matrix. For example, consider the SPO1 mode of FPU: The equations of motion

$$\ddot{x}_j(t) = x_{j+1} - 2x_j + x_{j-1} + \beta\Big((x_{j+1} - x_j)^3 - (x_j - x_{j-1})^3\Big), \quad j = 1, \ldots, N, \tag{3.29}$$

for fixed boundary conditions collapse to a single second order ODE:

$$\ddot{\hat{x}}(t) = -2\hat{x}(t) - 2\beta\hat{x}^3(t). \tag{3.30}$$

Its solution is well-known in terms of Jacobi elliptic functions

$$\hat{x}(t) = \mathscr{C}\mathrm{cn}(\mu t, \kappa^2), \tag{3.31}$$

where

$$\mathscr{C}^2 = \frac{2\kappa^2}{\beta(1 - 2\kappa^2)}, \quad \mu^2 = \frac{2}{1 - 2\kappa^2}, \tag{3.32}$$

and κ^2 is the modulus of the cn elliptic function.

Its stability is analyzed setting $x_j = \hat{x}_j + y_j$ and keeping up to linear terms in y_j to obtain the variational equations

$$\ddot{y}_j = (1 + 3\beta\hat{x}^2)(y_{j-1} - 2y_j + y_{j+1}), \quad j = 1, \ldots, N, \tag{3.33}$$

where $y_0 = y_{N+1} = 0$. We now separate these variational equations into N uncoupled Lamé equations

$$\ddot{z}_j(t) + 4(1 + 3\beta\hat{x}^2)\sin^2\left(\frac{\pi j}{2(N + 1)}\right)z_j(t) = 0, \quad j = 1, \ldots, N, \tag{3.34}$$

where the z_j variations are simple linear combinations of the y_j's (see Exercise 3.3). Changing variables to $u = \mu t$, this equation takes the form

$$z_j''(u) + 2(1 + 4\kappa^2 - 6\kappa^2 \text{sn}^2(u, \kappa^2))\sin^2\left(\frac{\pi j}{2(N + 1)}\right)z_j(u) = 0, \quad j = 1, \ldots, N, \tag{3.35}$$

where primes denote differentiation with respect to u. Equation 3.35 is an example of Hill's equation [240, 250, 344]

$$z''(u) + Q(u)z(u) = 0, \tag{3.36}$$

where $Q(u)$ is a T-periodic function, i.e. $Q(u) = Q(u + T)$ with $T = 2K$ and $K = K(\kappa^2)$ is the elliptic integral of the first kind (see (2.54)). In Exercise 3.3 we ask the reader to repeat the steps of the above analysis to analyze the stability of the other NNMs of the FPU and BEC systems mentioned above. In particular, we ask you to try to think of other such nonlinear modes, for which, due to symmetries, the N equations of motion reduce to a single nonlinear oscillator. If not, can you think of other NNMs (based on symmetry arguments), which reduce to the solution of $n = 2, 3, \ldots$, coupled oscillators? This is a topic we shall examine in detail when we discuss *bushes of orbits* in Chap. 4.

Thus, we now have two ways for studying the stability of these fundamental periodic orbits [13, 16]: One is to compute numerically the fundamental solutions of (3.34) and determine whether the eigenvalues of the monodromy matrix $M(T, 0)$ lie on the unit circle or not. In fact, we don't need to do this for all N of these equations. It suffices to study only the variation $z_m(u)$ (in the SPO1 case it is $m = (N - 1)/2$), which is the *first* to become unbounded as κ^2 (or the energy E) increases.

The other method is to apply the theory of Hill's equation [240, 250, 344] to the crucial $j = m$ Lamé equation (3.35) and determine the stability of the NNM according to the condition

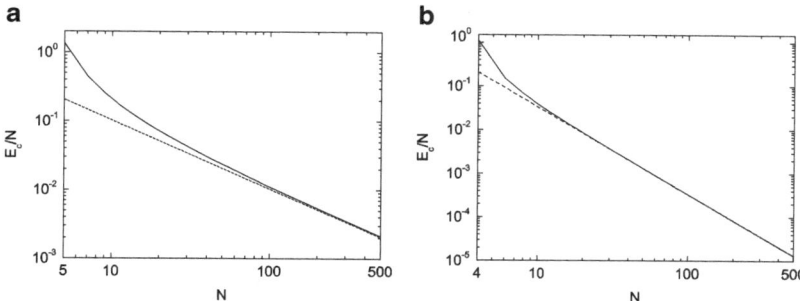

Fig. 3.2 (**a**) The *solid curve* corresponds to the energy per particle E_c/N, for $\beta = 1$, of the first destabilization of the SPO1 nonlinear mode of the FPU system (3.15) with fixed boundary conditions, obtained by the numerical evaluation of the monodromy matrix, while the *dashed line* corresponds to the function $\propto 1/N$. (**b**) Same as in (**a**) with the *solid curve* depicting the first destabilization of the OPM periodic solution of the FPU system (3.15). Here, however, the *dashed line* corresponds to the function $\propto 1/N^2$. Note that both axes are logarithmic (after [52])

$$\left| 1 - 2\sin^2 \left(K(\kappa^2) \sqrt{a_0} \right) \det \left(D(0) \right) \right| = \begin{cases} < 1, & \text{stable mode} \\ > 1, & \text{unstable mode} \end{cases}, \quad (3.37)$$

where $D(0)$ is an *infinite* Hill's matrix, whose entries are given in terms of the Fourier coefficients of the NNM and whose value is well approximated by rapidly convergent $n \times n$ approximants [13, 16]. Both ways lead to the interesting result that the critical energy threshold values for the first destabilization of the SPO1 solution satisfies $E_c/N \propto 1/N$, while for the OPM solution (3.22) we find $E_c/N \propto 1/N^2$, as shown in Fig. 3.2a, b respectively.

What does all this mean? Let us try to find out.

3.2.1   An Analytical Criterion for "Weak" Chaos

As we discovered from the above analysis, the NNMs of the FPU and BEC Hamiltonians studied so far experience a first destabilization at energy densities of the form

$$\frac{E_c}{N} \propto N^{-\alpha}, \quad \alpha = 1, \text{ or, } 2, \quad N \to \infty. \quad (3.38)$$

This means that for fixed N, some of these fundamental periodic solutions become unstable at much lower energy than others. We may, therefore, expect that those that destabilize at lower energies will have *smaller* chaotic regions around them, as the greater part of the constant energy surface is still occupied by tori of quasiperiodic motion. Could we then perhaps argue that near NNMs characterized by the exponent $\alpha = 2$ in (3.38) one would find a "weaker" form of chaos than in the $\alpha = 1$ case?

This indeed appears to be true at least for the FPU Hamiltonian model. As was recently shown in [128], the energy threshold for the destabilization of the low $q = 1, 2, 3, \ldots$, nonlinear modes, representing continuations of the corresponding linear model (see (3.28)), satisfies the analytical formula

$$\frac{E_c}{N} \approx \frac{\pi^2}{6\beta N(N+1)}, \tag{3.39}$$

and is therefore of the type $\alpha = 2$ in (3.38). Remarkably enough this local loss of stability coincides with the "weak" chaos threshold shown in [109, 110] to have global consequences regarding the dynamics of the system as a whole, as it is associated with the breakup of the famous *FPU recurrences*! In Chap. 6 we shall examine this very important phenomenon in detail. For the moment, let us simply point out that the destabilization of individual NNMs occurring at low energies appears to be somehow related to a transition from a "weak" to a "stronger" type of chaos in the full N particle chain.

Interestingly enough, it was later discovered [13, 16] that the energy threshold (3.39) for the low q modes, also coincides with the instability threshold of our SPO2 mode which corresponds to $q = 2(N + 1)/3$! In Fig. 3.3 we compare the approximate formula (dashed line) with our destabilization threshold for SPO2 obtained by the monodromy matrix analysis (solid line) for $\beta = 0.0315$ and find excellent agreement especially in the large N limit.

We may, therefore, arrive at the following conclusions based on the above results: Linear stability (or instability) of periodic solutions is certainly a local property and can only be expected to reveal how orbits behave in a limited region of phase space. And yet, we find that if these periodic solutions belong to the class of nonlinear continuations of linear normal modes, their stability character may have important consequences for the global dynamics of the Hamiltonian system. In particular, if the exponent of their first destabilization threshold in (3.38) is $\alpha = 2$ they are connected with the onset of "weak" chaos as a result of the breakdown of FPU

Fig. 3.3 The *solid curve* corresponds to the energy $E_{2u}(N)$ of the first destabilization of the SPO2 mode of the FPU system (3.15) with fixed boundary conditions and $\beta = 0.0315$ obtained by the numerical evaluation of the eigenvalues of the monodromy matrix. The *dashed line* corresponds to the approximate formula (3.38) for the $q = 3$ nonlinear normal solution (after [13])

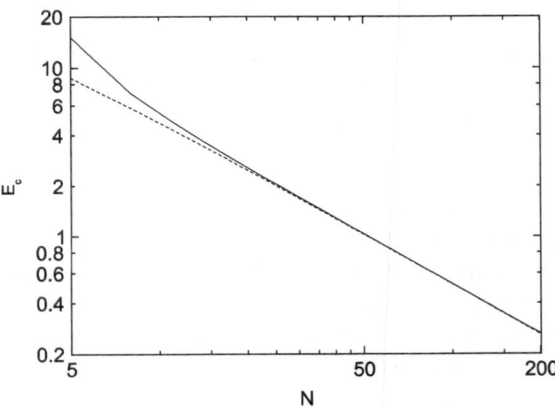

recurrences. On the other hand, if $\alpha = 1$ as in the case of the SPO1 mode, they arise in much wider chaotic domains and have orbits in their neighborhood which evolve from "weak" to "strong" chaos passing through quasi-periodic states of varying complexity, with different *statistical* properties, as we describe in detail in Chap. 8.

3.3 Lyapunov Characteristic Exponents and "Strong" Chaos

3.3.1 Lyapunov Spectra and Their Convergence

Let us now study the chaotic behavior in the neighborhood of our unstable SPOs, starting with the well-known method of the evaluation of the spectrum of Lyapunov characteristic exponents (LCEs) of a Hamiltonian dynamical system,

$$L_i, \quad i = 1, \ldots, 2N, \quad L_1 \equiv L_{\max} > L_2 > \ldots > L_{2N}. \tag{3.40}$$

The LCEs measure the rate of exponential divergence of initially nearby orbits in the phase space of the dynamical system as time approaches infinity. In Hamiltonian systems, the LCEs come in pairs of opposite sign, so their sum vanishes, $\sum_{i=1}^{2N} L_i = 0$, and two of them are always equal to zero corresponding to deviations along the orbit under consideration. If at least one of them (the largest one) $L_1 \equiv L_{\max} > 0$, the orbit is chaotic, i.e. almost all nearby orbits diverge exponentially in time, while if $L_{\max} = 0$ the orbit is stable (linear divergence of initially nearby orbits). The numerical algorithm we use here for the computation of all LCEs is the one proposed in [31, 32]. A detailed review of the theory of LCEs and the numerical techniques used for their evaluation can be found in [310].

In theory, $L_i \equiv L_i(\mathbf{x}(t))$ for a given orbit $\mathbf{x}(t)$ expresses the limit for $t \to \infty$ of a quantity of the form

$$K_t^i = \frac{1}{t} \ln \frac{\| \mathbf{w}_i(t) \|}{\| \mathbf{w}_i(0) \|}, \tag{3.41}$$

$$L_i = \lim_{t \to \infty} K_t^i \tag{3.42}$$

where $\mathbf{w}_i(0)$ and $\mathbf{w}_i(t)$, $i = 1, \ldots, 2N - 1$ are infinitesimal deviation vectors from the given orbit $\mathbf{x}(t)$ (at times $t = 0$ and $t > 0$ respectively) that are orthogonal to the vector *tangent* to the orbit (since the LCE in the direction along the orbit is zero). The time evolution of $\mathbf{w}_i$ is given by solving the variational equations of the system, i.e. the linearized equations about the orbit, assuming that the limit of (3.42) exists and converges to the same L_i, for almost all choices of initial deviations $\mathbf{w}_i(0)$.

In practice, however, the above computation is more involved [31, 32]: Since the exponential growth of $\mathbf{w}_i(t)$ is observed for short time intervals, one stops the evolution of $\mathbf{w}_i(t)$ after some time T_1, records the computed $K_{T_1}^i$, orthonormalizes

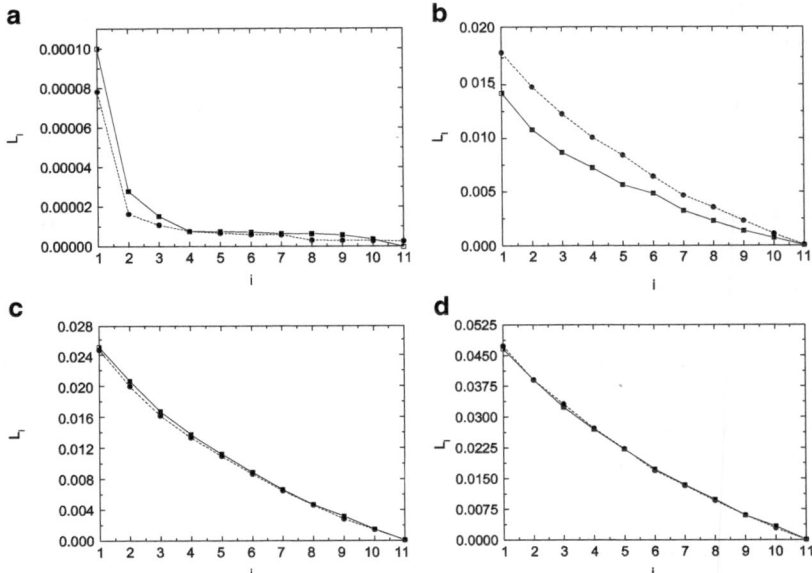

Fig. 3.4 (**a**) Lyapunov spectra of neighboring orbits of the SPO1 (*solid lines*) and SPO2 (*dashed lines*) modes of the FPU model (3.15) with fixed boundary conditions, for $N = 11$, at (**a**) energies $E = 1.94$ (SPO1) and $E = 0.155$ (SPO2), where both modes have just destabilized, (**b**) energy $E = 2.1$ for both SPOs, where the spectra are still distinct, (**c**) and convergence of the Lyapunov spectra of neighboring orbits of the SPOs at energy $E = 2.62$ where both of them are unstable. (**d**) Coincidence of Lyapunov spectra continues at energy $E = 5$ (after [13])

the vectors $\mathbf{w}_i(t)$ and repeats the calculation for the next time interval T_2 (with $t = 0$ replaced by $t = T_1$), etc. obtaining finally L_i as an average over many such T_j, $j = 1, 2, \ldots, n$ as

$$L_i = \frac{1}{n} \sum_{j=1}^{n} K_{T_j}^i, \quad n \to \infty. \tag{3.43}$$

For fixed N it has been found that as E increases, the Lyapunov spectrum (3.40) for all the unstable NNMs studied in the previous section appears to fall on a smooth curve [13, 16].

Let us examine this in more detail by plotting the Lyapunov spectra of two neighboring orbits of the SPO1 and SPO2 modes, of the FPU system with fixed boundary conditions, for $N = 11$ dof and energy values $E_1 = 1.94$ and $E_2 = 0.155$ respectively, where both SPOs have just destabilized (see Fig. 3.4a). Here, the maximum Lyapunov exponents L_1, are very small ($\approx 10^{-4}$) and the corresponding Lyapunov spectra are quite distinct.

Raising now the energy to $E = 2.1$, we observe in Fig. 3.4b that the Lyapunov spectra of the two SPOs are closer to each other, but still quite different. At $E = 2.62$, however, we see that the two spectra have nearly converged to the same exponentially decreasing function

$$L_i(N) \propto e^{-\alpha \frac{i}{N}}, \quad i = 1, 2, \ldots, N, \tag{3.44}$$

and their maximal Lyapunov exponents are virtually the same. The α exponents for the SPO1 and SPO2 are found to be approximately 2.3 and 2.32 respectively. Figure 3.4d also shows that this coincidence of Lyapunov spectra persists at higher energies.

3.3.1.1 Lyapunov Spectra and the Thermodynamic Limit

Let us now choose our initial conditions near unstable SPOs of our Hamiltonians and try to determine some important *statistical* properties of the dynamics in the so-called *thermodynamic limit*, where we let E and N grow indefinitely keeping E/N constant. In particular, we will compute the spectrum of the Lyapunov exponents of the FPU and BEC systems starting near the OPM solutions (3.22) and (3.23) for energies where these orbits are unstable. We thus find that the Lyapunov spectra are well approximated by smooth curves of the form $L_i \approx L_1 e^{-\alpha i/N}$, for both systems, with $\alpha \approx 2.76$, $\alpha \approx 3.33$ respectively (see Fig. 3.5).

The above exponential formula is found to hold quite well, up to $i = K(N) \approx 3N/4$. For the remaining exponents, the spectrum is seen to obey different decay laws, which are not easy to determine. In fact, such Lyapunov spectra provide important *invariants* of the dynamics, in the sense that, in the thermodynamic limit, we can use them to evaluate the average of the positive LCEs, yielding the so-called *Kolmogorov-Sinai entropy* $h_{KS}(N)$ [165, 267] for each system and find that it is a constant characterized by the value of the maximum Lyapunov exponent (MLE) L_1 and the exponent α appearing in them.

Thus, we plot in Fig. 3.6 the Kolmogorov-Sinai entropy (solid curves), which is defined as the sum of the $N - 1$ positive Lyapunov exponents,

$$h_{KS}(N) = \sum_{i=1}^{N-1} L_i(N), \quad L_i(N) > 0. \tag{3.45}$$

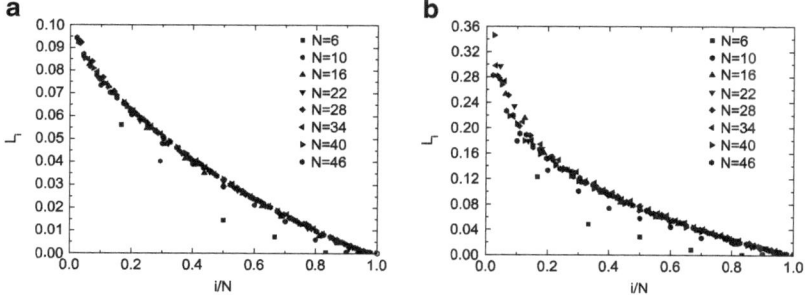

Fig. 3.5 The spectrum of positive LCEs (3.40) of (**a**) the OPM (3.22) of the FPU Hamiltonian (3.15) for fixed $E/N = 3/4$, (**b**) the OPM (3.23) of the BEC Hamiltonian (3.19) for fixed $E/N = 3/2$. For both models we consider periodic boundary conditions and i runs from 1 to N (after [16])

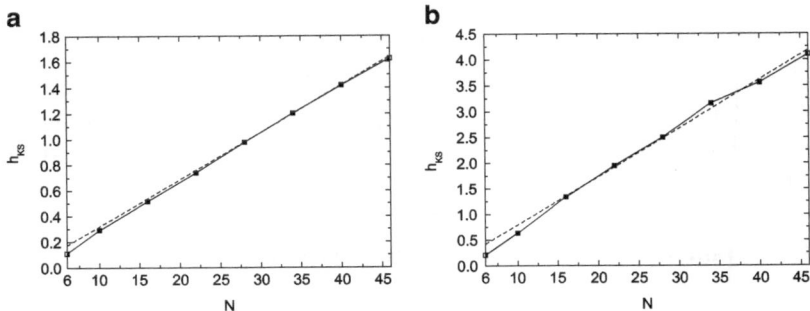

Fig. 3.6 The $h_{KS}(N)$ entropy near the OPM of (**a**) the FPU Hamiltonian for fixed $E/N = 3/4$ (*solid curve*) and the approximate formula (*dashed curve*), and (**b**) of the BEC Hamiltonian for fixed $E/N = 3/2$ (*solid curve*) and the approximate formula (*dashed curve*) (after [16])

In this way, we find, for both Hamiltonians, that $h_{KS}(N)$ is an *extensive* thermodynamic quantity as it is clearly seen to grow linearly with N ($h_{KS}(N) \propto N$), demonstrating that in their chaotic regions the FPU and BEC Hamiltonians behave as ergodic systems of BG statistical mechanics.

Finally, assuming that the exponential law $L_i \approx L_1 e^{-\alpha i/N}$ is valid for all $i = 1, \ldots, N-1$, we approximate the sum of the positive Lyapunov exponents L_i, and calculate the $h_{KS}(N)$ entropy from (3.45) as

$$h_{KS}(N) \propto L_{max} \frac{1}{-6 + e^{\frac{2.76}{N}}}, \quad L_{max} \approx 0.099, \tag{3.46}$$

for the OPM of the FPU model and

$$h_{KS}(N) \propto L_{max} \frac{1}{-1 + e^{\frac{3.33}{N}}}, \quad L_{max} \approx 0.34, \tag{3.47}$$

for the OPM of the BEC Hamiltonian. In Fig. 3.6, we have plotted (3.46) and (3.47) with dashed curves and obtain nearly straight lines with the same slope as the data computed by the numerical evaluation of the $h_{KS}(N)$, from (3.45). The reader is also advised to see [229] where analogous results are obtained for the Lyapunov spectra in the thermodynamic limit and the computation of $h_{KS}(N)$ in the FPU-β model.

3.4 Distinguishing Order from Chaos

The reader must have realized by now that if we wish to make serious progress in the study of the dynamics of Hamiltonian systems we must be able to develop accurate and efficient tools for distinguishing between order and chaos, both locally and globally. This means that these tools must be able to: (a) characterize correctly the time evolution of any given set of initial conditions and (b) approximately identify *regimes* of ordered vs. chaotic motion on a constant $(2N - 1)$-dimensional energy surface in the phase space $\mathbb{R}^{2N}$.

"But, surely", you will argue, "we already possess such tools. They are the Lyapunov exponents, whose usefulness was already evident in the results of the previous section". This is indeed true and we shall frequently mention these exponents in the chapters that follow. We must note, however, that powerful as they may be they also have a serious drawback: Their values vary significantly in time and may only be used in the long time limit when the exponents have converged with satisfactory accuracy.

Furthermore, it is well-known that a positive MLE not only implies chaotic behavior, it also *quantifies* it: As is commonly observed, the larger the MLE value the "stronger" the chaotic properties, i.e. the faster the divergence of nearby orbits. What happens, however, when we are very close to a region of ordered motion, where the MLE converges very slowly to a non-zero value, if it does so at all? And how do we know if it is not exactly zero and we are in fact following a quasiperiodic rather than a chaotic orbit?

It is for this reason that many researchers have developed over the last two decades, alternative approaches to characterize orbits of Hamiltonian systems as ordered or chaotic. These methods can be divided in two major categories, the ones which are based on the evolution of deviation vectors from a given orbit, like the computation of MLE, and the ones which rely on the analysis of the particular orbit under study.

Among other chaoticity detectors belonging to the same category with the evaluation of the MLE, are the fast Lyapunov indicator (FLI) [135, 137, 138] and its variants [26, 27], the smaller alignment index (SALI) [309, 313, 314] and its generalization, the so-called generalized alignment index (GALI) [247, 315, 316], the mean exponential growth of nearby orbits (MEGNO) [95, 96], the relative Lyapunov indicator (RLI) [296, 297], as well as methods based on the study of spectra of quantities related to the deviation vectors like the stretching numbers [136, 231, 340], the helicity angles (the angles of deviation vectors with a fixed direction) [102], the twist angles (the differences of two successive helicity angles) [103], or the study of the differences between such spectra [217, 341].

In the category of methods based on the analysis of a time series constructed by the coordinates of the orbit under study, one may list the frequency map analysis of Laskar [211–214], the '0-1' test [153, 154], the method of the low frequency spectral analysis [204, 342], the 'patterns method' [305], the recurrence plots technique [354, 355] and the information entropy index [258]. One could also refer to several ideas presented by various authors that could be used to distinguish order from chaos, like the differences appearing for ordered and chaotic orbits in the time evolution of their correlation dimension [134], the time averages of kinetic energies related to the virial theorem [171] and the statistical properties of series of time intervals between successive intersections of orbits with a PSS [181].

To our knowledge, a systematic and detailed comparative study of the efficiency and reliability of the various chaos detection techniques is not presently available. Still, a number of comparisons between some of the existing methods have been performed on various examples of dynamical systems [28, 239, 309, 314, 324].

Of all these methods, we have chosen to dwell on two indicators that the authors of this book have found most convenient and useful in their studies: the SALI and the GALI. These indicators are derived from a detailed analysis of the variational equations describing the tangent space of the orbits as time evolves. They are accurate and efficient in the sense that they: (a) correctly identify the chaotic nature of the orbits more rapidly than other methods and, perhaps more importantly, (b) successfully characterize quasiperiodic motion providing also the number of dimensions of the torus on which it lies. In the next subsections we discuss briefly the SALI and GALI indicators and return to examine them more carefully in later chapters, when we address more delicate issues regarding the complexity of Hamiltonian dynamics.

Before proceeding, however, it is instructive to point out a crucial difference between Hamiltonian (more generally conservative) and dissipative systems. As the reader has realized by now, chaotic regions and tori of ordered motion in Hamiltonian (and also conservative) systems are distributed in very complicated ways and are present down to infinitesimally small scales. Thus, the indicators mentioned above find their greatest usefulness in the conservative case, where the distinction between order and chaos is often a very delicate issue. When dissipation is added, the detailed features of the dynamics are smoothed out and all orbits converge either to *simple* (fixed points, periodic orbits, isolated tori) or *strange* attractors [160, 346]. In such cases, the identification of chaotic vs. ordered motion becomes a simple matter, which is usually settled by simply computing Lyapunov exponents and in particular the MLE.

3.4.1 The SALI Method

The SALI method was originally developed to distinguish between ordered and chaotic orbits in symplectic maps (see Sect. 5.1 for the definition of a symplectic map) and Hamiltonian systems [309, 313, 314], and was soon applied by many researchers to a number of important examples [55, 57, 72, 233, 234, 242, 245, 262, 269, 319, 320, 322, 325]. To compute this indicator, one follows simultaneously the time evolution of a reference orbit along with two deviation vectors with initial conditions $\mathbf{w}_1(0)$, $\mathbf{w}_2(0)$, normalizing them from time to time to 1, as follows

$$\hat{\mathbf{w}}_i(t) = \frac{\mathbf{w}_i(t)}{\|\mathbf{w}_i(t)\|}, \quad i = 1, 2. \tag{3.48}$$

The SALI is then defined as

$$\text{SALI}(t) = \min\left\{\|\hat{\mathbf{w}}_1(t) + \hat{\mathbf{w}}_2(t)\|, \|\hat{\mathbf{w}}_1(t) - \hat{\mathbf{w}}_2(t)\|\right\}. \tag{3.49}$$

In references [309, 314] it has been shown that, in the case of *chaotic orbits*, the deviation vectors $\hat{\mathbf{w}}_1$, $\hat{\mathbf{w}}_2$ eventually become aligned in the direction of the MLE and SALI(t) falls exponentially to zero as

$$\text{SALI}(t) \propto e^{-(L_1 - L_2)t}, \tag{3.50}$$

L_1, L_2 being the two largest LCEs. In the case of *ordered motion*, on the other hand, the orbit lies on a torus and eventually the vectors $\hat{\mathbf{w}}_1$, $\hat{\mathbf{w}}_2$ fall on the tangent space of the torus, following a t^{-1} time dependence. In this case, the SALI oscillates about values that are different from zero [309, 313], i.e.

$$SALI \approx const. > 0, \quad t \to \infty. \tag{3.51}$$

Thus, the different behavior of the index for ordered and chaotic orbits allows us to clearly distinguish between the two cases.

3.4.2 The GALI Method

The GALI is an efficient chaos detection technique introduced in [315] as a generalization of SALI. This generalization consists in the fact that GALI uses information of more than two deviation vectors from the reference orbit, leading to a faster and clearer distinction between regular and chaotic motion than SALI. The method has been applied successfully to different dynamical systems for the discrimination between order and chaos, as well as for the detection of quasiperiodic motion on low dimensional tori [14, 63, 93, 242–244, 246, 316].

The Generalized Alignment Index of order k (GALI$_k$), $2 \leq k \leq 2N$, is determined through the evolution of k initially linearly independent deviation vectors $\mathbf{w}_i(0)$. As in the case of SALI, deviation vectors $\mathbf{w}_i(t)$ are normalized from time to time to avoid overflow problems, but their directions are left intact. Thus, according to [315] GALI$_k$ is defined as the volume of the k-parallelepiped having as edges the k unitary deviation vectors $\hat{\mathbf{w}}_i(t) = \mathbf{w}_i(t)/\|\mathbf{w}_i(t)\|$, $i = 1, 2, \ldots, k$, determined through the wedge product of these vectors as

$$\text{GALI}_k(t) = \|\hat{\mathbf{w}}_1(t) \wedge \hat{\mathbf{w}}_2(t) \wedge \cdots \wedge \hat{\mathbf{w}}_k(t)\|, \tag{3.52}$$

where $\| \cdot \|$ denotes the usual norm. From this definition it is evident that if at least two of the deviation vectors become linearly *dependent*, the wedge product in (3.52) becomes zero and the GALI$_k$ vanishes.

In the case of a chaotic orbit, *all* deviation vectors tend to become linearly dependent, aligning in the direction defined by the MLE, and GALI$_k$ tends to zero exponentially following the law [315]

$$\text{GALI}_k(t) \propto e^{-[(L_1 - L_2) + (L_1 - L_3) + \cdots + (L_1 - L_k)]t}, \tag{3.53}$$

where $L_1, \ldots, L_k$ are the k largest LCEs.

In the $\mathbb{R}^{2N}$ phase space of an N dof Hamiltonian flow or a $2N$D map, regular orbits lie on s-dimensional tori, with $2 \leq s \leq N$ for Hamiltonian flows, and $1 \leq s \leq N$ for maps. For such orbits, all deviation vectors tend to fall on the s-dimensional tangent space of the torus on which the motion lies. Thus, if we start with $k \leq s$ general deviation vectors they will remain linearly independent on the s-dimensional tangent space of the torus, since there is no particular reason for them to become aligned. As a consequence, GALI_k remains practically constant and different from zero for $k \leq s$. On the other hand, GALI_k tends to zero for $k > s$, since some deviation vectors will eventually become linearly dependent. In particular, the generic behavior of GALI_k for quasiperiodic orbits lying on s-dimensional tori is given by [93, 315, 316]

$$\text{GALI}_k(t) \propto \begin{cases} \text{constant if } 2 \leq k \leq s \\ \frac{1}{t^{k-s}} & \text{if } s < k \leq 2N - s \\ \frac{1}{t^{2(k-N)}} & \text{if } 2N - s < k \leq 2N \end{cases} . \tag{3.54}$$

We note that these estimates are valid only when the corresponding conditions are satisfied. For example, in the case of 2D maps the only possible torus is an ($s =$)one-dimensional invariant curve, whose tangent space is also one-dimensional. Thus, the behavior of GALI_2 (which is the only possible index in that case) is given by the third case of (3.54), i.e. $\text{GALI}_2 \propto 1/t^2$, since the first two are not applicable. From (3.54) we also deduce that the behavior of GALI_k for the usual case of quasiperiodic orbits lying on an N-dimensional torus is given by

$$\text{GALI}_k(t) \propto \begin{cases} \text{constant if } 2 \leq k \leq N \\ \frac{1}{t^{2(k-N)}} & \text{if } N < k \leq 2N \end{cases} . \tag{3.55}$$

Exercises

Exercise 3.1. (a) Use the definition of Ω in (3.1), its properties listed in (3.2) and (3.3) to show that $\det \Omega = 1$ and $\det M = \pm 1$. Then prove that the determinant of the $2N \times 2N$ symplectic matrix M is exactly 1.
Hint: You may find it helpful to use the theorem of polar factorization of Linear Algebra (see [176], p. 188) to show that $\det M > 0$.

(b) Finally, show that the eigenvalues of the matrix M are expressed as the set of inverse pairs: $\mu_1, \mu_2, \ldots, \mu_N, \mu_{N+1} = 1/\mu_1, \ldots, \mu_{2N} = 1/\mu_N$.

Exercise 3.2. Consider the n-dimensional map $\mathbf{x}_{k+1} = P\mathbf{x}_k$, $k = 0, 1, 2, \ldots$, where P is a constant matrix. Assume that an invertible matrix S exists such that, in the new basis $\mathbf{x}_k = S\mathbf{y}_k$, $D = S^{-1}PS$ is a diagonal matrix whose elements are the eigenvalues of P, i.e. $D = \text{diag}(\mu_1, \ldots, \mu_n)$. Clearly, in this basis, the solution of the problem for any initial condition $\mathbf{y}_0 = S^{-1}\mathbf{x}_0$ is $\mathbf{y}_k = D^k\mathbf{y}_0$, $k = 0, 1, 2, \ldots$. Show that if all eigenvalues $|\mu_i| \leq 1$, $i = 1, 2, \ldots, n$, the map has only bounded solutions $\mathbf{y}_k$, while if there is (at least) one of them satisfying $|\mu_j| > 1$ the general solution is unbounded.

Exercise 3.3. Perform the linear transformation that uncouples (3.33) into (3.34), as described in Sect. 3.2, for the SPO1 mode under fixed boundary conditions (hint: You will have to diagonalize a tri-diagonal matrix). Next, using similar techniques, carry out the stability analysis for the other SPOs (or NNMs) of the FPU and BEC Hamiltonians listed at the beginning of Sect. 3.2 and derive the associated system of uncoupled Lamé equations. Are there any other such NNMs for which the N equations of motion reduce to the solution of a single nonlinear oscillator? Are there NNMs whose equations reduce to the solution of $n = 2, 3$, etc. coupled nonlinear oscillators?

Problems

Problem 3.1. Let us revisit the cubic Duffing oscillator of Exercise 2.4, in the undamped case $\alpha = 0$:

$$\dot{q} = p, \quad \dot{p} = q(1 + \gamma \sin(\Omega t)) - q^3 \tag{3.56}$$

(note the different signs in the second equation). Choose a small value of the driving amplitude, e.g. $\gamma = 0.01$ with $\Omega = 2$ and solve numerically the equations of motion to determine the periodic solutions $\hat{q}_i(t)$, $\hat{p}_i(t)$, $i = 1, 2$ crossing the PSS (2.41) at points P_1, P_2 close to the equilibrium points $(\pm 1, 0)$ respectively of the unforced $\gamma = 0$ case. For small enough γ, these should be π-periodic orbits of (3.56) that are stable, in the sense that they represent elliptic points of the Poincaré map (2.42).

(a) Based on your knowledge of these solutions, solve numerically the two-dimensional variational equations about them for initial conditions $(1, 0)$ and $(0, 1)$ to determine the 2×2 monodromy matrix $M(\pi, 0)$ (note that due to the simplicity of this problem the dimensionality of the monodromy matrix coincides with that of the Poincaré map). What are the eigenvalues of $M(\pi, 0)$ and what do you conclude from them?

(b) Start increasing the value of γ and repeat the previous calculation of $M(\pi, 0)$. Can you locate the first critical γ_c value at which the above two π periodic solutions become unstable?

Problem 3.2. Consider the non-integrable Hénon-Heiles system (2.36) with Hamiltonian

$$H(x, y, p_x, p_y) = \frac{1}{2}(p_x^2 + p_y^2) + \frac{1}{2}(x^2 + y^2) + x^2 y - \frac{1}{3}y^3 = E. \tag{3.57}$$

(a) Show that this system has an exact periodic solution of the form $\hat{x}(t) = 0$, $\hat{y}(t) = A\mathrm{cn}(\mu t, \kappa^2)$, with $A > 0$, where cn is the Jacobi elliptic cosine function with period T that depends on the modulus κ and the constants A, κ, μ satisfy certain relations.

(b) Write the four-dimensional variational system of equations about this solution and solve them numerically to determine the associated monodromy matrix $M(T, 0)$. Examining its eigenvalues, show that for small enough values of A this periodic solution is linearly stable.

(c) Increasing the value of A repeat the previous calculation of $M(T, 0)$ and locate the first critical value of $A = A_1$ (and corresponding value of the energy $E = E_1$) at which the periodic solution becomes unstable. What happens as you keep increasing A (and the energy E)? Hint: You should find a sequence of bifurcations at $A_k > A_{k-1}(E_k > E_{k-1})$, $k = 2, 3, \ldots$ that converges to $A_k \to 1(E_k \to 1/6)$.

Problem 3.3. Continuing the work of Exercise 3.3 pursue further the stability analysis of the OPM and IPM modes of the BEC Hamiltonian (3.19). Show that the variational equations cannot be uncoupled here as was possible for the NNMs of the FPU Hamiltonian. However, the mathematical form of the OPM and IPM solutions and the variational equations of the BEC Hamiltonian are simple to write down. Solve these variational equations for various choices of $N = 3, 4, 5, \ldots$, and find the eigenvalues of the associated monodromy matrix $M(T, 0)$ to determine the stability properties of the OPM and IPM solutions in terms of the values of the energy and norm integrals, E and F. What do you observe as the values of these integrals increase? Hint: Show that the IPM orbit remains stable even for large values of E and F, while for the OPM orbit consult Problem 3.4.

Problem 3.4. Integrate the variational equations and evaluate the monodromy matrix $M(T, 0)$ associated with the OPMs of the FPU and BEC Hamiltonians to show that they become unstable, as a pair of eigenvalues of $M(T, 0)$ exit the unit circle on the real axis: For FPU this happens at -1 (period-doubling bifurcation) and for BEC at $+1$ (pitchfork bifurcation). Fix N and examine the chaotic regions that arise about these orbits, as more and more pairs of eigenvalues exit the unit circle, by computing the Lyapunov exponents in the close vicinity of these NNMs. What do you observe?

Chapter 4
Normal Modes, Symmetries and Stability

Abstract The present chapter studies nonlinear normal modes (NNMs) of coupled oscillators from an altogether different perspective. Focusing entirely on periodic boundary conditions and using the Fermi Pasta Ulam β (FPU$-\beta$) and FPU$-\alpha$ models as examples, we demonstrate the importance of *discrete symmetries* in locating and analyzing exactly a class of NNMs called one-dimensional "bushes", depending on a single periodic function $\hat{q}(t)$. Using group theoretical arguments one can similarly identify n-dimensional bushes described by $\hat{q}_1(t), \ldots, \hat{q}_n(t)$, which represent quasiperiodic orbits characterized by n incommensurate frequencies. Expressing these solutions as linear combinations of single bushes, it is possible to simplify the linearized equations about them and study their stability analytically. We emphasize that these results are not limited to monoatomic particle chains, but can apply to more complicated molecular structures in two and three spatial dimensions, of interest to solid state physics.

4.1 Normal Modes of Linear One-Dimensional Hamiltonian Lattices

As the reader recalls, we began our discussion of N dof Hamiltonian systems in Chap. 2 by analyzing the case of two coupled linear oscillators of equal mass m and spring constant k described by (2.14) under fixed boundary conditions. Indeed, by a simple canonical transformation of variables (2.17), we were able to *uncouple* the two equations of motion into what we called their normal mode variables and obtain the complete solution of the problem as a linear combination of normal mode oscillations with frequencies $\omega_1 = \omega$, $\omega_2 = \omega\sqrt{3}$, $\omega = \sqrt{k/m}$. How would we proceed if we were to perform the same analysis to N such coupled linear oscillators? Good question.

T. Bountis and H. Skokos, *Complex Hamiltonian Dynamics*,
Springer Series in Synergetics, DOI 10.1007/978-3-642-27305-6_4,
© Springer-Verlag Berlin Heidelberg 2012

Let us note first that the Hamiltonian function of this system

$$H = \frac{1}{2m} \sum_{j=1}^{N} p_j^2 + \frac{k}{2} \sum_{j=0}^{N} (q_{j+1} - q_j)^2 = E, \qquad (4.1)$$

leads to the equations of motion

$$\frac{d^2 q_j}{dt^2} = \omega^2 (q_{j-1} - 2q_j + q_{j+1}), \quad j = 1, 2, ..., N \qquad (4.2)$$

imposing again fixed boundary conditions $q_0(t) \equiv q_{N+1}(t) \equiv 0$, $p_0(t) \equiv p_{N+1}(t) \equiv 0$, with $\omega^2 = k/m$. This may be called an "ordered", or translationally invariant particle chain (one-dimensional lattice), since all masses and spring constants are equal. In Chap. 7 we will discuss the very important case of a *disordered* lattice where translational invariance is broken by a random selection of the parameters of the system. For the time being, however, it suffices to take with no loss of generality $\omega = 1$.

How can we uncouple equations (4.2) into a set of N single harmonic oscillators? The case $N = 2$ was analyzed in Chap. 2 and easily led to the transformation (2.17). Here, we are dealing with a system of N ODEs (4.2) that, when written in matrix form, involves a tridiagonal matrix S on its right hand side. Since we are looking for normal mode oscillations with N frequencies ω_q, we proceed to diagonalize this matrix introducing new coordinates called normal mode variables Q_q, P_q by the transformation

$$q_j = \sqrt{\frac{2}{N+1}} \sum_{q=1}^{N} Q_q \sin\left(\frac{qj\pi}{N+1}\right), \quad p_j = \sqrt{\frac{2}{N+1}} \sum_{q=1}^{N} P_q \sin\left(\frac{qj\pi}{N+1}\right),$$
$$(4.3)$$

in terms of which (4.1) reduces to the Hamiltonian of N uncoupled harmonic oscillators

$$H_2 = \frac{1}{2} \sum_{q=1}^{N} P_q^2 + \omega_q^2 Q_q^2, \qquad (4.4)$$

whose frequencies

$$\omega_q = 2 \sin\left(\frac{q\pi}{2(N+1)}\right), \quad 1 \leq q \leq N \qquad (4.5)$$

are related to the eigenvalues λ_q of S by the simple formula $\lambda_q = \omega_q^2$ (see (3.28) and Exercise 4.1).

Before studying this case further, let us also write down the analogous result for the case of periodic boundary conditions $q_j(t) \equiv q_{j+N}(t)$, $p_j(t) \equiv p_{j+N}(t)$, $j = 1, ..., N$. As we ask you to demonstrate in Exercise 4.2, the reduction of (4.1) to the Hamiltonian (4.4) of N uncoupled oscillators is achieved through the transformation

$$Q_q = \frac{1}{\sqrt{N}} \sum_{j=1}^{N} \left(\sin \frac{2\pi q j}{N} + \cos \frac{2\pi q j}{N} \right) q_j,$$

$$P_q = \frac{1}{\sqrt{N}} \sum_{j=1}^{N} \left(\sin \frac{2\pi q j}{N} + \cos \frac{2\pi q j}{N} \right) p_j,$$

(4.6)

and the normal mode frequencies now take the form

$$\omega_q = 2 \sin \left(\frac{q\pi}{N} \right), \quad 1 \leq q \leq N.$$

(4.7)

As we remark in Exercises 4.1 and 4.2, knowing the normal mode spectra in the above two cases, allows us to examine the linear (in)dependence (or commensurability properties) of these frequencies. Remember what happened in the case of $N = 2$ such oscillators in Chap. 2 under fixed boundary conditions: The two normal mode frequencies were rationally independent and the motion was in general quasiperiodic! Does the same hold for all N? And what about the periodic boundary condition case, where $\sin(q\pi/N) = \sin((N - q)\pi/N)$, holds for $q = 1, 2, \ldots, [N/2]$ (with $[\cdot]$ denoting the integer part of the argument)? These are important issues to which we will return later in this chapter.

4.2   Nonlinear Normal Modes (NNMs) and the Problem of Continuation

Now that we have understood the importance of normal modes in the linear case let us turn to the more difficult situation of coupled systems of nonlinear oscillators. In particular, we shall focus on our old friend the FPU$-\beta$ Hamiltonian (3.15), which we already met in Chap. 3 as

$$H = \frac{1}{2} \sum_{j=1}^{N} p_j^2 + \sum_{j=0}^{N} \frac{1}{2}(q_{j+1} - q_j)^2 + \frac{1}{4}\beta(q_{j+1} - q_j)^4 = E,$$

(4.8)

representing an N-particle chain with quadratic and quartic nearest neighbor interactions. In Sect. 3.2, we studied some of its SPOs, under fixed and periodic boundary conditions, and found that their linear stability properties had important consequences for the global dynamics of the system. In particular, in the case of fixed boundaries, we showed that the first destabilization of solutions SPO1 and SPO2, corresponding to normal modes $q = (N + 1)/2$ and $q = 2(N + 1)/3$ respectively, was related to the onset of "strong" and "weak" chaos as the total energy E is increased.

Clearly, these SPOs are linear modes, which continue to exist as the parameter β in (4.8) is turned on and our oscillator system becomes nonlinear! Is that true? Of course it is. Remember what we said in Sect. 3.2 about Lyapunov's Theorem 1.3, according to which all normal modes of the linear $\beta = 0$ system can be rigorously continued to the nonlinear lattice, for fixed boundary conditions. This is because the linear frequencies (4.5) can be shown in many cases to be mutually rationally independent (see Exercise 4.1). Unfortunately, Lyapunov's theorem does not apply to the FPU problem under periodic boundary conditions.

So what do we do in that case? Don't worry. We will construct its NNMs later in this chapter using discrete symmetries of the equations of motion and discuss their stability properties in detail. For the time being let us complete our discussion of the NNMs of the FPU$-\beta$ system under fixed boundaries. This is indeed a very important problem as these NNMs are intimately connected with the historical paradox of the FPU recurrences, which we shall discuss in detail in Chap. 6.

Note that substituting (4.3) into (4.8) allows us to write the FPU$-\beta$ Hamiltonian in the form $H = H_2 + H_4$ in which the quadratic part corresponds to N uncoupled harmonic oscillators, as in (4.4) above. On the other hand, the quartic part of the Hamiltonian becomes

$$H_4 = \frac{\beta}{8(N+1)} \sum_{q,l,m,n=1}^{N} C_{q,l,m,n} \omega_q \omega_l \omega_m \omega_n Q_q Q_l Q_m Q_n, \qquad (4.9)$$

where the coefficients $C_{q,l,m,n}$ take non-zero values only for particular combinations of the indices q, l, m, n, namely

$$C_{q,l,m,n} = \sum_{\pm} \left(\delta_{q \pm l \pm m \pm n, 0} - \delta_{q \pm l \pm m \pm n, \pm 2(N+1)} \right) \qquad (4.10)$$

in which all possible combinations of the $\pm$ signs arise, with $\delta_{ij} = 1$ for $i = j$ and $\delta_{ij} = 0$ for $i \neq j$. Thus, in the new canonical variables, the equations of motion are expressed as

$$\ddot{Q}_q + \omega_q^2 Q_q = -\frac{\beta}{2(N+1)} \sum_{l,m,n=1}^{N} C_{q,l,m,n} \omega_q \omega_l \omega_m \omega_n Q_l Q_m Q_n. \qquad (4.11)$$

If $\beta = 0$, the individual harmonic energies $E_q = (P_q^2 + \omega_q^2 Q_q^2)/2$ are preserved since they constitute a set of N integrals in involution. When $\beta \neq 0$, however, the harmonic energies become functions of time and only the total energy E, (4.8), is conserved. One may thus define a specific energy of the system as $\varepsilon = E/N$, while the average harmonic energy of each mode over a time interval $0 \leq t \leq T$ is given by the integral $\bar{E}_q(T) = \frac{1}{T} \int_0^T E_q(t) dt$.

As we discuss in Chap. 6, in classical FPU experiments, one starts with the total energy distributed among a small subset of linear modes. Then, as every physics student knows, equilibrium statistical mechanics predicts that, due to nonlinear interactions, after a short time interval the energy of all non-excited modes will grow and a kind of *equipartition* will occur with the energy being shared equally by all modes, i.e.

$$\lim_{T \to \infty} \bar{E}_q(T) = \varepsilon \ , \ q = 1, \ldots, N. \tag{4.12}$$

But this does not happen in the FPU system! At least for a range of low E values, one finds that the total energy returns periodically to the originally excited modes, yielding the famous FPU recurrences which persist even when T becomes very large. The paradox lies in the fact that such deviations from equipartition were not expected to occur in a nonlinear and non-integrable Hamiltonian system as the FPU. Is it still a paradox to date? Not any longer, we claim. As will be explained in Chap. 6, the dynamics of FPU recurrences can be understood in terms of exponential localization of periodic and quasiperiodic solutions in q-modal space.

4.3   Periodic Boundary Conditions and Discrete Symmetries

Let us recall from the above discussion that, when periodic boundary conditions are imposed, i.e. $q_j = q_{j+N}$, $p_j = p_{N+j}$, $j = 1, 2, \ldots, N$, the linear mode spectrum of the FPU-β particle chain becomes degenerate for all N and Lyapunov's theorem cannot be invoked. How do we proceed to study existence and stability of NNMs in that case? As we explain below, this is a situation where the identification of the system's discrete symmetries turns out to be very helpful in the analysis of the dynamics.

In the present chapter we shall demonstrate how one may use the powerful techniques of group theory to establish the existence of such NNMs for a variety of mechanical systems, including particle chains in one dimension, as well as certain two-dimensional and three-dimensional structures of interest to solid state physics. These NNMs are simple periodic orbits, which we shall call *one-dimensional bushes*. We will then show how one can combine such periodic solutions to form multi-dimensional bushes of NNMs and exploit symmetries to simplify the variational equations about them and study their (linear) stability. Ultimately, of course, one would like to be able to obtain *invariant manifolds* on which the dynamics of the system is as simple as possible. This is not easy to do in general, but if the equations of motion possess discrete symmetry groups, one can single out some of these manifolds using group theoretical methods developed in [81, 82, 295] (see also [64] for a recent review).

4.3.1   NNMs as One-Dimensional Bushes

Let us illustrate the main steps of the bush theory on the FPU-β Hamiltonian system (4.8). The ODEs describing the longitudinal vibrations of the FPU-β chain can be written in the form

$$\ddot{q}_i = f(q_{i+1} - q_i) - f(q_i - q_{i-1}), \qquad i = 1, \ldots, N, \tag{4.13}$$

where $q_i(t)$ is the displacement of the ith particle from its equilibrium state at time t, while the force $f(\Delta q)$ depends on the spring deformation Δq as follows:

$$f(\Delta q) = \Delta q + \beta(\Delta q)^3. \tag{4.14}$$

We also assume that $\beta > 0$ and impose periodic boundary conditions. Thus, we may study the dynamics of this chain by attempting to solve (4.13) for the "configuration" vector

$$\mathbf{X}(t) = \{q_1(t), q_2(t), \ldots, q_N(t)\}, \tag{4.15}$$

whose components are the individual particle displacements.

As is well-known, it is pointless to try to obtain this vector as a general solution of (4.13). We, therefore, concentrate on studying special solutions represented by the NNMs described in the previous section. In particular, let us begin with the following simple periodic solution

$$\mathbf{X}(t) = \{\hat{q}(t), -\hat{q}(t), \hat{q}(t), -\hat{q}(t), \ldots, \hat{q}(t), -\hat{q}(t)\}, \tag{4.16}$$

which is easily seen to exist in the FPU$-\beta$ chain with an even number of particles ($N \bmod 2 = 0$). This solution is called the OPM or "π-mode" and is fully determined by only one arbitrary function $\hat{q}(t)$ as we already discovered in Chap. 3. It represents an example of the type of NNMs introduced in [286] and expresses an exact dynamical state which can be written in the form

$$\mathbf{X}(t) = \hat{q}(t)\{1, -1, |1, -1, | \ldots, |1, -1\}. \tag{4.17}$$

Hence, the following question naturally arises: Are there any other such exact NNMs in the FPU-β chain? Some examples of such modes are already known in the literature [13, 203, 218, 272, 283, 303, 304, 351], under various terminologies. Below we list these exact states in detail

$$\mathbf{X}(t) = \hat{q}(t)\{1, 0, -1, |1, 0, -1, |, \ldots\}, \quad \omega^2 = 3, \quad (N \bmod 3 = 0), \tag{4.18}$$

$$\mathbf{X}(t) = \hat{q}(t)\{1, -2, 1, |1, -2, 1, |, \ldots\}, \quad \omega^2 = 3, \quad (N \bmod 3 = 0), \tag{4.19}$$

$$\mathbf{X}(t) = \hat{q}(t)\{0, 1, 0, -1, |0, 1, 0, -1, |, \ldots\}, \quad \omega^2 = 2, \quad (N \bmod 4 = 0), \tag{4.20}$$

$$\mathbf{X}(t) = \hat{q}(t)\{1, 1, -1, -1, |1, 1, -1, -1|, \ldots\}, \quad \omega^2 = 2, \quad (N \bmod 4 = 0), \tag{4.21}$$

$$\mathbf{X}(t) = \hat{q}(t)\{0, 1, 1, 0, -1, -1, |0, 1, 1, 0, -1, -1|, \ldots\}, \quad \omega^2 = 1,$$
$$(N \bmod 6 = 0). \tag{4.22}$$

Let us comment on certain properties of the above NNMs: Note that in (4.17)–(4.22) we have identified for each of these NNMs a primitive cell (divided by vertical lines), whose number of elements m is called the multiplicity number. These numbers have the values: $m = 2$ for (4.17), $m = 3$ for (4.18) and (4.19), $m = 4$ for (4.20) and (4.21) and $m = 6$, for (4.22). Observe also that, due to the periodicity of the boundary conditions, the FPU-β chain in its equilibrium state is invariant under translation by the inter-particle distance a, while in its vibrational states it is invariant under a translation by ma. Moreover, NNMs (4.17)–(4.22) differ not only by translational symmetry (ma), but also by some additional symmetry transformations which are discussed below.

Each of the NNMs (4.17)–(4.22) depends on only one function $\hat{q}(t)$ and is, therefore, said to describe a *one-dimensional* "dynamical domain", borrowing the term from the theory of phase transitions in crystals. For example, consider the mode (4.18): It is easy to check that cyclic permutations of each primitive cell produces other modes, which differ only by the position of their stationary particles. As a consequence, these NNMs possess equivalent dynamics and, in particular, turn out to share the same stability properties. Thus, we only need to study one representative member of each set.

Many aspects of existence and stability of NNMs have been discussed in the literature [13, 66, 85, 87, 203, 218, 272, 283, 298, 303, 304, 351]. What is important from our point of view is to pose certain fundamental questions concerning these NNMs, which we shall proceed to answer in the sections that follow:

(Q1) Is the list (4.17)–(4.22) of NNMs for the FPU-β chain complete? Indeed, at first sight, it seems that many other NNMs exist, for example, modes whose multiplicity number m is different from those listed above.

(Q2) What kind of NNMs arise in nonlinear chains with different interactions than those of the FPU-β chain? In most papers (see [203, 272, 303, 304, 351]) the NNMs listed above are discussed by analyzing dynamical equations connected only with FPU-β inter-particle interactions.

(Q3) Do there exist NNMs for Hamiltonian systems that are more complicated than monoatomic chains? For example, can one pose this question for diatomic nonlinear chains (with particles having alternating masses), two-dimensional (2D) lattices, or 3D crystal structures?

(Q4) Can one construct exact *multi-dimensional bushes* in nonlinear N-particle Hamiltonian systems and study their stability by locating the unstable manifolds of orbits characterized by a number of different frequencies? As we shall discover in subsequent sections, there exist interesting domains of regular motion depending on more than one frequency and characterized by families of quasiperiodic functions, which form s-dimensional tori, with $s \geq 2$.

4.3.2 Higher-Dimensional Bushes and Quasiperiodic Orbits

Let us consider the following exact solution of the FPU-β chain with N mod $6 = 0$:

$$\mathbf{X}(t) = \{0, \hat{q}_1(t), \hat{q}_2(t), 0, -\hat{q}_2(t), -\hat{q}_1(t) | \ldots | 0, \hat{q}_1(t), \hat{q}_2(t), 0, -\hat{q}_2(t), -\hat{q}_1(t) |\}.$$
(4.23)

This bush has a multiplicity number $m = 6$ and is defined by two functions, $\hat{q}_1(t)$ and $\hat{q}_2(t)$, satisfying a system of two nonlinear autonomous ODEs. Thus, it is an example of what we call a two-dimensional bush that represents an exact quasiperiodic motion involving two frequencies.

We remark at this point that, in the framework of bush theory, the problem of finding invariant manifolds of exact solutions in a physical system with discrete symmetry can be solved without any information about the inter-particle forces. However, the explicit form of the bush dynamical equations *does* depend on the inter-particle interactions of the system under consideration. In the sections that follow, we will show that the theory of bushes gives definite and quite general answers to the above questions Q1–Q3 in nonlinear systems with discrete symmetries. We shall then proceed to answer question Q4, using the analytical and numerical results described in Chap. 6.

More specifically, we shall demonstrate that it is possible to use Poincaré-Linstedt series expansions to construct exact quasiperiodic solutions with $s \geq 2$ frequencies, which belong to the lower part of the normal mode spectrum, with $q = 1, 2, 3, \ldots$. Thus, we will be able to establish the important property that the energies E_q excited by these low-dimensional, so-called q-tori, are exponentially localized in q-space [94]. The relevance of this fact becomes apparent when we study the well-known phenomenon of FPU recurrences, which are remarkably persistent in FPU chains (with fixed or periodic boundary conditions), at low values of the total energy E.

We will also study in Chap. 6 the stability of these q-tori as accurately as possible, using methods that do not rely on Floquet theory. In particular, we shall exploit the approach of the GALI$_k$ indicators $k = 2, \ldots, 2N$, described in Chaps. 3 and 5, to estimate the destabilization threshold at which q-tori are destroyed, leading to the breakdown of FPU recurrences and the eventual onset of energy equipartition among all modes.

4.4 A Group Theoretical Study of Bushes

Let us begin with the case of NNMs, which represent one-dimensional bushes of orbits that help to determine all symmetry groups of the equations describing the vibrations of a given mechanical system. The set of these groups constitutes the *parent* symmetry group of all transformations that leave the given system of equations invariant. Let us consider, for example, the dynamical equations (4.13) with (4.14) for the FPU-β chain with periodic boundary conditions and an even number of particles. Clearly, these remain invariant under the action of an operator $\hat{a}$ that shifts it by the lattice spacing a. This operator generates the translational group

$$T = \{\hat{e}, \hat{a}, \hat{a}^2, \ldots, \hat{a}^{N-1}\}, \quad \hat{a}^N = \hat{e}, \tag{4.24}$$

where $\hat{e}$ is the identity element and N is the order of the cyclic group T. The operator $\hat{a}$ induces a cyclic permutation of all particles of the chain and, therefore, acts on the configuration vector $\mathbf{X}(t)$ of (4.15) as follows

$$\hat{a}\mathbf{X}(t) \equiv \hat{a}\{q_1(t), q_2(t), \ldots, q_{N-1}(t), q_N(t)\} =$$
$$\{q_N(t), q_1(t), q_2(t), \ldots, q_{N-1}(t)\}. \tag{4.25}$$

Another important element of the symmetry group of transformations of the monoatomic chain is the *inversion* $\hat{i}$ with respect to the center of the chain, which acts on the vector $\mathbf{X}(t)$ in the following way

$$\hat{i}\mathbf{X}(t) \equiv \hat{i}\{q_1(t), q_2(t), \ldots, q_{N-1}(t), q_N(t)\} =$$
$$\{-q_N(t), -q_{N-1}(t), \ldots, -q_2(t), -q_1(t)\}. \tag{4.26}$$

The complete set of symmetry transformations, of course, includes also all products $\hat{a}^k\hat{i}$ of the pure translations $\hat{a}^k$ ($k = 1, 2, \ldots, N - 1$) with the inversion $\hat{i}$ and forms the so-called dihedral group D which can be written as a direct sum of the two cosets T and $T \cdot \hat{i}$

$$D = T \oplus T \cdot \hat{i}. \tag{4.27}$$

This is a non-Abelian group induced by two *generators* ($\hat{a}$ and $\hat{i}$) through the following generating relations

$$\hat{a}^N = \hat{e}, \quad \hat{i}^2 = \hat{e}, \quad \hat{i}\hat{a} = \hat{a}^{-1}\hat{i}. \tag{4.28}$$

Clearly, operators $\hat{a}$ and $\hat{i}$ induce the following changes of variables

$$\begin{aligned} \hat{a} &: q_1(t) \to q_N(t), q_2(t) \to q_1(t), \ldots, q_N(t) \to q_{N-1}(t); \\ \hat{i} &: q_1(t) \leftrightarrow -q_N(t), q_2(t) \leftrightarrow -q_{N-1}(t), q_3(t) \leftrightarrow -q_{N-2}(t), \ldots. \end{aligned} \tag{4.29}$$

It is now straightforward to check that upon acting on (4.13) with transformation (4.29) the system is transformed to an equivalent form. Moreover, since (4.13) are invariant under the actions of $\hat{a}$ and $\hat{i}$, they are also invariant with respect to all products of these two operators and, therefore, the dihedral group D is indeed a symmetry group of (4.13) for a monoatomic chain with arbitrary inter-particle interactions. As a consequence, the dihedral group D can be considered as a parent symmetry group for *all* monoatomic nonlinear chains as, for example, the FPU-α chain, whose inter-particle interactions are characterized by the force

$$f(\Delta q) = \Delta q + \alpha(\Delta q)^2. \tag{4.30}$$

Of course, since in the case of the FPU-β chain the force is an *odd* function of its argument, the FPU-β chain possesses a *higher* symmetry group than the FPU-α model. Indeed, let us introduce the operator $\hat{u}$ which changes the signs of all atomic displacements without their transposition

$$\hat{u}\mathbf{X} \equiv \hat{u}\{q_1(t), q_2(t), \ldots, q_{N-1}(t), q_N(t)\} = \{-q_1(t), -q_2(t), \ldots,$$
$$- q_{N-1}(t), -q_N(t)\}. \tag{4.31}$$

It can be easily checked that the operator $\hat{u}$ generates a transformation of all the variables $q_i(t)$, $i = 1, \ldots, N$ in (4.13), (4.14), which leads to an equivalent form of these equations. Therefore, the operator $\hat{u}$ and all its products with elements of the dihedral group D belong to the full symmetry group of the FPU-β chain. Clearly, the operator $\hat{u}$ *commutes* with all the elements of the dihedral group D and thus we can consider the group

$$G = D \oplus D \cdot \hat{u} \tag{4.32}$$

as the parent symmetry group of the FPU-β chain. The group G contains twice as many elements as the dihedral group D and, therefore, possesses a greater number of subgroups.

Finally, let us note that it is sufficient for our purposes to define any symmetry group by listing only its *generators* which we denote by square brackets, for example, we write $T[\hat{a}]$, $D[\hat{a}, \hat{i}]$, $G[\hat{a}, \hat{i}, \hat{u}]$. The complete set of group elements is written in curly brackets, as in (4.24).

4.4.1 Subgroups of the Parent Group and Bushes of NNMs

Let us consider now a specific configuration vector $\mathbf{X}^{(j)}(t)$, see (4.15), which determines a displacement pattern at time t, and let us act on it successively by the operators $\hat{g}$ that correspond to all the elements of a parent group G. The full set G_j of elements of the group G under which $\mathbf{X}^{(j)}(t)$ turns out to be invariant generates a certain subgroup of G ($G_j \subset G$). We then call $\mathbf{X}^{(j)}(t)$ *invariant* under the action of the subgroup G_j of the parent group G and use it to determine the bush of NNMs corresponding to this subgroup.

Thus, in the framework of this approach, one must list all the subgroups of the parent group G to obtain all the bushes of NNMs of different types. This can be done by standard group theoretical methods. In [87] a simple crystallographic technique was developed for singling out all the subgroups of the parent group of any monoatomic chain, following the approach of a more general method [80, 83, 84]. Here, we demonstrate how one can obtain bushes of NNMs if the subgroups are already known.

Let us consider the subgroups G_j of the dihedral group D. Each G_j contains its own translational subgroup $T_j \subset T$, where T is the full translational group (4.24). If N is divisible by 4, say, there exists a subgroup $T_4 = [\hat{a}^4]$ of the group

$T = [\hat{a}]$. If a vibrational state of the chain possesses the symmetry group $T_4 = [\hat{a}^4] \equiv \{\hat{e}, \hat{a}^4, \hat{a}^8, \ldots, \hat{a}^{N-4}\}$, the displacements of atoms that lie at a distance $4a$ from each other in the equilibrium state turn out to be equal, since the operator $\hat{a}^4$ leaves the vector $\mathbf{X}(t)$ invariant.

For example, for the case $N = 12$, the operator $\hat{a}^4$ permutes the coordinates of $\mathbf{X} = \{q_1, q_2, \ldots, q_{12}\}$ taken in quadruplets $(q_i, q_{i+1}, q_{i+2}, q_{i+3})$, $i = 1, 5, 9$, while from equation $\hat{a}^4 \mathbf{X}(t) = \mathbf{X}(t)$ one deduces $q_i = q_{i+4}$, $i = 1, 2, 3, 4$. Thus, the vector $\mathbf{X}(t)$ contains three times the quadruplets q_1, q_2, q_3, q_4, where $q_i(t)$ ($i = 1, 2, 3, 4$) are arbitrary functions of time and can be written as follows

$$\mathbf{X}(t) = \{ q_1(t), q_2(t), q_3(t), q_4(t) \mid q_1(t), q_2(t), q_3(t), q_4(t) \mid q_1(t), q_2(t), q_3(t), q_4(t) \}. \tag{4.33}$$

In other words, the complete set of atomic displacements can be divided into $N/4$ (in our case, $N/4 = 3$) *identical* subsets, which are called *extended primitive cells*. In the bush (4.33), the extended primitive cell contains four atoms and the vibrational state of the whole chain is described by three such cells. Thus, the extended primitive cell for the vibrational state with the symmetry group $T_4 = [\hat{a}^4]$ has size equal to $4a$, which is four times larger than the primitive cell of the chain at the equilibrium state.

It is essential that some symmetry elements of the dihedral group D disappear as a result of the symmetry reduction $D = [\hat{a}, \hat{i}] \rightarrow T_4 = [\hat{a}^4]$. There are four other subgroups of the dihedral group D, corresponding to the same translational subgroup $T_4 = [\hat{a}^4]$, namely

$$[\hat{a}^4, \hat{i}], \quad [\hat{a}^4, \hat{a}\hat{i}], \quad [\hat{a}^4, \hat{a}^2\hat{i}], \quad [\hat{a}^4, \hat{a}^3\hat{i}]. \tag{4.34}$$

Each of these subgroups possesses two generators, namely $\hat{a}^4$ and an inversion element $\hat{a}^k\hat{i}$ ($k = 0, 1, 2, 3$). Note that the $\hat{a}^k\hat{i}$ differ from each other by the position of the center of inversion. Subgroups $[\hat{a}^4, \hat{a}^k\hat{i}]$ with $k > 3$ are equivalent to those listed in (4.34), since the second generator $\hat{a}^k\hat{i}$ can be multiplied from the left by $\hat{a}^{-4}$, representing the inverse element with respect to the generator $\hat{a}^4$. Thus, there exist only five subgroups of the dihedral group (with N mod $4 = 0$) constructed on the basis of the translational group $T_4 = [\hat{a}^4]$: This one and the four listed in (4.34).

Now, let us examine the bushes corresponding to the subgroups (4.34). The subgroup $[\hat{a}^4, \hat{i}]$ consists of the following six elements

$$\hat{e}, \hat{a}^4, \hat{a}^8, \hat{i}, \hat{a}^4\hat{i}, \hat{a}^8\hat{i} \equiv \hat{i}\hat{a}^4. \tag{4.35}$$

The invariance of $\mathbf{X}(t)$ with respect to this group can be written as follows:

$$\hat{a}^4 \mathbf{X}(t) = \mathbf{X}(t), \quad \hat{i}\mathbf{X}(t) = \mathbf{X}(t), \tag{4.36}$$

while the invariance of the vector $\mathbf{X}(t)$ under the action of the group generators $[\hat{a}^4]$ and $[\hat{i}]$ guarantees its invariance under all elements of this group.

As explained above, the equation $\hat{a}^4 \mathbf{X}(t) = \mathbf{X}(t)$ is satisfied by the vector $\mathbf{X}(t)$ (see (4.33)), while $\hat{i}\mathbf{X}(t) = \mathbf{X}(t)$ also holds, from which we obtain the following relations: $q_1(t) = -q_4(t)$, $q_2(t) = -q_3(t)$. Therefore, for $N = 12$, the invariant vector $\mathbf{X}(t)$ of the group $[\hat{a}^4, \hat{i}]$ can be written in the form

$$\mathbf{X}(t) = \{q_1(t), q_2(t), -q_2(t), -q_1(t) | q_1(t), q_2(t), -q_2(t), -q_1(t)$$

$$|q_1(t), q_2(t), -q_2(t), -q_1(t)\}, \quad (4.37)$$

where $q_1(t)$ and $q_2(t)$ are arbitrary functions of time.

Thus, the subgroup $[\hat{a}^4, \hat{i}]$ of the dihedral group D generates a *two-dimensional* bush of NNMs. The explicit form of the differential equations governing the two variables $q_1(t)$ and $q_2(t)$ can now be obtained by substitution of the ansatz (4.37) into the FPU-β equations (4.13), (4.14). We shall, hereafter, denote the bush (4.37) in the form

$$B[\hat{a}^4, \hat{i}] = |q_1, q_2, -q_2, -q_1|, \quad (4.38)$$

showing the atomic displacements in only one extended primitive cell and omitting the argument t in the variables $q_1(t)$, $q_2(t)$.

Proceeding in a similar manner, we obtain bushes of NNMs for the other three groups listed in (4.34)

$$B[\hat{a}^4, \hat{a}\hat{i}] = |0, q, 0, -q|, \quad (4.39)$$

$$B[\hat{a}^4, \hat{a}^2\hat{i}] = |q_1, -q_1, q_2, -q_2|, \quad (4.40)$$

$$B[\hat{a}^4, \hat{a}^3\hat{i}] = |q, 0, -q, 0|. \quad (4.41)$$

More generally, we conclude that for sufficiently large extended primitive cells it will not be possible to find enough symmetry elements to give rise to NNMs, since the bushes of the corresponding displacement patterns are multi-dimensional. For this reason, there exists only a very specific number of bushes for any fixed dimension beyond the (one-dimensional) NNMs!

4.4.2 Bushes in Modal Space and Stability Analysis

Recall now that our vectors $\mathbf{X}(t)$, giving rise to bushes of NNMs, are defined in the configuration space $\mathbb{R}^N$. If we now introduce in this space a basis set of vectors $\{\boldsymbol{\varphi}_1, \boldsymbol{\varphi}_2, \ldots, \boldsymbol{\varphi}_N\}$, we can represent the dynamical regime of our mechanical system by a linear combination of the vectors $\boldsymbol{\varphi}_j$ with time dependent coefficients $v_j(t)$ as follows

$$\mathbf{X}(t) = \sum_{j=1}^{N} v_j(t)\boldsymbol{\varphi}_j, \quad (4.42)$$

where the functions $v_j(t)$ entering this decomposition may be thought of as new dynamical variables.

Let us observe that every term in the sum (4.42) has the form of a NNM, whose basis vector $\boldsymbol{\varphi}_j$ determines a displacement pattern, while the functions $v_j(t)$ determine the time evolution of the atomic displacements. Because of this interpretation, we can consider a given dynamical regime $\mathbf{X}(t)$ as a bush of NNMs. In fact, we may also speak of *root* modes and *secondary* modes of a given bush (see below).

Note that each term $v_j(t)\boldsymbol{\varphi}_j$ in the sum (4.42) is *not*, in general, a solution of the dynamical equations of the considered mechanical system, while a specific linear combination of a number of these modes can represent such a solution and thus describe an exact dynamical regime. Sometimes, for brevity, we will use the word "mode" not only for the term $v_j(t)\boldsymbol{\varphi}_j$, but also for $v_j(t)$ itself.

We now write our Hamiltonian as the sum of a kinetic energy and a potential energy $V(\mathbf{X})$ part, and assume that $V(\mathbf{X})$ can be decomposed into a Taylor series with respect to the atomic displacements $q_i(t)$ from their equilibrium positions and all terms whose orders are higher than 2 are neglected. As a result, Newton's equations

$$m_i \ddot{q}_i = -\frac{\partial V}{\partial q_i}, \quad (i = 1, \ldots, N) \tag{4.43}$$

are *linear* differential equations, with constant coefficients. As we discussed in Chap. 2, each normal mode is a particular solution to (4.43) of the form

$$\mathbf{X}(t) = \mathbf{c}\cos(\omega t + \varphi_0), \tag{4.44}$$

where the N-dimensional constant vector $\mathbf{c} = \{c_1, c_2, \ldots, c_N\}$ and the constant phase ϕ_0 are determined by the initial displacements of all particles from their equilibrium state and ω is the normal mode frequency.

Since $V(\mathbf{X})$ has a quadratic form, substituting (4.44) into (4.43) and eliminating $\cos(\omega t + \varphi_0)$ from the resulting equations reduces the problem of finding the normal modes to the task of evaluating the eigenvalues and eigenvectors of the matrix K with coefficients

$$k_{ij} = \left.\frac{\partial^2 V}{\partial q_i \partial q_j}\right|_{\mathbf{X}=0}, \quad i, j = 1, \ldots, N. \tag{4.45}$$

Since the the $N \times N$ matrix K is real and symmetric, it possesses N eigenvectors $\mathbf{c}_j$ ($j = 1, \ldots, N$) and N eigenvalues ω_j^2. The complete collection of these vectors, i.e. the N normal coordinates, can be used as the basis of our configuration space, hence we may write

$$\mathbf{X}(t) = \sum_{i=1}^{N} \mu_j(t)\mathbf{c}_j, \tag{4.46}$$

where $\mathbf{X}(t) = \{q_1(t), q_2(t), \ldots, q_N(t)\}$, while $\mu_j(t)$ are the new dynamical variables that replace the old variables $q_i(t)$ ($i = 1, \ldots, N$).

If the transformation to normal coordinates is used in the absence of degeneracies, the corresponding system of linear ODEs leads to a set of uncoupled harmonic oscillators

$$\ddot{\mu}_j(t) + \omega_j^2 \mu_j(t) = 0, \quad j = 1, \ldots, N,$$

with the well-known solution

$$\mu_j(t) = a_j \cos\left(\omega_j t + \varphi_{0j}\right), \tag{4.47}$$

where a_j and φ_{0j} are arbitrary constants.

Note the distinction we make between a *normal coordinate*, represented by the eigenvector $\mathbf{c}_j$, and a normal mode, referring to the product of the vector $\mathbf{c}_j$ and the time-periodic function $\mu_j(t) = \cos\left(\omega_j t + \varphi_{0j}\right)$.

Let us begin by considering individual NNMs, representing one-dimensional bushes of the FPU-β Hamiltonian. To study bushes of NNMs in monoatomic chains, we shall choose the complete set of normal coordinates $\boldsymbol{\varphi}_k$ as the basis of our configuration space. Here, we use the normal coordinates in the form presented in [272]:

$$\boldsymbol{\varphi}_k = \left\{ \frac{1}{\sqrt{N}} \left[\sin\left(\frac{2\pi k}{N} n\right) + \cos\left(\frac{2\pi k}{N} n\right) \right] \Big| n = 1, \ldots, N \right\}, \quad k = 0, \ldots, N-1, \tag{4.48}$$

where the subscript k refers to the mode and the subscript n refers to the specific particle. The vectors $\boldsymbol{\varphi}_k$, $k = 0, 1, 2, \ldots, N - 1$, form an orthonormal basis, in which we can expand the set of atomic displacements $\mathbf{X}(t)$ corresponding to a given bush as follows

$$\mathbf{X}(t) = \sum_{k=0}^{N-1} v_k(t) \boldsymbol{\varphi}_k \tag{4.49}$$

(see (4.42)). For example, one obtains in this way the following expressions for the bushes $B[\hat{a}^4, \hat{i}]$ and $B[\hat{a}^4, \hat{a}^2\hat{i}]$ (see (4.38) and (4.40))

$$B[\hat{a}^4, \hat{i}] : \mathbf{X}(t) = \{ q_1(t), q_2(t), -q_1(t), -q_2(t) \mid q_1(t), q_2(t),$$
$$- q_1(t), -q_2(t) \mid \ldots \}$$
$$= \mu(t) \boldsymbol{\varphi}_{N/2} + v(t) \boldsymbol{\varphi}_{3N/4}, \tag{4.50}$$

$$B[\hat{a}^4, \hat{a}^2\hat{i}] : \mathbf{X}(t) = \widetilde{\mu}(t) \boldsymbol{\varphi}_{N/2} + \widetilde{v}(t)\text{'}_{N/4}. \tag{4.51}$$

From the complete basis (4.48) only the vectors

$$\boldsymbol{\varphi}_{N/2} = \frac{1}{\sqrt{N}}(-1, 1, -1, 1, -1, 1, -1, 1, -1, 1, -1, 1, \ldots), \tag{4.52}$$

$$\boldsymbol{\varphi}_{N/4} = \frac{1}{\sqrt{N}}(1, -1, -1, 1, 1, -1, -1, 1, 1, -1, -1, 1, \ldots), \tag{4.53}$$

$$\varphi_{3N/4} = \frac{1}{\sqrt{N}}(-1, -1, 1, 1, -1, -1, 1, 1, -1, -1, 1, 1, \ldots), \qquad (4.54)$$

contribute to the two-dimensional bushes (4.50), (4.51), which are equivalent to each other and constitute examples of dynamical domains. For the bush $B[\hat{a}^4, \hat{i}]$, we find the following relations between the old variables $q_1(t)$, $q_2(t)$ (corresponding to the configuration space) and the new ones $\mu(t)$, $\nu(t)$ (corresponding to the modal space)

$$\mu(t) = -\frac{\sqrt{N}}{2}[q_1(t) - q_2(t)],$$
$$\nu(t) = -\frac{\sqrt{N}}{2}[q_1(t) + q_2(t)]. \qquad (4.55)$$

Thus, each of the above bushes consists of two modes. One of these modes is the root mode ($\varphi_{3N/4}$ for the bush $B[\hat{a}^4, \hat{i}]$ and $\varphi_{N/4}$ for the bush $B[\hat{a}^4, \hat{a}^2\hat{i}]$), while the other mode $\varphi_{N/2}$ is the secondary mode. Indeed, according to [81,82,294] the symmetry of the secondary modes must be higher or equal to the symmetry of the root mode. In our case, as we deduce from (4.53), the translational symmetry of the mode $\varphi_{N/4}$ is $\hat{a}^4$ (acting by this element on (4.53) produces the same displacement pattern), while the translational symmetry of the mode $\varphi_{N/2}$ is $\hat{a}^2$, which has twice as many symmetry elements as that of $\varphi_{N/4}$ (see (4.52) and (4.53)). Note that the full symmetry of the modes $\varphi_{N/4}$ and $\varphi_{N/2}$ is represented by $[\hat{a}^4, \hat{a}^2\hat{i}]$ and $[\hat{a}^2, \hat{i}]$, respectively.

Let us now return to the Hamiltonian of the FPU system written as follows

$$H = T + V = \frac{1}{2}\sum_{n=1}^{N}\dot{q}_n^2 + \frac{1}{2}\sum_{n=1}^{N}(q_{n+1} - q_n)^2 + \frac{\gamma}{p}\sum_{n=1}^{N}(q_{n+1} - q_n)^p. \qquad (4.56)$$

Here $p = 3$, $\gamma = \alpha$ for the FPU-α chain, and $p = 4$, $\gamma = \beta$ for the FPU-β chain, while T and V are the kinetic and potential energies, respectively. We assume again periodic boundary conditions.

Let us consider the set of atomic displacements corresponding to the two-dimensional bush $B[\hat{a}^4, \hat{i}]$

$$\mathbf{X}(t) = \{\, x, y, -y, -x \mid x, y, -y, -x \mid x, y, -y, -x \mid \ldots \,\}, \qquad (4.57)$$

where we rename $q_1(t)$ and $q_2(t)$ from (4.50) as $x(t)$ and $y(t)$, respectively. Substituting the particle displacements from (4.57) into the Hamiltonian (4.56), choosing in this case the FPU-α chain, we obtain the kinetic and potential energies

$$T = \frac{N}{4}(\dot{x}^2 + \dot{y}^2), \qquad (4.58)$$

$$V = \frac{N}{4}(3x^2 - 2xy + 3y^2) + \frac{N\alpha}{2}(x^3 + x^2y - xy^2 - y^3). \qquad (4.59)$$

These expressions are valid for an arbitrary FPU-α chain with N mod $4 = 0$. The size of the extended primitive cell for the vibrational state (4.57) is equal to $4a$ and, therefore, when calculating the energies T and V, we may restrict ourselves to summing over only one such cell. In the present case, Newton's equations become

$$\begin{cases} \ddot{x} + (3x - y) + \alpha(3x^2 + 2xy - y^2) = 0, \\ \ddot{y} + (3y - x) + \alpha(x^2 - 2xy - 3y^2) = 0. \end{cases} \tag{4.60}$$

Let us emphasize that these equations do not depend on the number N of the particles in the chain, only N mod $4 = 0$ must hold.

Equations 4.60 are written in terms of the particle displacements $x(t)$ and $y(t)$. From them, it is easy to obtain the dynamical equations for the bush in terms of the normal modes $\mu(t)$ and $v(t)$. Using the relations (4.55) between the old and new variables, we find the following equations for the bush $B[\hat{a}^4, \hat{i}]$ in the modal space

$$\ddot{\mu} + 4\mu - \frac{4\alpha}{\sqrt{N}} v^2 = 0, \tag{4.61}$$

$$\ddot{v} + 2v - \frac{8\alpha}{\sqrt{N}} \mu v = 0. \tag{4.62}$$

Thus, the Hamiltonian for the bush $B[\hat{a}^4, \hat{i}]$, considered as a two-dimensional dynamical system, can be written in the modal space as follows

$$H[\hat{a}^4, \hat{i}] = \frac{1}{2}(\dot{\mu}^2 + \dot{v}^2) + (2\mu^2 + v^2) - \frac{4\alpha}{\sqrt{N}} \mu v^2. \tag{4.63}$$

The stability of bushes of modes was discussed, in general, in [81,82,294], while in the case of the FPU chains it was considered in [85,87]. Following these papers, we also discuss here the stability of a given bush of normal modes with respect to its interactions. Note that there is an essential difference between the interactions of the modes that belong and those that do not belong to a given bush: In the former case, we speak of a "force interaction" and in the latter of a "parametric interaction" (see [81,82]). Let us illustrate this with an example.

As was shown above, the two-dimensional bush $B[\hat{a}^4, \hat{i}]$ for the FPU-α chain is described by (4.61) and (4.62). These equations admit a special solution of the form

$$\mu(t) \neq 0, \quad v(t) \equiv 0. \tag{4.64}$$

which can be excited by imposing the initial conditions: $\mu(t_0) = \mu_0 \neq 0$, $\dot{\mu}(t_0) = 0$, $v(t_0) = 0$, $\dot{v}(t_0) = 0$. Substitution of (4.64) into (4.61) produces the dynamical equation of the one-dimensional bush $B[\hat{a}^2, \hat{i}]$ (see [87]) consisting of only one mode $\mu(t)$

$$\ddot{\mu} + 4\mu = 0, \tag{4.65}$$

with the simple solution

$$\mu(t) = \mu_0 \cos(2t), \tag{4.66}$$

taking (with no loss of generality) the initial phase to be equal to zero. Substituting (4.66) into (4.62), we obtain

$$\ddot{v} + \left[2 - \frac{8\alpha\mu_0}{\sqrt{N}} \cos(2t)\right] v = 0, \tag{4.67}$$

which is easily transformed into the standard form of the Mathieu equation [6]

$$\ddot{v} + [a - 2q\cos(2t)]v = 0. \tag{4.68}$$

As is well-known, there exist domains of stable and unstable motion of the Mathieu equation (4.68) in the (a, q) plane of its parameters [6]. The one-dimensional bush $B[\hat{a}^2, \hat{i}]$ is stable for sufficiently small amplitudes μ_0 of the mode $\mu(t)$, but becomes unstable when $\mu_0 > 0$ is increased. This phenomenon, similar to the well-known parametric resonance, takes place at μ_0 values which lie within the domains of unstable motion of Mathieu's equation (4.67).

The loss of stability of the dynamical regime (4.64) (representing the bush $B[\hat{a}^2, \hat{i}]$), manifests itself in the exponential growth of the mode $v(t)$, which was identically zero for the vibrational state (4.64) and oscillatory for the stable regime of (4.68). As a result of $v(t) \neq 0$, the *dimension* of the original one-dimensional bush $B[\hat{a}^2, \hat{i}]$ increases and the bush is transformed into the two-dimensional bush $B[\hat{a}^4, \hat{i}]$. This is accompanied by a breaking of the symmetry of the vibrational state (the symmetry of the bush $B[\hat{a}^2, \hat{i}]$ is twice as high as that of the bush $B[\hat{a}^4, \hat{i}]$).

In general, we may view a given bush as a stable dynamical object if the complete collection of its modes (and hence also its dimension) do not change in time. All other modes of the system possess zero amplitudes and are therefore called "sleeping" modes. If we increase the intensity of bush vibrations, some sleeping modes (because of parametric interactions with the active modes [81, 82]) can lose their stability and become excited. In this situation, we speak of the *loss of stability* of the original bush, since the dimension of the vibrational state (the number of active modes) becomes larger, while its symmetry becomes lower. Thus, as a consequence of stability loss, the original bush transforms into another bush of higher dimension.

Note, however, that the above destabilization of the (one-dimensional) bush $B[\hat{a}^2, \hat{i}]$ with respect to its transformation into the (two-dimensional) bush $B[\hat{a}^4, \hat{i}]$, falls within the framework of Floquet theory and the analysis of a simple Mathieu's equation. In the case of a more general perturbation of the N particle chain, one must examine the stability of the bush $B[\hat{a}^4, \hat{i}]$ with respect to all other modes, as well. This cannot be done using Floquet theory, since the variations that need to be studied are not about a periodic but a quasiperiodic orbit with two rationally independent frequencies. In Chap. 6, the stability analysis of a multi–dimensional bush is

described theoretically and is performed efficiently and accurately employing the
GALI method described in Chap. 5.

Indeed, as we shall show in Chap. 6, it is possible that tori continue to exist
at energies *higher* than the thresholds (obtained by Floquet theory) where their
"parent" modes become linearly unstable! This ensures stability over a wider
domain than just an infinitesimal neighborhood of the simple periodic orbits
representing the NNM's of the FPU chain.

4.5 Applications to Solid State Physics

As is well-known, the dynamics of Hamiltonian systems is a broad field with a wide
range of applications. Of course, in this book the emphasis is mainly placed on
coupled oscillators of N dof that one encounters primarily in classical mechanics.
However, as we have already seen in Chap. 3, with regard to N–particle one-
dimensional lattices, when N becomes arbitrarily large, it is possible to examine
certain very important issues of interest to statistical mechanics. Still, there are
some basic questions concerning these Hamiltonian systems that may be posed more
generally. One such question addresses the existence and stability of NNMs, which
are the most natural states to consider in certain problems of atomic or molecular
vibrations.

In the previous sections of this chapter, we described a convenient way to study
NNMs, as bushes of orbits in systems which possess discrete symmetries, using
some basic concepts of group theory. From this viewpoint, one realizes that this
approach might perhaps be useful for the study of more general mechanical systems
occurring in solid state physics. This expectation indeed turns out to be fulfilled,
as we will discuss in this section, in the examples of a square molecule in two
dimensions and an octahedral molecule in three dimensions.

4.5.1 Bushes of NNMs for a Square Molecule

Let us consider, therefore, a square molecule represented by a mechanical system
whose equilibrium state is shown in Fig. 4.1. The four atoms of this molecule are
shown as filled circles at the vertices of the square, while the number of every atom
and its (x, y) coordinates are also given in the figure.

If we suppose that the four atoms can oscillate about their equilibrium positions
only in the (x, y) plane, we immediately realize that this model can be described
by a Hamiltonian system possessing eight dof. Furthermore, we will not assume
at this stage any specific type of inter-particle interactions, so that we may treat
bushes of NNMs of this system as purely geometrical objects. The equilibrium
configurations of our molecule possess a symmetry group denoted by C_{4v} below,

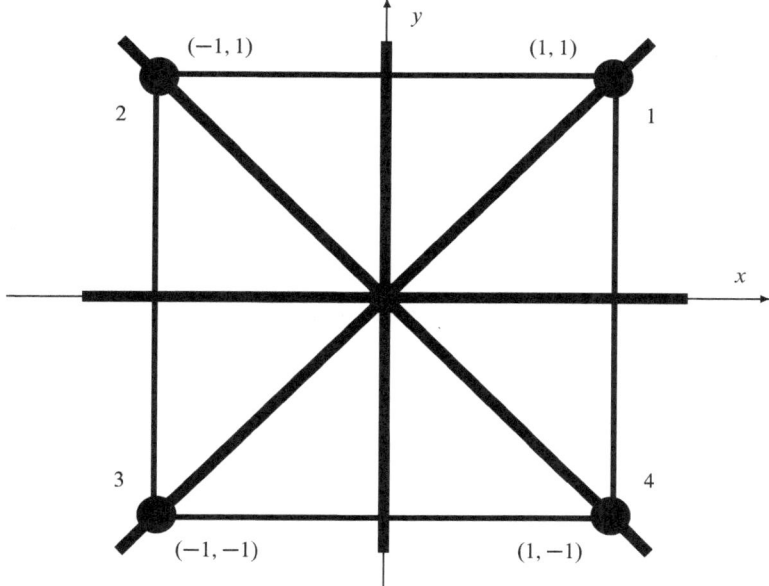

Fig. 4.1 The model of a square molecule with one atom at each of its four vertices (after [64])

according to which, at equilibrium, the system is invariant under the action of the following transformations

1. Rotations by the angles $0°$, $90°$, $180°$, $270°$ about a z axis orthogonal to the plane of Fig. 4.1 passing through the center of the square. We denote these rotations by g_1, g_2, g_3, g_4, respectively.
2. Reflections across four mirror planes orthogonal to the plane of the molecule containing the z axis and shown by bold lines in Fig. 4.1. Two of them are "coordinate" planes (g_5, g_7) and the other two are "diagonal" planes (g_6, g_8).

The above symmetry elements can be explicitly defined as follows

$$g_1(x, y) = (x, y), \quad g_2(x, y) = (-y, x), \quad g_3(x, y) = (-x, -y), \quad g_4(x, y) = (y, -x)$$
$$g_5(x, y) = (-x, y), \quad g_6(x, y) = (-y, -x), \quad g_7(x, y) = (x, -y), \quad g_8(x, y) = (y, x).$$
$$\tag{4.69}$$

Thus, the symmetry group of the square molecule, C_{4v}, contains eight elements $g_1, \ldots, g_8$, determined by (4.69). This group is non-Abelian since, e.g., $g_2 \cdot g_8 = g_5$, while $g_8 \cdot g_2 = g_7$. According to Lagrange's theorem, the order of any subgroup is a divisor of the order of the full group. Therefore, for the case of the group $G = C_{4v}$ with the order $m = 8$, there exist only four subgroups G_j with order equal to $m = 1, 2, 4, 8$ as follows

$$
\begin{aligned}
m = 1 : \quad & G_1 = \{g_1\} = C_1 \\
m = 2 : \quad & G_2 = \{g_1, g_3\} = C_2 \\
& G_3 = \{g_1, g_5\} \text{ and } G_3' = \{g_1, g_7\} = C_s^c \\
& G_4 = \{g_1, g_6\} \text{ and } G_4' = \{g_1, g_8\} = C_s^d \\
m = 4 : \quad & G_5 = \{g_1, g_2, g_3, g_4\} = C_4 \\
& G_6 = \{g_1, g_3, g_5, g_7\} = C_{2v}^c \\
& G_7 = \{g_1, g_3, g_6, g_8\} = C_{2v}^d \\
m = 8 : \quad & G_8 = \{g_1, g_2, g_3, g_4, g_5, g_6, g_7, g_8\} = C_{4v}.
\end{aligned}
\tag{4.70}
$$

Let us suppose now that the equilibrium state of our molecule is stable under arbitrary infinitesimal displacements of the atoms in the (x, y) plane. Moreover, we will also assume that this state is isolated in the sense that within a finite size neighborhood around it there are no other equilibrium states.

Next, we shall consider planar vibrations, i.e. vibrations of the molecule in the plane of its equilibrium configuration. Let us excite a vibrational regime of our molecule by displacing the atoms from their equilibrium positions in a specific manner. As a result of such displacements, the initial configuration of the molecule will have a well defined symmetry described by one of the subgroups of the group G_{4v} listed in (4.70). Indeed, the first configuration in Fig. 4.2 with atoms displaced arbitrarily corresponds to the symmetry group $G_1 = C_1$. In this figure, we depict by arrows the atomic displacements and by thick lines the resulting instantaneous configurations of the molecule.

Note that it is essential for our analysis that the symmetry of the instantaneous configuration of the mechanical system be preserved, during the vibrational motions. More precisely, if an atomic pattern at time t_0 has a symmetry element g, this element cannot disappear spontaneously at any $t > t_0$. This proposition, which we call the "symmetry preservation theorem", can be proved rigorously by examining the Hamiltonian equations of motion. Let us also note, on the other hand, that spontaneous lowering of the symmetry can occur when the dynamical regime loses its stability leading most frequently to the appearance of another bush of higher dimension. This important phenomenon, which may be regarded as the dynamical analogue of a phase transition was discussed in the previous sections of this chapter.

Thus, all possible vibrational regimes of the square molecule can be classified according to eight subgroups of the group $G = C_{4v}$. The different configurations of the vibrating molecule, as depicted in Fig. 4.2, are the following: an arbitrary quadrangle $G_1 = C_1$, Fig. 4.2a, a rotating and pulsating square $G_5 = C_4$, Fig. 4.2b, a parallelogram $G_2 = C_2$, Fig. 4.2c, a rectangle $G_6 = C_{2v}^c$, Fig. 4.2d, a trapezoid $G_3 = C_s^c$, Fig. 4.2e, a rhombus $G_7 = C_{2v}^d$, Fig. 4.2f, a deltoid $G_4 = C_s^d$, Fig. 4.2g, or a pulsating square $G_8 = C_{4v}$, Fig. 4.2h. All these configurations vary in size as time progresses, but the *type* of the corresponding quadrangle does not change. Observe also that these different types of vibrational regimes of the molecule are described by different numbers of dof. Let us discuss this in more detail:

Each of the eight types of vibrational regimes in Fig. 4.2 corresponds to a certain bush of NNMs. For example, the dynamical regime representing a pulsating square

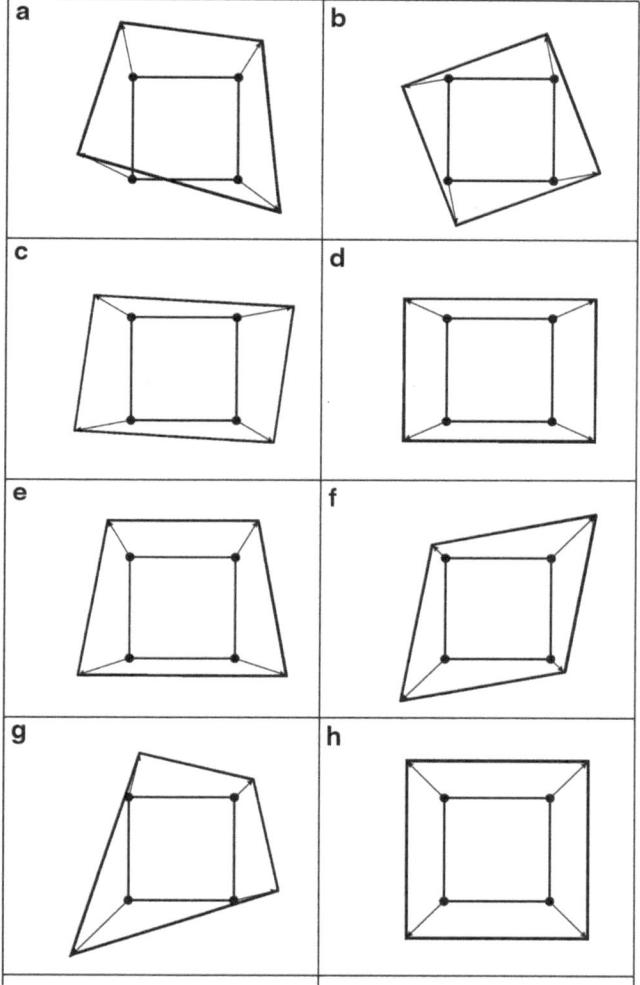

Fig. 4.2 Different vibrational regimes (bushes of NNMs) of the square molecule (after [64])

$G = C_{4v}$, Fig. 4.2h, can be characterized by only one dof, which can be represented either by the edge of the square or the displacement of a certain atom from its equilibrium position along the corresponding diagonal. Thus, such a vibrational regime is described by a one-dimensional bush consisting only of the so-called "breathing" NNM.

On the other hand, the rhombus-like vibration $G_7 = C_{2v}^d$, Fig. 4.2f, and rectangle-like vibration $G_6 = C_{2v}^c$, Fig. 4.2d, are characterized by two dof: The lengths of the diagonals in the former case and the lengths of the adjacent edges in the latter. Thus, both of these vibrational regimes are described by two-dimensional bushes of modes. Similarly, one can check that a trapezoid-type vibration $G = C_s^c$, Fig. 4.2e, corresponds to a three-dimensional bush, the deltoid-type vibration $G_4 = C_s^d$, Fig. 4.2g, to a four-dimensional bush and the vibration with arbitrary quadrangle $G_1 = C_1$, Fig. 4.2a, is represented by a five-dimensional bush of vibrational modes. Note, finally, that vibrations belonging to the $G_5 = C_4$ group of Fig. 4.2b are described by a two-dimensional bush consisting of one rotating and one pulsating mode.

4.5.2 Bushes of NNMs for a Simple Octahedral Molecule

We end this section by describing the occurrence of bushes in a three-dimensional mechanical system consisting of a molecule with six atoms, whose interactions are described by an isotropic pair-particle potential $V(r)$ depending only on the distance between two particles. Here, we suppose that, at equilibrium, these particles form a regular octahedron with edge a_0 as shown in Fig. 4.3. In a Cartesian coordinate system four particles of the octahedron lie in the (x, y) plane and form a square with edge a_0, while the other two particles are located on the z-axis and are called "top particle" and "bottom particle" with respect to the direction of the z-axis. The distance between these particles is also equal to a_0.

At equilibrium, this system possesses the point symmetry group O_h, whose possible bushes of vibrational modes have been listed in [86]. We may suppose, for example, that $u(r)$ is the well-known Lennard-Jones potential

$$u(r) = \frac{1}{r^{12}} - \frac{1}{r^6}, \tag{4.71}$$

as was done in [86], where the stability of the bushes in the octahedral molecule was also studied. In the present discussion, however, we shall consider $u(r)$ as an arbitrary pair-potential. Let us now write the potential energy of our system in its vibrational state in the form

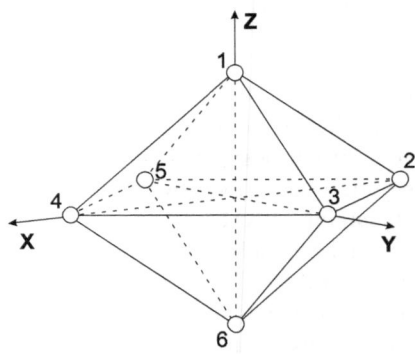

Fig. 4.3 Model of an octahedral molecule consisting of six atoms (after [64])

$$V(\mathbf{X}) = \sum_{\substack{i,j \\ (i<j)}} u(r_{ij}), \tag{4.72}$$

where r_{ij} is the distance between the ith and jth particles. Here, the N-dimensional vector $\mathbf{X} = \{x_1(t), x_2(t), \ldots, x_N(t)\}$ in (4.72) determines the displacements of all particles at an arbitrary instant t, while $N = 18$ is the number of dof of the octahedral mechanical system.

The configuration vector $\mathbf{X}(t) = \{x_1(t), x_2(t), \ldots, x_{18}(t)\}$ determines all the dynamical variables $x_i(t)$ as follows: The first three components of this vector correspond to the x, y and z displacements of particle 1, the next three components correspond to x, y, z displacements of particle 2, etc., as numbered in Fig. 4.3. The dynamics of this system is described by Newton's equations

$$\ddot{x}_i = -\frac{\partial V}{\partial x_i}, \quad i = 1, 2, \ldots, 18,$$

with all particles having mass equal to unity. Each bush of modes corresponds to a certain subgroup G_j of the symmetry group $G = O_h$ of the mechanical system at equilibrium. This means that the vibrational state described by the bush possesses the symmetry group $G_j \subset G$ and, therefore, there exist certain restrictions on the dynamical variables $x_i(t)$. As a result, the number m of dof corresponding to the given bush (i.e. its dimension), is *smaller* than the total number of dynamical variables of the system.

Indeed, at any time t, the atom configuration in a vibrational state represents a polyhedron characterized by the symmetry group $G_j \subset O_h$. Thus, instead of the old variables $x_i(t)$, we introduce m new variables $y_j(t)$ ($j = 1, \ldots, m$), which completely determine this polyhedron. Next, we employ the well-known Lagrange method (see e.g. [209]) for obtaining the dynamical equations in terms of the new variables $y_j(t)$. For this purpose, we introduce the Lagrange function $L = T - V$, where T is the kinetic and V the potential energy expressed as functions of the new variables y_j and new momenta $\dot{y}_j$ ($j = 1, \ldots, m$). The Euler-Lagrange equations in these variables read

$$\frac{\mathrm{d}}{\mathrm{d}t}\left(\frac{\partial L}{\partial \dot{y}_j}\right) - \frac{\partial L}{\partial y_j} = 0, \quad j = 1, \ldots, m, \tag{4.73}$$

where m is the dimension of the considered bush. We will not present here the tedious derivation of these equations for the octahedral molecule (see e.g. [86], where dynamical equations for bushes are obtained by an appropriate MAPLE program). Rather, we will give the final results for the bushes $B[O_h]$, $B[D_{4h}]$ and $B[C_{4v}]$. Here, in the square brackets next to the bush symbol B, we give the symmetry group of the corresponding bush of vibrational modes. The symmetry groups of the above bushes satisfy the following group-subgroup relations

$$C_{4v} \subset D_{4h} \subset O_h. \tag{4.74}$$

The bushes $B[O_h]$, $B[D_{4h}]$, and $B[C_{4v}]$ have one, two and three dimensions, respectively, while their geometrical features can be immediately identified from the symmetry groups that correspond to them. Indeed, the one-dimensional bush $B[O_h]$ consists of only one ("breathing") mode: The appropriate nonlinear dynamical regime describes the evolution of a regular octahedron whose edge $a = a(t)$ periodically changes in time. The two-dimensional bush $B[D_{4h}]$ describes a dynamical regime with two dof. The symmetry group $G = D_{4h}$ of this bush contains the fourfold axis symmetry coinciding with the z coordinate axis and the mirror plane coinciding with the x, y plane. This symmetry group severely restricts the shape of the polyhedron describing our mechanical system in the vibrational state. Indeed, the presence of the fourfold axis demands that the quadrangle in the x, y plane be a square. For the same reason, the four edges connecting the particles in the (x, y) plane at the vertices of the above square with the top particle lying on the z axis must have the same length, denoted here by $b(t)$.

Similarly, let the length of the edges connecting the bottom particle on the z axis with any of the four particles in the (x, y) plane be denoted by $c(t)$. In the case of the bush $B[D_{4h}]$, $b(t) = c(t)$ for any t, due to the horizontal mirror plane in the group $G = D_{4h}$. However, for the three-dimensional bush $B[C_{4v}]$, this mirror plane is absent and, therefore, $b(t) \neq c(t)$. In Fig. 4.4 we illustrate the instantaneous configuration of our mechanical system vibrating according to the bush $B[C_{4v}]$.

Let us also introduce the heights, $h_1(t)$ and $h_2(t)$ for the perpendicular distances, respectively, of the top and bottom vertices of our polyhedron from the (x, y) plane. We can now write the dynamical equations of the above bushes in terms of the purely geometrical variables $a(t)$, $b(t)$, $c(t)$, $h_1(t)$ and $h_2(t)$. Choosing $a(t)$ and $h(t) \equiv h_1(t) \equiv h_2(t)$ as suitable variables for describing the two-dimensional bush $B[D_{4h}]$ and $a(t), h_1(t)$ and $h_2(t)$ as appropriate for describing the three-dimensional bush $B[C_{4v}]$, we express the potential energy for our bushes of vibrational modes as follows

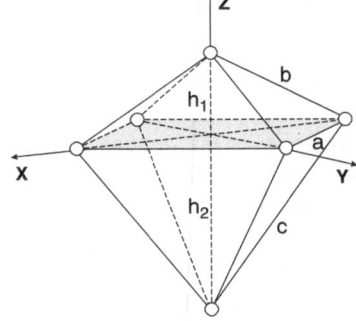

Fig. 4.4 The distorted octahedron shown here illustrates the positions of the particles for the bush $B[C_{4v}]$ at a fixed instant in time (after [64])

$$B[O_h] \; : V(a) = 12u(a) + 3u(\sqrt{2}a),$$

$$B[D_{4h}] : V(a,h) = 4u(a) + 2u(\sqrt{2}a) + 8u\left(\sqrt{h^2 + \tfrac{a^2}{2}}\right) + u(2h),$$

$$B[C_{4v}] : V(a,h_1,h_2) = 4u(a) + 2u(\sqrt{2}a) + 4u(b) + 4u(c) + u(h_1 + h_2),$$

where $b = \sqrt{\frac{a^2}{2} + \left(\frac{5}{4}h_1 - \frac{1}{4}h_2\right)^2}$, $c = \sqrt{\frac{a^2}{2} + \left(\frac{5}{4}h_2 - \frac{1}{4}h_1\right)^2}$.

Using then Euler-Lagrange equations, we obtain the following dynamical equations for the above bushes of vibrational modes

$$B[O_h] :$$
$$\ddot{a} = -4u'(a) - \sqrt{2}u'(\sqrt{2}a);$$
$$B[D_{4h}] :$$
$$\ddot{a} = -2u'(a) - \sqrt{2}u'(\sqrt{2}a) - 2u'(b)\tfrac{a}{b},$$
$$\ddot{h} = -4u'(b)\tfrac{h}{b} - u'(2h); \tag{4.75}$$
$$B[C_{4v}] :$$
$$\ddot{a} = -2u'(a) - \sqrt{2}u'(\sqrt{2}a) - u'(b)\tfrac{a}{b} - u'(c)\tfrac{a}{c},$$
$$\ddot{h}_1 = -u'(b)\tfrac{5h_1 - h_2}{b} - u'(h_1 + h_2),$$
$$\ddot{h}_2 = -u'(c)\tfrac{5h_2 - h_1}{c} - u'(h_1 + h_2).$$

Thus, we have derived dynamical equations for our bushes of vibrational modes in terms of variables having an explicit geometrical meaning. Each bush describes a certain nonlinear dynamical regime corresponding to a vibrational state of the considered system, so that at any fixed time the configuration of this system is represented by a definite polyhedron characterized by the symmetry group G of the given bush. The above dynamical equations for the bushes $B[O_h]$, $B[D_{4h}]$ and $B[C_{4v}]$ are *exact*, but rather complicated to solve. One may thus consider approximate equations, which can be obtained by expanding the potential energy in Taylor series (near the equilibrium state) keeping only the lowest order terms in the expansion. Using then the first few leading terms in such decompositions, it turns out that bushes belonging to systems of different physical nature, with different symmetry groups and different structures, can produce equivalent dynamical equations. More precisely, the equations for many bushes can be written in the same form, and this fact leads to the idea of "classes of dynamical universality" mentioned in [81, 82, 294, 295].

Exercises

Exercise 4.1. Write the equations of motion of N coupled harmonic oscillators under *fixed* boundary conditions (4.2) as a system of ODEs whose rhs is expressed in terms of a tridiagonal matrix S. Using the eigenvectors and eigenvalues of this matrix, perform a basis transformation that diagonalizes S and change to new variables Q_q, given by (4.3), where the eigenvalues of S, $\lambda_q = \omega_q^2$ provide the N normal mode frequencies (4.5). Plot the frequency spectrum ω_q, $q = 1, 2, \ldots, N$ as a function of q. What can you say about the linear independence of these frequencies for general values of N? Can you find N for which they are linearly dependent? Hint: Consider the cases $N + 1$ is a prime or a (positive integer) power of 2.

Exercise 4.2. Repeat the analysis of Exercise 4.1 for N coupled harmonic oscillators under *periodic* boundary conditions and derive expressions (4.6) and (4.7). Note that the matrix S is not exactly tridiagonal in this case, as it contains non-zero entries in its $S_{1,N}$ and $S_{N,1}$ elements. Still it can be diagonalized and and its eigenvalues provide the normal mode frequencies ω_q in the same manner. Now plot these frequencies vs. q and comment on their commensurability (linear dependence) properties.

Problems

Problem 4.1. Consider an FPU-β chain with N=12 atoms and periodic boundary conditions. Study the stability of all six NNMs listed in (4.18)–(4.22) by straightforward integration of the nonlinear dynamical equations of the FPU-β model for different values of their amplitudes. Check that these NNMs are stable for small values of the initial amplitude A. Find maximal values of A for which the dynamical regime corresponding to the considered NNM loses its stability. Compare your results with those presented in [87] in the form of universal stability diagrams, which can be used to analyse the stability of NNMs in the FPU-β chain with an arbitrary number of atoms.

Problem 4.2. Consider the plane square molecule discussed in Sect. 4.5.1, whose atoms interact via the Lennard-Jones potential (4.71). Note that in dimensionless variables you may take $A = B = 1$. Using a mathematical package like MAPLE or MATHEMATICA, find the potential energy of this molecule in the harmonic approximation and obtain all linear normal modes, as well as their frequencies by finding eigenvalues and eigenvectors of the matrix (4.45). Show that the square equilibrium state turns out to be unstable with respect to the rhombus distortion (the eigenfrequency of the mode with rhombus symmetry turns out to be an imaginary number!). Hint: additional information can be found in Appendix B of [87], and in [86].

Problem 4.3. Place an additional atom to the center of the square molecule of Sect. 4.5.1, whose interaction with other four atoms is described by Lennard-Jones potential with coefficients A, B different from those used in Problem 4.2. Show that one can choose such values of A and B for this potential so that the square configuration of the molecule becomes stable, as all normal mode frequencies turn out to be real. Hint: Consult Appendix B of [81], and [86].

Problem 4.4. Study the nonlinear dynamics of the molecule of Problem 4.3 by integrating the corresponding dynamical equations, using rhombus linear normal coordinate as initial displacements of the atoms at the edges of square. Decompose the configuration vector $\mathbf{X}(t)$ according to (4.42), with φ_k being normal coordinates of the square molecule. Check that only two normal coordinates enter this decomposition: the rhombus-like and square-like (all other linear normal coordinates do not contribute). Thus, you have found one example of a two-dimensional bush. Can you find in this way another example of a two-dimensional bush of NNMs? Hint: Consult Appendix B of [81].

Problem 4.5. Choosing initial conditions corresponding to other linear normal coordinates, find all bushes of NNMs for the square molecule depicted in Fig. 4.2. Find also two-dimensional, three-dimensional, etc. bushes for the FPU-β chain. Hint: Consult the results presented in [81] and [82].

Problem 4.6. Following the approach outlined in Sect. 4.5.2 and the references listed therein, show that the analysis of the vibrations of the octahedral molecule depicted in Figs. 4.3 and 4.4, based on the Euler-Lagrange equations (4.73) leads to the dynamical equations for the corresponding bushes listed in (4.75). How would you go about solving these equations, if $u(r)$ is the Lennard-Jones Potential (4.71)? What kind of oscillatory solutions do you find? Hint: Consult results presented in [86].

Chapter 5
Efficient Indicators of Ordered and Chaotic Motion

Abstract This chapter opens with an introduction to the *variational equations*, derived by the linearization of the ordinary differential equations of a Hamiltonian system, and the equations of the *tangent map*, obtained by linearizing the difference equations of a symplectic map. We then focus on the particularly efficient strategy of *alignment indicators* (called SALI and GALI), which employ more than one deviation vector to uncover the various degrees of chaotic behavior of an orbit, as well as the dimensionality of the torus in the case of quasiperiodic motion. We summarize the derivation of asymptotic formulas of these indicators for long times and provide extensive numerical evidence demonstrating their successful application to many examples of Hamiltonian systems and symplectic maps of interest to mathematical physics.

5.1 Variational Equations and Tangent Map

As the reader must have undoubtedly realized, one of the principal aims of this book is to explore in some detail the concepts of *local* vs. *global* stability of motion in multi-dimensional Hamiltonian systems. In particular, we would like to understand the interrelationship between these concepts and investigate their role in revealing important properties of Hamiltonian dynamics in a variety of physical applications.

Global methods, you might rightfully argue, are very rare and quite difficult to apply to systems whose phase space motion varies dramatically from domain to domain. On the other hand, local methods should not be confined only to the analysis of small neighborhoods around fixed points and periodic orbits. As we will find out later in this book, the existence and stability of certain types of tori of quasiperiodic motion and the location and size of "strongly" vs. "weakly" chaotic domains are of crucial importance to many problems of Hamiltonian dynamics.

Thus, even though we will continue to use local methods to study the flow in the tangent space of different orbits of Hamiltonian systems, we shall exploit a new approach that: (1) analyzes as completely as possible the 'geometrical' properties of

T. Bountis and H. Skokos, *Complex Hamiltonian Dynamics*,
Springer Series in Synergetics, DOI 10.1007/978-3-642-27305-6_5,
© Springer-Verlag Berlin Heidelberg 2012

this space, and (2) is rapid and efficient enough to be simultaneously applied to many initial conditions, yielding, as fast as possible, a global portrait of the dynamics. It is this kind of approach that we turn to in the present chapter.

In Chap. 3 we referred to some particular chaos indicators, namely the maximum Lyapunov exponent (MLE), the smaller alignment index (SALI) and the generalized alignment index (GALI), which rely on the evolution of deviation vectors from a given orbit. In this chapter, we will discuss in more detail the behavior of the SALI and the GALI, emphasizing the usefulness of these indices in distinguishing accurately and efficiently between ordered and chaotic motion.

Let us discuss first, for reasons of consistency, the variational equations governing the evolution of deviation vectors in continuous and discrete time, which we have already met repeatedly in previous chapters. For this purpose, let us consider an *autonomous Hamiltonian system* of N dof having a Hamiltonian function

$$H(q_1, q_2, \ldots, q_N, p_1, p_2, \ldots, p_N) = h = \text{constant}, \tag{5.1}$$

where q_i and p_i, $i = 1, 2, \ldots, N$ are the generalized coordinates and conjugate momenta respectively. An orbit in the $2N$-dimensional phase space $\mathscr{S}$ of this system is defined by the vector

$$\mathbf{x}(t) = (q_1(t), q_2(t), \ldots, q_N(t), p_1(t), p_2(t), \ldots, p_N(t)), \tag{5.2}$$

with $x_i = q_i$, $x_{i+N} = p_i$, $i = 1, 2, \ldots, N$. The time evolution of this orbit is governed by *Hamilton's equations of motion*, which in matrix form are given by

$$\dot{\mathbf{x}} = \left[\frac{\partial H}{\partial \mathbf{p}} \ -\frac{\partial H}{\partial \mathbf{q}} \right]^{\mathrm{T}} = \Omega \cdot \nabla H, \tag{5.3}$$

with $\mathbf{q} = (q_1(t), q_2(t), \ldots, q_N(t))$, $\mathbf{p} = (p_1(t), p_2(t), \ldots, p_N(t))$, and

$$\nabla H = \left[\frac{\partial H}{\partial q_1} \ \frac{\partial H}{\partial q_2} \ \cdots \ \frac{\partial H}{\partial q_N} \ \frac{\partial H}{\partial p_1} \ \frac{\partial H}{\partial p_2} \ \cdots \ \frac{\partial H}{\partial p_N} \right]^{\mathrm{T}} \tag{5.4}$$

with $(^{\mathrm{T}})$ denoting the transpose matrix, as we have already seen in Chap. 3. The matrix Ω has the following block form

$$\Omega = \begin{bmatrix} 0_N & I_N \\ -I_N & 0_N \end{bmatrix}, \tag{5.5}$$

(see (3.1) of Sect. 3.1), with I_N being the $N \times N$ identity matrix and 0_N being the $N \times N$ matrix with all its elements equal to zero.

An initial deviation vector $\mathbf{w}(0) = (\delta x_1(0), \delta x_2(0), \ldots, \delta x_{2N}(0))$ from an orbit $\mathbf{x}(t)$ evolves in the *tangent space* $\mathscr{T}_{\mathbf{x}}\mathscr{S}$ of $\mathscr{S}$ according to the *variational equations*

$$\dot{\mathbf{w}} = \left[\Omega \cdot \nabla^2 H(\mathbf{x}(t)) \right] \cdot \mathbf{w} =: B(t) \cdot \mathbf{w}, \tag{5.6}$$

with $\nabla^2 H(\mathbf{x}(t))$ the Hessian matrix of Hamiltonian (5.1) calculated on the reference orbit $\mathbf{x}(t)$, i.e.

$$\nabla^2 H(\mathbf{x}(t))_{i,j} = \left. \frac{\partial^2 H}{\partial x_i \partial x_j} \right|_{\mathbf{x}(t)} , \quad i, j = 1, 2, \ldots, 2N. \tag{5.7}$$

Equations 5.6 are a set of linear differential equations with respect to $\mathbf{w}$, having time dependent coefficients since matrix $B(t)$ depends on the particular reference orbit, which is a function of time t.

In order to follow the evolution of a deviation vector, the variational equations (5.6) have to be integrated simultaneously with the Hamilton's equations of motion (5.3), since matrix $\nabla^2 H(t)$ depends on the particular reference orbit $\mathbf{x}(t)$, which is a solution of (5.3). Any general purpose numerical integration algorithm can be used for the integration of the whole set of (5.3) and (5.6). On the other hand, *symplectic integrators* are often preferred when integrating Hamiltonian dynamical systems. A thorough discussion of such methods can be found in [162] (see also [349, 350]). In [312]. So it was shown that it is possible to integrate the Hamilton's equations of motion *and* the corresponding variational equations using the so-called *"tangent map (TM) method"*, a scheme based on symplectic integration techniques.

According to this method, a symplectic integrator is used to approximate the solution of (5.3) by the repeated action of a symplectic map S, while the corresponding tangent map TS, is used for the integration of (5.6). A simple and systematic technique to construct TS was also presented in [312]. The TM method was proved to be very efficient and superior to other commonly used numerical schemes, both with respect to its accuracy and its speed, for low- [146, 312] and high-dimensional Hamiltonian systems [147].

Let us also consider discrete time ($t = n \in \mathbb{N}$) conservative dynamical systems defined by $2N$D symplectic map F. A *symplectic map* is an area-preserving map whose *Jacobian matrix*

$$M = \nabla F(\mathbf{x}) = \frac{\partial F}{\partial \mathbf{x}} = \begin{bmatrix} \frac{\partial F_1}{\partial x_1} & \frac{\partial F_1}{\partial x_2} & \cdots & \frac{\partial F_1}{\partial x_{2N}} \\ \frac{\partial F_2}{\partial x_1} & \frac{\partial F_2}{\partial x_2} & \cdots & \frac{\partial F_2}{\partial x_{2N}} \\ \vdots & \vdots & \vdots & \vdots \\ \frac{\partial F_{2N}}{\partial x_1} & \frac{\partial F_{2N}}{\partial x_2} & \cdots & \frac{\partial F_{2N}}{\partial x_{2N}} \end{bmatrix}, \tag{5.8}$$

satisfies

$$M^{\mathrm{T}} \cdot \Omega \cdot M = \Omega, \tag{5.9}$$

(see (3.3)). The evolution of an orbit in the $2N$-dimensional space $\mathscr{S}$ of the map is governed by the difference equation

$$\mathbf{x}(n + 1) \equiv \mathbf{x}_{n+1} = F(\mathbf{x}_n). \tag{5.10}$$

In this case, the evolution of a deviation vector $\mathbf{w}(n) \equiv \mathbf{w}_n$, with respect to a reference orbit $\mathbf{x}_n$, is given by the corresponding *tangent map*

$$\mathbf{w}(n+1) \equiv \mathbf{w}_{n+1} = \frac{\partial F}{\partial \mathbf{x}}(\mathbf{x}_n) \cdot \mathbf{w}_n. \tag{5.11}$$

5.2   The SALI Method

The SALI method was introduced in [309] and has been applied successfully to detect regular and chaotic motion in Hamiltonian flows as well as symplectic maps [55, 57, 72, 233, 234, 242, 245, 262, 269, 319, 320, 322, 325]. The basic idea behind the success of the SALI method (which essentially distinguishes it from the computation of MLE) is the introduction of one additional deviation vector with respect to a reference orbit. Indeed, by considering the relation between two deviation vectors (instead of one deviation vector and the reference orbit), one is able in the case of chaotic orbits to circumvent the difficulty of the slow convergence of Lyapunov exponents to non-zero values as $t \to \infty$.

As has already been mentioned in Sect. 3.4.1, in order to compute the SALI one follows simultaneously the time evolution of a reference orbit along with two deviation vectors with initial conditions $\mathbf{w}_1(0)$, $\mathbf{w}_2(0)$. Since we are only interested in the directions of these two vectors we normalize them, from time to time, keeping their norm equal to 1, setting

$$\hat{\mathbf{w}}_i(t) = \frac{\mathbf{w}_i(t)}{\|\mathbf{w}_i(t)\|}, \quad i = 1, 2, \tag{5.12}$$

where $\| \cdot \|$ is the Euclidean norm and the hat ($^\wedge$) over a vector denotes that it is of unit magnitude. The SALI is then defined as

$$\mathrm{SALI}(t) = \min\left\{ \|\hat{\mathbf{w}}_1(t) + \hat{\mathbf{w}}_2(t)\|, \|\hat{\mathbf{w}}_1(t) - \hat{\mathbf{w}}_2(t)\| \right\}, \tag{5.13}$$

whence, it is evident that $\mathrm{SALI}(t) \in [0, \sqrt{2}]$. $\mathrm{SALI} = 0$ indicates that the two deviation vectors have become aligned in the same direction (and are equal or opposite to each other); in other words, they are linearly dependent.

In the case of chaotic orbits, the deviation vectors $\hat{\mathbf{w}}_1$, $\hat{\mathbf{w}}_2$ eventually become aligned in the direction of MLE, and $\mathrm{SALI}(t)$ falls exponentially to zero. An analytical study of SALI's behavior for chaotic orbits was carried out in [314] where it was shown that

$$\mathrm{SALI}(t) \propto e^{-(L_1 - L_2)t}, \tag{5.14}$$

with L_1, L_2 being the two largest LCEs.

In the case of regular motion, on the other hand, the orbit lies on a torus and the vectors $\hat{\mathbf{w}}_1$, $\hat{\mathbf{w}}_2$ eventually fall on its tangent space (following a t^{-1} time evolution) having in general different directions as there is no reason for them to become aligned. This behavior is due to the fact that for regular orbits the norm of a deviation vector increases linearly in time along the flow. Thus, our normalization procedure brings about a decrease of the magnitude of the coordinates perpendicular to the torus at a rate proportional to t^{-1} and so $\hat{\mathbf{w}}_1$, $\hat{\mathbf{w}}_2$ eventually fall on the tangent space of the torus. In this case, the SALI oscillates about values that are different from zero (for more details see [313]), so that

$$SALI \approx const. > 0, \quad t \to \infty. \tag{5.15}$$

A small comment is necessary here. In the case of 2D maps the torus is actually an invariant curve and its tangent space is one-dimensional. Thus, in this case, the two unit deviation vectors eventually become linearly dependent and SALI becomes zero following a power law, SALI $\propto 1/t^2$. This is, of course, different than the exponential decay of SALI for chaotic orbits and thus SALI can distinguish easily between the two cases even in 2D maps [309]. We note that the analytical derivation of SALI's behavior for ordered and chaotic orbits will be given in Sect. 5.3.1 below.

Following [309, 314], let us now demonstrate the behavior of the SALI by considering some typical ordered and chaotic orbits of low-dimensional dynamical systems. Initially we consider an ordered (point A in Fig. 5.1a) and a chaotic orbit (point B in Fig. 5.1a) of the 2D map

$$\begin{aligned} x_1' &= x_1 + x_2 \\ x_2' &= x_2 - \nu \sin(x_1 + x_2) \end{aligned} \quad (\text{mod } 2\pi), \tag{5.16}$$

for $\nu = 0.5$, with initial conditions $x_1 = 2$, $x_2 = 0$ and $x_1 = 3$, $x_2 = 0$, respectively. The evolution of the finite time Lyapunov characteristic number K_t^1, having as limit for $t = n \to \infty$ the MLE (3.42), for both orbits is shown in Fig. 5.1b where

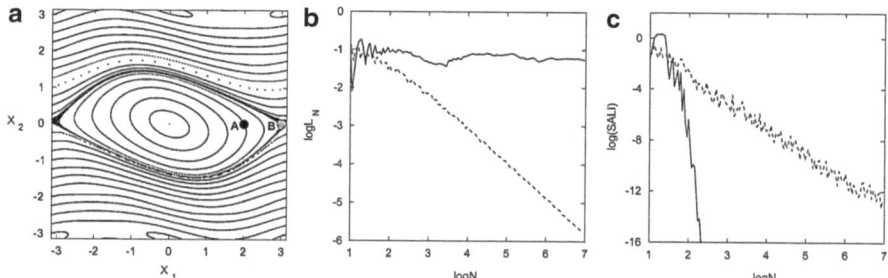

Fig. 5.1 (**a**) Phase plot of the 2D map (5.16) for $\nu = 0.5$. The initial conditions of the ordered orbit A ($x_1 = 2$, $x_2 = 0$) and the chaotic orbit B ($x_1 = 3$, $x_2 = 0$) are marked by *black* and *light-grey filled circles* respectively. The evolution of the finite time Lyapunov characteristic number K_t^1 (denoted by L_N), and the smaller alignment index SALI, with respect to the number n (denoted by N) for orbits A (*dashed curves*) and B (*solid curves*) is plotted in (**b**) and (**c**) respectively (after [309])

it is denoted by L_N. K_t^1 of orbit A (dashed line) decreases as the number n of iterations (which is denoted by N in Fig. 5.1b) increases, following a power law, reaching the value $K_t^1 \approx 1.6 \times 10^{-6}$ after 10^7 iterations. On the other hand, K_t^1 of the chaotic orbit B (solid line), after some fluctuations seems to stabilize near a constant value and becomes $K_t^1 \approx 5 \times 10^{-2}$ after 10^7 iterations.

As we have already explained, since the map (5.16) is 2D any two deviation vectors tend to coincide or become opposite, both for ordered and chaotic orbits. In particular, the SALI of the ordered orbit A decreases as n increases, following a power law (SALI $\propto n^{-2}$) and it becomes SALI $\approx 10^{-13}$ after 10^7 iterations, which means that the two deviation vectors practically coincide (Fig. 5.1c—dashed curve). On the other hand, the SALI of the chaotic orbit B decreases exponentially fast reaching the limit of accuracy of the computer (10^{-16}) after about 200 iterations (Fig. 5.1c—solid curve). After that time the two vectors are identical since their coordinates are represented by the same (or opposite) numbers in the computer. So, the SALI can distinguish between ordered and chaotic motion in a 2D map, since it decreases to zero following completely different time rates.

Let us now study the case of the 4D map

$$
\begin{aligned}
x_1' &= x_1 + x_2 \\
x_2' &= x_2 - \nu \sin(x_1 + x_2) - \mu[1 - \cos(x_1 + x_2 + x_3 + x_4)] \\
x_3' &= x_3 + x_4 \\
x_4' &= x_4 - \kappa \sin(x_3 + x_4) - \mu[1 - \cos(x_1 + x_2 + x_3 + x_4)]
\end{aligned}
\quad (\text{mod } 2\pi), \quad (5.17)
$$

for $\nu = 0.5$, $\kappa = 0.1$ and $\mu = 10^{-3}$ by considering the orbits C and D with initial conditions $x_1 = 0.5$, $x_2 = 0$, $x_3 = 0.5$, $x_4 = 0$ and $x_1 = 3$, $x_2 = 0$, $x_3 = 0.5$, $x_4 = 0$ respectively. The projections on the $x_1 x_2$ plane of some thousands of consequents of these orbits, as well as their initial conditions are shown in Fig. 5.2a. From this figure we can easily guess that orbit C is ordered, since its

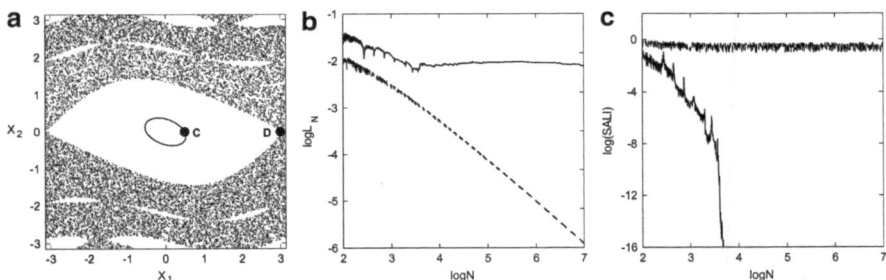

Fig. 5.2 (a) Projections on the $x_1 x_2$ plane of the orbits with initial conditions $x_1 = 0.5$, $x_2 = 0$, $x_3 = 0.5$, $x_4 = 0$ (ordered orbit C) and $x_1 = 3$, $x_2 = 0$, $x_3 = 0.5$, $x_4 = 0$ (chaotic orbit D) of the 4D map (5.17) for $\nu = 0.5$, $\kappa = 0.1$, $\mu = 10^{-3}$. The initial conditions of the orbits are marked by *filled circles*. The evolution of K_t^1 (denoted by L_N), and the SALI, with respect to the number n of iterations (denoted by N), for the ordered orbit C (*dashed curves*) and for the chaotic orbit D (*solid curves*) is plotted in (b) and (c) respectively (after [309])

projection on the $x_1 x_2$ plane forms a closed curve, while orbit D is chaotic, since it produces scattered points on the $x_1 x_2$ plane. This guess is verified from the evolution of K_t^1 (denoted by L_N), shown in Fig. 5.2b.

In Fig. 5.2c we show the evolution of the SALI for these two orbits. For the ordered orbit C the SALI remains almost constant, fluctuating around SALI ≈ 0.28, while it decreases abruptly in the case of the chaotic orbit D, reaching the limit of accuracy of the computer (10^{-16}) after about 4.7×10^3 iterations. After that time the coordinates of the two vectors are represented by opposite numbers in the computer, and the chaotic nature of the orbit is determined beyond any doubt. On the other hand, the computation of the MLE (Fig. 5.2b) is not able to definitely characterize the orbit as chaotic after the same number of iterations, since it is not yet clear if K_t^1 will tend to a positive value.

Let us now verify the validity of (5.14) for two simple Hamiltonian systems with two and three dof: the well-known Hénon-Heiles system [164], having the Hamiltonian function (2.36)

$$H_2 = \frac{1}{2}(p_x^2 + p_y^2) + \frac{1}{2}(x^2 + y^2) + x^2 y - \frac{1}{3}y^3, \tag{5.18}$$

and the three dof Hamiltonian

$$H_3 = \frac{1}{2}(p_x^2 + p_y^2 + p_z^2) + \frac{1}{2}(Ax^2 + By^2 + Cz^2) - \epsilon x z^2 - \eta y z^2, \tag{5.19}$$

studied in [100, 101]. We keep the parameters of the two systems fixed at the energies $H_2 = 0.125$ and $H_3 = 0.00765$, with $A = 0.9$, $B = 0.4$, $C = 0.225$, $\epsilon = 0.56$ and $\eta = 0.2$.

We recall that two dof Hamiltonian systems have only one positive LCE, L_1, since the second largest is $L_2 = 0$. So, (5.14) becomes for such systems

$$\mathrm{SALI}(t) \propto e^{-L_1 t}. \tag{5.20}$$

In Fig. 5.3a we plot in log-log scale K_t^1 (denoted by L_t) as a function of time t for a chaotic orbit of the system (5.18). K_t^1 remains different from zero, which implies the chaotic nature of the orbit. Following its evolution for a sufficiently long time interval to achieve reliable estimates ($t \approx 10,000$) we obtain $L_1 \approx 0.047$. In Fig. 5.3b we plot the SALI for the same orbit (solid curve) using linear scale for the time t. Again we conclude that the orbit is chaotic as SALI $\approx 10^{-16}$ for $t \approx 800$. If (5.20) is valid, the slope of the SALI in Fig. 5.3b should be given approximately by $-L_1/\ln 10$, because $\log(\mathrm{SALI})$ is a linear function of t. From Fig. 5.3b we see that a line having precisely this slope with $L_1 = 0.047$ (dashed line) approximate quite accurately the computed values of the SALI. We note here that a MAPLE algorithm for the computation of the SALI for orbits of the Hénon-Heiles system is presented in Appendix B.

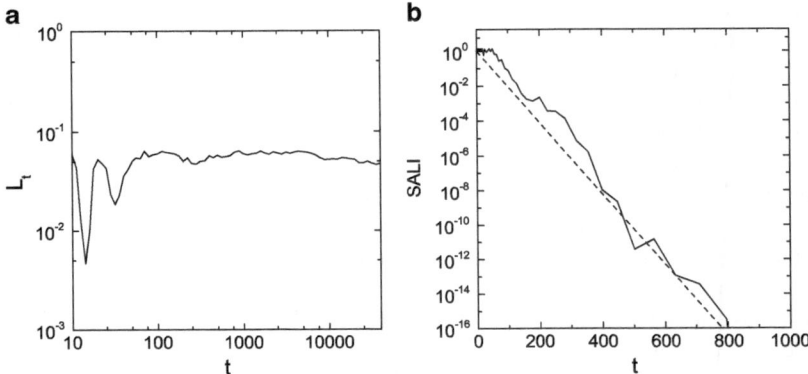

Fig. 5.3 (**a**) The evolution of K_t^1 (denoted by L_t) for the chaotic orbit with initial condition $x = 0$, $y = -0.25$, $p_x \simeq 0.42081$, $p_y = 0$ of the 2D system (5.18). (**b**) The SALI of the same orbit (*solid curve*) and a function proportional to $e^{-L_1 t}$ (*dashed curve*) for $L_1 = 0.047$. Note that the t-axis is linear (after [309])

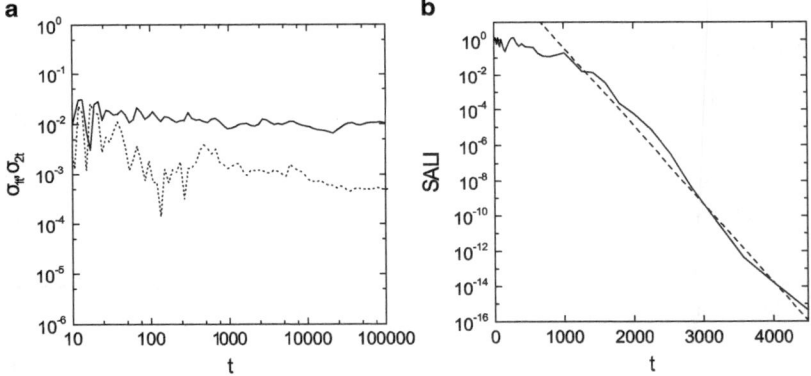

Fig. 5.4 (**a**) The evolution of the two finite time Lyapunov characteristic numbers K_t^1, K_t^2 (denoted by σ_{1t} and σ_{2t} respectively) for the chaotic orbit with initial condition $x = 0$, $y = 0$, $z = 0$, $p_x = 0$, $p_y = 0$, $p_z \simeq 0.123693$ of the 3D system (5.19). (**b**) The SALI of the same orbit (*solid curve*) and a function proportional to $e^{-(L_1 - L_2)t}$ (*dashed curve*) for $L_1 = 0.0107$, $L_2 = 0.0005$. Note that the t-axis is linear (after [309])

Chaotic orbits of Hamiltonian systems with three dof generally have two positive Lyapunov exponents, L_1 and L_2. So, for approximating the behavior of the SALI by (5.14), both L_1 and L_2 are needed. We compute L_1, L_2 for a chaotic orbit of the 3D system (5.19) as the long time estimates of some appropriate quantities, K_t^1, K_t^2, by applying the method proposed by Benettin et al. [32] (see also [310] for more details). The results are presented in Fig. 5.4a. The computation is carried out until K_t^1 and K_t^2 stop having high fluctuations and approach some non-zero values (since the orbit is chaotic), which could be considered as good approximations of their

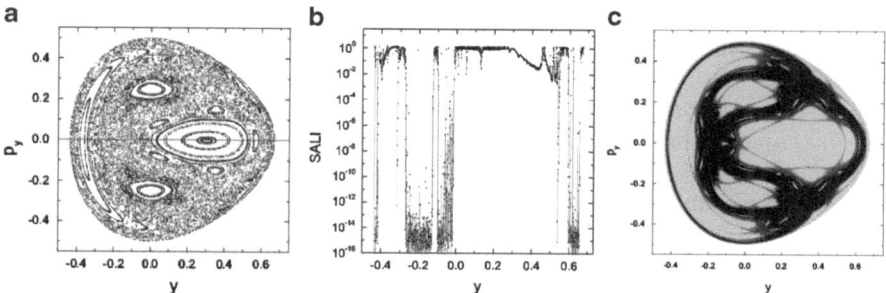

Fig. 5.5 (**a**) The $x = 0$ PSS of the 2D Hénon-Heiles system (5.18). The axis $p_y = 0$ is also plotted. (**b**) The values of the SALI at $t = 1,000$ for orbits with initial conditions on the $p_y = 0$ line of panel (**a**) as a function of the y coordinate of the initial condition. (**c**) Regions of different values of the SALI on the PSS of panel (**a**) at $t = 1,000$. The initial conditions are colored black if their SALI $\leqslant 10^{-12}$, deep gray if $10^{-12} <$ SALI $\leqslant 10^{-8}$, gray if $10^{-8} <$ SALI $\leqslant 10^{-4}$ and light gray if SALI $> 10^{-4}$ (after [309])

limits L_1, L_2. Actually for $t \approx 10^5$ we have $K_t^1 \approx 0.0107$, $K_t^2 \approx 0.0005$. Using these values as good approximations of L_1, L_2 we see in Fig. 5.4b that the slope of the SALI (solid curve) is well reproduced by (5.14) (dashed curve).

Exploiting the advantages of the SALI we can use it to efficiently identify ordered and chaotic orbits in large regions of phase space where large scale ordered and chaotic regions are both present. As an example, we consider the two dof system (5.18), whose PSS for $x = 0$ is presented in Fig. 5.5a. Regions of ordered motion, around stable periodic orbits, are seen to coexist with chaotic regions filled by scattered points. In order to demonstrate the effectiveness of the SALI method, we first consider orbits whose initial conditions lie on the line $p_y = 0$. In particular we take 5,000 equally spaced initial conditions on this line and compute the value of the SALI for each one. The results are presented in Fig. 5.5b where we plot the SALI as a function of the y coordinate of the initial condition of these orbits for $t = 1,000$. The data points are line connected, so that the changes of the SALI values are clearly visible.

Note that there are intervals where the SALI has large values (e.g. larger than 10^{-4}), which correspond to ordered motion in the island of stability crossed by the $p_y = 0$ line in Fig. 5.5a. There also exist regions where the SALI has very small values (e.g. smaller than 10^{-12}) denoting that in these regions the motion is chaotic. These intervals correspond to the regions of scattered points crossed by the $p_y = 0$ line in Fig. 5.5a. Although most of the initial conditions give large ($>10^{-4}$) or very small ($\leqslant 10^{-12}$) values for the SALI, there also exist initial conditions that have intermediate values of the SALI ($10^{-12} <$ SALI $\leqslant 10^{-4}$). These initial conditions correspond to *sticky* chaotic orbits, remaining for long time intervals at the borders of islands, whose chaotic nature will be revealed later on (see also Chap. 8). Note that it is not easy to define a threshold value, such that if the SALI is smaller, this reliably signifies chaoticity. Nevertheless, numerical experiments in several systems show that in general a good guess for this value could be $\lesssim 10^{-4}$.

In Fig. 5.5b, around $y \approx -0.1$ there exists a group of points inside a big chaotic region having SALI $> 10^{-4}$. These points correspond to orbits with initial conditions inside a small stability island, which is not easily visible in Fig. 5.5a. Also the point with $y = -0.2088$ has very high value of the SALI (>0.1) in Fig. 5.5b, while all its neighboring points have SALI $< 10^{-9}$. This point actually corresponds to an ordered orbit inside a tiny island of stability, which can be revealed only after a very high magnification of this region of the PSS. Thus, we conclude that the systematic application of the SALI method can reveal very fine details of the dynamics.

By carrying out the above analysis for points not only along a line but on the whole plane of the PSS, and giving to each point a color according to the value of the SALI, we can have a clear picture of the regions where chaotic or ordered motion occurs. The outcome of this procedure is presented in Fig. 5.5c where the values of the logarithm of the SALI are divided in four intervals. Initial conditions having different values of the SALI at $t = 1,000$ are plotted by different shades of gray: black if SALI $\leq 10^{-12}$, deep gray if $10^{-12} <$ SALI $\leq 10^{-8}$, gray if $10^{-8} <$ SALI $\leq 10^{-4}$ and light gray if SALI $> 10^{-4}$. Thus, in Fig. 5.5c we clearly distinguish between light gray regions, where the motion is ordered and black regions, where it is chaotic. At the borders between these regions we find deep gray and gray points, which correspond to sticky chaotic orbits. It is worth-mentioning that in Fig. 5.5c we can see small islands of stability inside the large chaotic sea, which are not visible in the PSS of Fig. 5.5a, like the one for $y \approx -0.1$, $p_y \approx 0$.

As an example of SALI's usefulness for investigating dynamical systems we present the application of the index to a model of a simplified accelerator ring with linear frequencies (tunes) q_x, q_y, having a localized thin sextupole magnet. The evolution of a charged particle in this ring is modeled by the 4D symplectic map

$$
\begin{pmatrix} x_1' \\ x_2' \\ x_3' \\ x_4' \end{pmatrix} = \begin{pmatrix} \cos\omega_1 & -\sin\omega_1 & 0 & 0 \\ \sin\omega_1 & \cos\omega_1 & 0 & 0 \\ 0 & 0 & \cos\omega_2 & -\sin\omega_2 \\ 0 & 0 & \sin\omega_2 & \cos\omega_2 \end{pmatrix} \times \begin{pmatrix} x_1 \\ x_2 + x_1^2 - x_3^2 \\ x_3 \\ x_4 - 2x_1 x_3 \end{pmatrix}, \quad (5.21)
$$

where x_1 (x_3) denotes the initial deflection from the ideal circular orbit in the horizontal (vertical) direction before the particle enters the magnetic element, and x_2 (x_4) is the associated momentum [57, 59]. Primes denote positions and momenta after one turn in the ring. The parameters ω_1 and ω_2 are related to the accelerator's tunes q_x and q_y by $\omega_1 = 2\pi q_x$ and $\omega_2 = 2\pi q_y$. In [57] the particular case of $q_x = 0.61803$, $q_y = 0.4152$ was considered and the SALI method was applied to construct phase space "charts", where regions of chaos and order were clearly

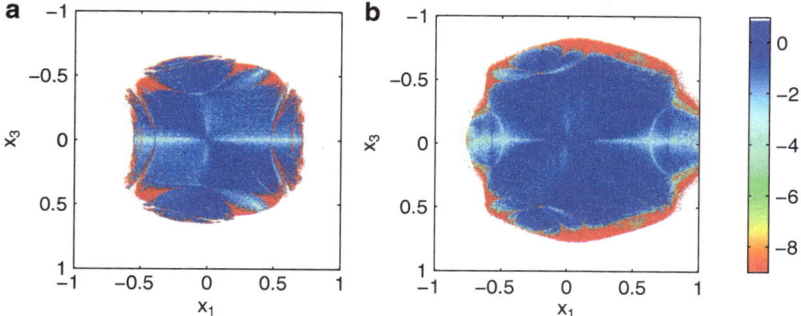

Fig. 5.6 Regions of different SALI values on (**a**) the (x_1, x_3) plane of the uncontrolled map (5.21), and (**b**) the controlled map constructed in [49]. $16,000$ uniformly distributed initial conditions in the square $(x_1, x_3) \in [-1, 1] \times [-1, 1]$, $x_2(0) = x_4(0) = 0$ are followed for 10^5 iterations, and they are colored according to their final $\log(\text{SALI})$ value. The *white colored* regions correspond to orbits that escape in less than 10^5 iterations (after [50])

Fig. 5.7 The percentages of regular (*solid curves*) and chaotic (*dashed curves*) orbits after $n = 10^5$ iterations of the uncontrolled map (5.21) (*black curves*), the controlled map of [49] (*blue curves*), in a four-dimensional hypersphere centered at $x_1 = x_2 = x_3 = x_4 = 0$, as a function of the hypersphere radius r (after [50])

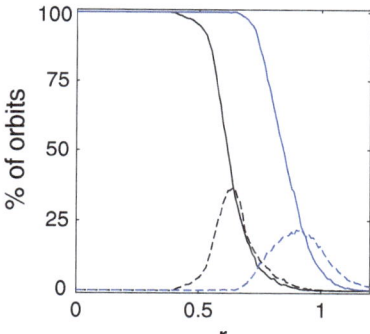

identified in Fig. 5.6, and for evaluating the percentages of ordered and chaotic orbits in a hypersphere centered at the origin $x_1 = x_2 = x_3 = x_4 = 0$ (black curves in Fig. 5.7). Later on, in [49, 50] chaos control techniques were used to increase the so-called *dynamic aperture* (i.e. the stability domain around the ideal circular orbit) of the system. The procedure was successful since the constructed controlled map has a larger stability region as can be seen from Figs. 5.6 and 5.7.

From all the results presented above we conclude that the SALI offers indeed an easy and efficient method for distinguishing the chaotic vs. ordered nature of orbits in a variety of problems, since it tends exponentially to zero in the case of chaotic orbits, while it fluctuates around non-zero values for regular trajectories of Hamiltonian systems and $2N$D symplectic maps with $N > 1$. In the case of 2D maps, the SALI tends to zero both for regular and chaotic orbits but with very different time rates, which allows us again to distinguish between the two cases

[309]. In particular the SALI tends to zero following an exponential law for chaotic orbits and decays to zero following a power law for regular orbits.

Thus, although the behavior of SALI in 2D maps is clearly understood, the fact remains that SALI does not always have the same behavior for regular orbits, as it may oscillate about a constant or decay to zero by a power law, depending on the dimensionality of the tangent space of the reference orbit. It is therefore interesting to ask whether this index can be generalized, so that different power laws may be found to characterize regular motion in higher dimensions.

Let us make one more remark concerning the behavior of SALI for chaotic orbits. Looking at (5.14), one might wonder what would happen in the case of a chaotic orbit whose two largest Lyapunov exponents L_1 and L_2 are equal or almost equal. Although this may not be common in generic Hamiltonian systems, such cases can be found in the literature. In one such example presented in [16], very close to a particular unstable periodic orbit of a 15 dof Hamiltonian system, the two largest Lyapunov exponents are nearly equal $L_1 - L_2 \approx 0.0002$. Even though, in that example, SALI still tends to zero at the rate indicated by (5.14), it is evident that the chaotic nature of an orbit cannot be revealed very fast by the SALI method. How can we overcome this problem? Can we define, for example, a generalized index that depends on *several* Lyapunov exponents (instead of only the two largest ones), as this might accelerate considerably the identification of chaotic motion? The answer is yes. Such a generalized index exists, and is discussed in detail in the next section.

5.3 The GALI Method

Observe first that seeking the minimum of the two positive quantities in (5.13) (which are bounded above by 2) is essentially equivalent to evaluating the product

$$P(t) = \|\hat{\mathbf{w}}_1(t) + \hat{\mathbf{w}}_2(t)\| \cdot \|\hat{\mathbf{w}}_1(t) - \hat{\mathbf{w}}_2(t)\| \,, \qquad (5.22)$$

at every value of t. Indeed, if the minimum of these two quantities is zero (as in the case of a chaotic reference orbit), so will be the value of $P(t)$. On the other hand, if it is not zero, $P(t)$ will be proportional to the constant about which this minimum oscillates (as in the case of regular motion). This suggests that, instead of computing the SALI(t) from (5.13), one might as well evaluate the "exterior" or "wedge" product of the two deviation vectors $\hat{\mathbf{w}}_1 \wedge \hat{\mathbf{w}}_2$ for which it holds

$$\|\hat{\mathbf{w}}_1 \wedge \hat{\mathbf{w}}_2\| = \frac{\|\hat{\mathbf{w}}_1 - \hat{\mathbf{w}}_2\| \cdot \|\hat{\mathbf{w}}_1 + \hat{\mathbf{w}}_2\|}{2} \,, \qquad (5.23)$$

and which represents the "area" of the parallelogram formed by the two deviation vectors (see Exercise 5.1). For the definition of the wedge product see the Appendix at the end of this chapter. Indeed, the 'wedge' product of two vectors can be generalized to represent the "volume" of a parallelepiped formed by the deviation vectors $\hat{\mathbf{w}}_1, \hat{\mathbf{w}}_2, \ldots, \hat{\mathbf{w}}_k$, $2 \leq k \leq 2N$, an N dof Hamiltonian system, or a $2N$D symplectic map.

Hence, the SALI is suitable for checking whether or not two normalized deviation vectors $\hat{\mathbf{w}}_1$, $\hat{\mathbf{w}}_2$ (having norm 1), eventually become linearly dependent, by falling in the same direction. The linear dependence of the two vectors is equivalent to the vanishing of the "area" of the parallelogram having as edges the two vectors. Generalizing this idea we now follow the evolution of k deviation vectors $\hat{\mathbf{w}}_1$, $\hat{\mathbf{w}}_2$, ..., $\hat{\mathbf{w}}_k$, with $2 \leq k \leq 2N$, and determine whether these eventually become linearly dependent, by checking if the "volume" of the parallelepiped having these vectors as edges goes to zero. This volume is equal to the norm of the wedge product of these vectors and is defined as the Generalized Alignment Index (GALI) of order k

$$\text{GALI}_k(t) = \|\hat{\mathbf{w}}_1(t) \wedge \hat{\mathbf{w}}_2(t) \wedge \cdots \wedge \hat{\mathbf{w}}_k(t)\| . \tag{5.24}$$

Let us see how one can actually evaluate this norm and consequently the GALI_k. All normalized deviation vectors $\hat{\mathbf{w}}_i$, $i = 1, 2, \ldots, k$, belong to the $2N$-dimensional tangent space of the Hamiltonian flow. Using as a basis of this space the usual set of orthonormal vectors

$$\hat{e}_1 = (1, 0, 0, \ldots, 0), \hat{e}_2 = (0, 1, 0, \ldots, 0), \ldots, \hat{e}_{2N} = (0, 0, 0, \ldots, 1), \tag{5.25}$$

any deviation vector $\hat{\mathbf{w}}_i$ can be written as a linear combination

$$\hat{\mathbf{w}}_i = \sum_{j=1}^{2N} w_{ij} \hat{e}_j , \quad i = 1, 2, \ldots, k, \tag{5.26}$$

where w_{ij} are real numbers satisfying $\sum_{j=1}^{2N} w_{ij}^2 = 1$. Thus, (5.26) gives

$$\begin{bmatrix} \hat{\mathbf{w}}_1 \\ \hat{\mathbf{w}}_2 \\ \vdots \\ \hat{\mathbf{w}}_k \end{bmatrix} = \begin{bmatrix} w_{11} & w_{12} & \cdots & w_{1\,2N} \\ w_{21} & w_{22} & \cdots & w_{2\,2N} \\ \vdots & \vdots & & \vdots \\ w_{k1} & w_{k2} & \cdots & w_{k\,2N} \end{bmatrix} \cdot \begin{bmatrix} \hat{e}_1 \\ \hat{e}_2 \\ \vdots \\ \hat{e}_{2N} \end{bmatrix} = W \cdot \begin{bmatrix} \hat{e}_1 \\ \hat{e}_2 \\ \vdots \\ \hat{e}_{2N} \end{bmatrix}, \tag{5.27}$$

and according to (5.24) the GALI_k is evaluated as

$$\text{GALI}_k = \|\hat{\mathbf{w}}_1 \wedge \hat{\mathbf{w}}_2 \wedge \cdots \wedge \hat{\mathbf{w}}_k\| = \left\{ \sum_{1 \leq i_1 < i_2 < \cdots < i_k \leq 2N} \begin{vmatrix} w_{1i_1} & w_{1i_2} & \cdots & w_{1i_k} \\ w_{2i_1} & w_{2i_2} & \cdots & w_{2i_k} \\ \vdots & \vdots & & \vdots \\ w_{ki_1} & w_{ki_2} & \cdots & w_{ki_k} \end{vmatrix}^2 \right\}^{1/2} , \tag{5.28}$$

where the sum is performed over all possible combinations of k indices out of $2N$.

In order to compute $GALI_k$, therefore, we follow the evolution of an orbit and of k initially linearly independent unit deviation vectors $\hat{\mathbf{w}}_i$, $i = 1, 2, \ldots, k$. From time to time we normalize these deviation vectors to unity, keeping their directions intact, and compute $GALI_k$ as the norm of their wedge product using (5.28).

Consequently, if $GALI_k(t)$ tends to zero, this would imply that the volume of the parallelepiped having the vectors $\hat{\mathbf{w}}_i$ as edges also shrinks to zero, as at least one of the deviation vectors becomes linearly dependent on the remaining ones. On the other hand, if $GALI_k(t)$ remains far from zero, as t grows arbitrarily, this would indicate the linear independence of the deviation vectors and the existence of a corresponding parallelepiped, whose volume is different from zero for all time.

Equation 5.28 is ideal for the theoretical determination of the asymptotic behavior of GALIs for chaotic and ordered orbits (see Sect. 5.3.1 below). However, from a practical point of view using (5.28) for the numerical evaluation of $GALI_k$ is not very efficient as it might require the computation of a large number of determinants. In [14, 316] a more efficient numerical technique for the computation of $GALI_k$, was proposed. This technique is based on the Singular Value Decomposition of the matrix W in (5.27), whose rows are the coordinates of the unit deviation vectors $\hat{\mathbf{w}}_i$, $i = 1, 2, \ldots, k$. In particular, it has been shown that $GALI_k$ is equal to the product of the singular values $z_i \geq 0$, $i = 1, 2, \ldots, k$ of W^T

$$GALI_k = \prod_{i=1}^{k} z_i, \qquad (5.29)$$

(see Exercise 5.2).

5.3.1 Theoretical Results for the Time Evolution of GALI

5.3.1.1 Exponential Decay of GALI for Chaotic Orbits

In order to investigate the dynamics in the vicinity of a chaotic orbit of an N dof Hamiltonian system or a $2N$D symplectic map, let us first recall some well-known properties of the LCEs of these systems, following for example [310]. Oseledec [259] has shown that the mean exponential rate of divergence $L(\mathbf{x}(0), \mathbf{w})$ from a reference orbit with initial condition $\mathbf{x}(0)$ given by

$$L(\mathbf{x}(0), \mathbf{w}) = \lim_{t \to \infty} \frac{1}{t} \ln \frac{\|\mathbf{w}(t)\|}{\|\mathbf{w}(0)\|}, \qquad (5.30)$$

exists and is finite. Furthermore, there is a $2N$-dimensional basis $\{\hat{u}_1, \hat{u}_2, \ldots, \hat{u}_{2N}\}$ of the system's tangent space so that $L(\mathbf{x}(0), \mathbf{w})$ takes one of the $2N$ (possibly non-distinct) values

$$L_i(\mathbf{x}(0)) = L(\mathbf{x}(0), \hat{u}_i), \quad i = 1, 2, \ldots, 2N \qquad (5.31)$$

which are the LCEs ordered in size as

$$L_1 \geq L_2 \geq \ldots \geq L_{2N}. \tag{5.32}$$

In addition, it has been proven in [32] that the LCEs are grouped in pairs of opposite sign, i.e. $L_i = -L_{2N-i+1}$, $i = 1, 2, \ldots, N$, while two of them are equal to zero ($L_N = L_{N+1} = 0$) in the case of autonomous Hamiltonian systems.

The time evolution of an initial deviation vector

$$\mathbf{w}(0) = \sum_{i=1}^{2N} c_i \, \hat{u}_i, \tag{5.33}$$

can be approximated by

$$\mathbf{w}(t) = \sum_{i=1}^{2N} c_i \, e^{d_i t} \hat{u}_i, \tag{5.34}$$

where c_i, d_i are real numbers depending on the specific phase space location through which the reference orbit passes. Thus, the quantities d_i, $i = 1, 2, \ldots, 2N$ may be thought of as "local Lyapunov exponents" having as limits for $t \to \infty$ the LCEs L_i, $i = 1, 2, \ldots, 2N$.

Assuming that, after a certain time interval, the d_i, $i = 1, 2, \ldots, 2N$, do not fluctuate significantly about their limiting values, we write $d_i \approx L_i$ and express the evolution of the deviation vectors $\mathbf{w}_i$ in the form

$$\mathbf{w}_i(t) = \sum_{j=1}^{2N} c_j^i \, e^{L_j t} \hat{u}_j. \tag{5.35}$$

Thus, if $L_1 > L_2$, a leading order estimate of the deviation vector's Euclidean norm (for t large enough), is given by

$$\|\mathbf{w}_i(t)\| \approx |c_1^i| e^{L_1 t}. \tag{5.36}$$

Consequently, the matrix W in (5.27) of coefficients of k normalized deviation vectors $\hat{\mathbf{w}}_i(t) = \mathbf{w}_i(t)/\|\mathbf{w}_i(t)\|$, $i = 1, 2, \ldots, k$, with $2 \leq k \leq 2N$, using as basis of the vector space the set $\{\hat{u}_1, \hat{u}_2, \ldots, \hat{u}_{2N}\}$ becomes

$$C(t) = \begin{bmatrix} s_1 & \frac{c_2^1}{|c_1^1|}e^{-(L_1-L_2)t} & \frac{c_3^1}{|c_1^1|}e^{-(L_1-L_3)t} & \ldots & \frac{c_{2N}^1}{|c_1^1|}e^{-(L_1-L_{2N})t} \\ s_2 & \frac{c_2^2}{|c_1^2|}e^{-(L_1-L_2)t} & \frac{c_3^2}{|c_1^2|}e^{-(L_1-L_3)t} & \ldots & \frac{c_{2N}^2}{|c_1^2|}e^{-(L_1-L_{2N})t} \\ \vdots & \vdots & \vdots & & \vdots \\ s_k & \frac{c_2^k}{|c_1^k|}e^{-(L_1-L_2)t} & \frac{c_3^k}{|c_1^k|}e^{-(L_1-L_3)t} & \ldots & \frac{c_{2N}^k}{|c_1^k|}e^{-(L_1-L_{2N})t} \end{bmatrix}, \tag{5.37}$$

with $s_i = \text{sign}(c_1^i)$ and $i = 1, 2, \ldots, k$.

Note that this matrix is *not identical* to the one appearing in (5.28) as it is not expressed with respect to the basis (5.25). Nevertheless, we shall proceed with our analysis using matrix $C(t)$, as we do not expect that the use of a different basis affects significantly our results since the sets $\{\hat{u}_i\}$ and $\{\hat{e}_i\}$, $i = 1, 2, \ldots, 2N$, are valid bases of the vector space, related by a non-singular transformation of the form $\left[\hat{u}_1 \ \hat{u}_2 \ldots \hat{u}_{2N} \right]^{\mathrm{T}} = T_c \cdot \left[\hat{e}_1 \ \hat{e}_2 \ldots \hat{e}_{2N} \right]^{\mathrm{T}}$, with T_c denoting the transformation matrix. So, by studying analytically the time evolution of the determinants of matrix $C(t)$, we expect to derive accurate approximations of the behavior of the GALI_k (5.24) for chaotic orbits. The validity of this approximation is numerically verified in Sect. 5.3.2.

From all the possible $k \times k$ determinants of $C(t)$ (5.37), the one that decreases the *slowest* is

$$
D_{1,2,3,\ldots,k} = \begin{vmatrix} s_1 \frac{c_2^1}{|c_1^1|} e^{-(L_1-L_2)t} & \cdots & \frac{c_{j_k}^1}{|c_1^1|} e^{-(L_1-L_{j_k})t} \\ s_2 \frac{c_2^2}{|c_1^2|} e^{-(L_1-L_2)t} & \cdots & \frac{c_{j_k}^2}{|c_1^2|} e^{-(L_1-L_{j_k})t} \\ \vdots & \vdots & \vdots \\ s_k \frac{c_2^k}{|c_1^k|} e^{-(L_1-L_2)t} & \cdots & \frac{c_{j_k}^k}{|c_1^k|} e^{-(L_1-L_{j_k})t} \end{vmatrix} =
$$

$$
= \begin{vmatrix} s_1 \frac{c_2^1}{|c_1^1|} & \cdots & \frac{c_{j_k}^1}{|c_1^1|} \\ s_2 \frac{c_2^2}{|c_1^2|} & \cdots & \frac{c_{j_k}^2}{|c_1^2|} \\ \vdots & \vdots & \vdots \\ s_k \frac{c_2^k}{|c_1^k|} & \cdots & \frac{c_{j_k}^k}{|c_1^k|} \end{vmatrix} e^{-\left[(L_1-L_2)+(L_1-L_3)+\cdots+(L_1-L_{j_k})\right]t} \Rightarrow
$$

$$
D_{1,2,3,\ldots,k} \propto e^{-\left[(L_1-L_2)+(L_1-L_3)+\cdots+(L_1-L_k)\right]t}. \tag{5.38}
$$

All other determinants of $C(t)$ tend to zero *faster* than $D_{1,2,3,\ldots,k}$ and we, therefore, conclude that the rate of decrease of GALI_k is dominated by (5.38), yielding the approximation

$$
\mathrm{GALI}_k(t) \propto e^{-\left[(L_1-L_2)+(L_1-L_3)+\cdots+(L_1-L_k)\right]t}. \tag{5.39}
$$

In the previous analysis we assumed that $L_1 > L_2$ so that the norm of each deviation vector can be well approximated by (5.36). If the first m LCEs, with $1 < m < k$, are equal, or very close to each other, i.e. $L_1 \simeq L_2 \simeq \cdots \simeq L_m$, (5.39) becomes

$$
\mathrm{GALI}_k(t) \propto e^{-\left[(L_1-L_{m+1})+(L_1-L_{m+2})+\cdots+(L_1-L_k)\right]t}, \tag{5.40}
$$

which still describes an exponential decay. However, for $k \leq m < N$ the GALI_k does not tend to zero as there exists at least one determinant of $C(t)$ that does not vanish. In this case, of course, one should increase the number of deviation vectors until an exponential decrease of GALI_k is achieved. The extreme situation that all $L_i = 0$ corresponds to motion on quasiperiodic tori, where all orbits are regular and is described below.

5.3.1.2 The Evaluation of GALI for Ordered Orbits

Ordered orbits of an N dof Hamiltonian system (or a $2N$D symplectic map) typically lie on N-dimensional tori found around stable periodic orbits. Such tori can be accurately described by N formal local integrals of motion in involution, so that the system appears locally integrable. This means that we could perform a local transformation to action-angle variables, considering as actions $J_1, J_2, \ldots, J_N$ the values of the N formal integrals, so that Hamilton's equations of motion, locally attain the form

$$\begin{aligned} \dot{J}_i &= 0 \\ \dot{\theta}_i &= \omega_i(J_1, J_2, \ldots, J_N) \end{aligned} \quad i = 1, 2, \ldots, N. \tag{5.41}$$

These can be easily integrated to give

$$\begin{aligned} J_i(t) &= J_{i0} \\ \theta_i(t) &= \theta_{i0} + \omega_i(J_{10}, J_{20}, \ldots, J_{N0})\, t \end{aligned} \quad i = 1, 2, \ldots, N, \tag{5.42}$$

where $J_{i0}, \theta_{i0}, i = 1, 2, \ldots, N$ are the initial conditions.

By denoting as $\xi_i, \eta_i, i = 1, 2, \ldots, N$, small deviations of J_i and θ_i respectively, the variational equations of system (5.41), describing the evolution of a deviation vector are

$$\begin{aligned} \dot{\xi}_i &= 0 \\ \dot{\eta}_i &= \sum_{j=1}^{N} \omega_{ij} \xi_j \end{aligned} \quad i = 1, 2, \ldots, N, \tag{5.43}$$

where

$$\omega_{ij} = \frac{\partial \omega_i}{\partial J_j}|_{\mathbf{J}_0} \quad i, j = 1, 2, \ldots, N, \tag{5.44}$$

and $\mathbf{J}_0 = (J_{10}, J_{20}, \ldots, J_{N0}) = $ constant, represents the N–dimensional vector of the initial actions. The solution of these equations is:

$$\begin{aligned} \xi_i(t) &= \xi_i(0) \\ \eta_i(t) &= \eta_i(0) + \left[\sum_{j=1}^{N} \omega_{ij} \xi_j(0) \right] t \end{aligned} \quad i = 1, 2, \ldots, N. \tag{5.45}$$

From (5.45) we see that an initial deviation vector $\mathbf{w}(0)$ with coordinates $\xi_i(0)$, $i = 1, 2, \ldots, N$, in the action variables and $\eta_i(0)$, $i = 1, 2, \ldots, N$ in the angles, i.e. $\mathbf{w}(0) = (\xi_1(0), \xi_2(0), \ldots, \xi_N(0), \eta_1(0), \eta_2(0), \ldots, \eta_N(0))$, evolves in time in such a way that its action coordinates remain constant, while its angle coordinates increase linearly in time. This behavior implies an almost linear increase of the norm of the deviation vector.

Using as a basis of the $2N$-dimensional tangent space of the Hamiltonian flow the $2N$ unit vectors $\{\hat{v}_1, \hat{v}_2, \ldots, \hat{v}_{2N}\}$, such that the first N of them, $\hat{v}_1, \hat{v}_2, \ldots, \hat{v}_N$, correspond to the N action variables and the remaining ones, $\hat{v}_{N+1}, \hat{v}_{N+2}, \ldots, \hat{v}_{2N}$, to the N conjugate angle variables, any unit deviation vector $\hat{w}_i$, $i = 1, 2, \ldots$, can be written as

$$\hat{w}_i(t) = \frac{1}{\|\mathbf{w}(t)\|} \left[\sum_{j=1}^{N} \xi_j^i(0)\, \hat{v}_j + \sum_{j=1}^{N} \left(\eta_j^i(0) + \sum_{k=1}^{N} \omega_{kj} \xi_j^i(0) t \right) \hat{v}_{N+j} \right]. \quad (5.46)$$

We point out that the quantities ω_{ij}, $i, j = 1, 2 \ldots, N$, in (5.44), depend only on the particular reference orbit and not on the choice of the deviation vector. We also note that the basis $\hat{v}_i$, $i = 1, 2, \ldots, 2N$ depends on the specific torus on which the motion occurs and is related to the usual vector basis $\hat{e}_i$, $i = 1, 2, \ldots, 2N$, (5.25), through a non-singular transformation. The basis $\{\hat{e}_1, \hat{e}_2, \ldots, \hat{e}_{2N}\}$ is used to describe the evolution of a deviation vector with respect to the original q_i, p_i $i = 1, 2, \ldots, N$ coordinates of the Hamiltonian system, while the basis $\{\hat{v}_1, \hat{v}_2, \ldots, \hat{v}_{2N}\}$ is used to describe the same evolution in action-angle variables, so that the equations of motion are the ones given by (5.41).

Let us now study the case of k, general, linearly independent unit deviation vectors $\{\hat{w}_1, \hat{w}_2, \ldots, \hat{w}_k\}$ with $2 \leq k \leq 2N$. As we have already explained, the k deviation vectors will eventually fall on the N-dimensional tangent space of the torus on which the motion occurs. In their final state, the deviation vectors will have coordinates only in the N-dimensional space spanned by $\hat{v}_{N+1}, \hat{v}_{N+2}, \ldots, \hat{v}_{2N}$. Now, if we start with $2 \leq k \leq N$ general deviation vectors there is no particular reason for them to become linearly dependent and their wedge product will be different from zero, yielding GALI$_k$ which are *not zero*. However, if we start with $N < k \leq 2N$ deviation vectors, some of them will necessarily become linearly dependent. Thus, in this case, their wedge product, as well as GALI$_k$ will be zero.

Let us examine in more detail the behavior of GALIs in these cases. The matrix $D(t)$ having as rows the coordinates of the deviation vectors with respect to the basis $\{\hat{v}_1, \hat{v}_2, \ldots, \hat{v}_{2N}\}$ has the form

$$D(t) = \frac{1}{\prod_{m=1}^{k} \|\mathbf{w}_m(t)\|}$$

$$\cdot \begin{bmatrix} \xi_1^1(0) & \cdots & \xi_N^1(0) & \eta_1^1(0) + \sum_{m=1}^{N} \omega_{1m}\xi_m^1(0)t & \cdots & \eta_N^1(0) + \sum_{m=1}^{N} \omega_{Nm}\xi_m^1(0)t \\ \xi_1^2(0) & \cdots & \xi_N^2(0) & \eta_1^2(0) + \sum_{m=1}^{N} \omega_{1m}\xi_m^2(0)t & \cdots & \eta_N^2(0) + \sum_{m=1}^{N} \omega_{Nm}\xi_m^2(0)t \\ \vdots & & \vdots & \vdots & & \vdots \\ \xi_1^k(0) & \cdots & \xi_N^k(0) & \eta_1^k(0) + \sum_{m=1}^{N} \omega_{1m}\xi_m^k(0)t & \cdots & \eta_N^k(0) + \sum_{m=1}^{N} \omega_{Nm}\xi_m^k(0)t \end{bmatrix},$$

$$(5.47)$$

where $i = 1, 2, \ldots, k$. Denoting by $\boldsymbol{\xi}_i^k$ and $\boldsymbol{\eta}_i^k$ the $k \times 1$ column matrices

$$\boldsymbol{\xi}_i^k = \left[\xi_i^1(0)\ \xi_i^2(0) \ldots \xi_i^k(0) \right]^{\mathrm{T}}, \quad \boldsymbol{\eta}_i^k = \left[\eta_i^1(0)\ \eta_i^2(0) \ldots \eta_i^k(0) \right]^{\mathrm{T}}, \quad (5.48)$$

the matrix $D(t)$ of (5.47) assumes the much simpler form

$$D(t) = \frac{1}{\prod_{i=1}^{k} M_i(t)} \cdot \left[\boldsymbol{\xi}_1^k \cdots \boldsymbol{\xi}_N^k \ \boldsymbol{\eta}_1^k + \sum_{i=1}^{N} \omega_{1i} \boldsymbol{\xi}_i^k t \cdots \boldsymbol{\eta}_N^k + \sum_{i=1}^{N} \omega_{Ni} \boldsymbol{\xi}_i^k t \right]$$

$$= \frac{1}{\prod_{i=1}^{k} M_i(t)} \cdot D^k(t). \tag{5.49}$$

So all determinants appearing in the definition of GALI_k have as a common factor the quantity $1/\prod_{i=1}^{k} M_i(t)$. This quantity decreases to zero according to the power law

$$\frac{1}{\prod_{i=1}^{k} M_i(t)} \propto \frac{1}{t^k}, \tag{5.50}$$

due to the fact that the norm of each deviation vector grows linearly with time t for long times. In order to determine the asymptotic time evolution of GALI_k, we search for the fastest increasing determinants of all the possible $k \times k$ minors of the matrix D^k in (5.49), as time t grows.

Let us start with k being less than or equal to the dimension of the tangent space of the torus, i.e. $2 \leq k \leq N$. The fastest increasing determinants in this case are the $N!/(k!(N-k)!)$ determinants, whose k columns are chosen among the last N columns of matrix D^k:

$$\Delta_{j_1, j_2, \ldots, j_k}^k = \left| \boldsymbol{\eta}_{j_1}^k + \sum_{i=1}^{N} \omega_{j_1 i} \boldsymbol{\xi}_i^k t \ \boldsymbol{\eta}_{j_2}^k + \sum_{i=1}^{N} \omega_{j_2 i} \boldsymbol{\xi}_i^k t \cdots \boldsymbol{\eta}_{j_k}^k + \sum_{i=1}^{N} \omega_{j_k i} \boldsymbol{\xi}_i^k t \right|, \tag{5.51}$$

with $1 \leq j_1 < j_2 < \ldots < j_k \leq N$. Using standard properties of determinants, we easily see that the time evolution of $\Delta_{j_1, j_2, \ldots, j_k}^k$ is mainly determined by the behavior of determinants of the form

$$\left| \omega_{j_1 m_1} \boldsymbol{\xi}_{m_1}^k t \ \omega_{j_2 m_2} \boldsymbol{\xi}_{m_2}^k t \cdots \omega_{j_k m_k} \boldsymbol{\xi}_{m_k}^k t \right| = t^k \prod_{i=1}^{k} \omega_{j_i m_i} \cdot \left| \boldsymbol{\xi}_{m_1}^k \ \boldsymbol{\xi}_{m_2}^k \cdots \boldsymbol{\xi}_{m_k}^k \right| \propto t^k, \tag{5.52}$$

where $m_i \in \{1, 2, \ldots, N\}$, $i = 1, 2, \ldots, k$, with $m_i \neq m_j$, for all $i \neq j$. Thus, from (5.50) to (5.52) we conclude that the contribution to the behavior of GALI_k of the determinants related to $\Delta_{j_1, j_2, \ldots, j_k}^k$ is to provide constant terms in (5.28). All other determinants appearing in the definition of GALI_k, not being of the form of $\Delta_{j_1, j_2, \ldots, j_k}^k$, contain at least one column from the first N columns of matrix D^k and introduce in (5.28) terms that grow at a rate *slower* than t^k, which will ultimately have no bearing on the behavior of $\text{GALI}_k(t)$. Thus, we get

$$\text{GALI}_k(t) \approx \text{constant for } 2 \leq k \leq N. \tag{5.53}$$

Next, we turn to the case of k deviation vectors with $N < k \leq 2N$. The fastest growing determinants are again those containing the last N columns of the matrix D^k, i.e.

$$\Delta^k_{j_1,j_2,\ldots,j_{k-N},1,2,\ldots,N} = \left| \boldsymbol{\xi}^k_{j_1} \cdots \boldsymbol{\xi}^k_{j_{k-N}} \; \boldsymbol{\eta}^k_1 + \sum_{i=1}^N \omega_{1i} \boldsymbol{\xi}^k_i t \cdots \boldsymbol{\eta}^k_N + \sum_{i=1}^N \omega_{Ni} \boldsymbol{\xi}^k_i t \right|,$$
(5.54)

with $1 \leq j_1 < j_2 < \ldots < j_{k-N} \leq N$. The first $k - N$ columns of $\Delta^k_{j_1,j_2,\ldots,j_{k-N},1,2,\ldots,N}$ are chosen among the first N columns of D^k which are time independent. So there exist $N!/((k-N)!(2N-k)!)$ determinants of the form (5.54), which can be written as a sum of simpler $k \times k$ determinants, each containing in the position of its last N columns $\boldsymbol{\eta}^k_i$, $i = 1, 2, \ldots, N$ and/or columns of the form $\omega_{ji} \boldsymbol{\xi}^k_i t$ with $i, j = 1, 2, \ldots, N$. We exclude the ones where $\boldsymbol{\xi}^k_i$, $i = 1, 2, \ldots, N$ appear more than once, since in that case the corresponding determinant is zero. Among the remaining determinants, the fastest increasing ones are those containing as many columns proportional to t as possible.

Since t is always multiplied by the $\boldsymbol{\xi}^k_i$, and such columns occupy the first $k - N$ columns of $\Delta^k_{j_1,j_2,\ldots,j_{k-N},1,2,\ldots,N}$, t appears at most $N - (k - N) = 2N - k$ times. Otherwise the determinant would contain the same $\boldsymbol{\xi}^k_i$ column at least twice and would be equal to zero. The remaining $k - (2N - k) - (k - N) = k - N$ columns are filled by the $\boldsymbol{\eta}^k_i$ each of which appears at most once. Thus, the time evolution of $\Delta^k_{j_1,j_2,\ldots,j_{k-N},1,2,\ldots,N}$ is mainly expressed by determinants of the form

$$\left| \boldsymbol{\xi}^k_{j_1} \cdots \boldsymbol{\xi}^k_{j_{k-N}} \; \boldsymbol{\eta}^k_{i_1} \cdots \boldsymbol{\eta}^k_{i_{k-N}} \; \omega_{i_{k-N+1}m_1} \boldsymbol{\xi}^k_{m_1} t \cdots \omega_{i_N m_{2N-k}} \boldsymbol{\xi}^k_{i_{2N-k}} t \right| \propto t^{2N-k},$$
(5.55)

with $i_l \in \{1, 2, \ldots, N\}$, $l = 1, 2, \ldots, N$, $i_l \neq i_j$, for all $l \neq j$ and $m_l \in \{1, 2, \ldots, N\}$, $l = 1, 2, \ldots, 2N - k$, $m_l \notin \{j_1, j_2, \ldots, j_{k-N}\}$, $m_l \neq m_j$, for all $l \neq j$. So determinants of the form (5.54) contribute to the time evolution of GALI_k by introducing terms proportional to $t^{2N-k}/t^k = 1/t^{2(k-N)}$. All other determinants appearing in the definition of GALI_k, not having the form of $\Delta^k_{j_1,j_2,\ldots,j_{k-N},1,2,\ldots,N}$, introduce terms that tend to zero faster than $1/t^{2(k-N)}$ since they contain more than $k - N$ time independent columns of the form $\boldsymbol{\xi}^k_i$, $i = 1, 2, \ldots, N$. Thus, GALI_k tends to zero following a power law of the form

$$\mathrm{GALI}_k(t) \propto \frac{1}{t^{2(k-N)}} \quad \text{for } N < k \leq 2N.$$
(5.56)

In summary, the behavior of GALI_k for ordered orbits lying on an N-dimensional torus in the $2N$-dimensional phase space of an N dof Hamiltonian system (or a $2N$D symplectic map) is given by

$$\mathrm{GALI}_k(t) \propto \begin{cases} \text{constant if } 2 \leq k \leq N \\ \dfrac{1}{t^{2(k-N)}} \quad \text{if } N < k \leq 2N \end{cases}.$$
(5.57)

We remark that SALI is practically equivalent to GALI_2 as we see from the definitions of SALI (5.13) and GALI (5.24), as well as (5.23). We also note that estimates (5.57) are valid only when the particular conditions of each case are satisfied. For example, in the case of 2D maps, where the only possible torus is an one-dimensional invariant curve, the tangent space is one-dimensional. Thus, the behavior of GALI$_2$ (which is the only possible index in this case) is given by the second branch of (5.57), i.e. GALI$_2 \propto 1/t^2$, since the first case of (5.57) is not applicable. The SALI $\propto t^{-2}$ behavior has actually been seen in Fig. 5.1c for an ordered orbit of the 2D map (5.16).

Let us finally turn our attention to ordered orbits of N dof Hamiltonian systems (or $2N$D symplectic maps), lying on s-dimensional tori with $2 \leq s \leq N$ for Hamiltonian flows, and $1 \leq s \leq N$ for maps. For such orbits, all deviation vectors tend to fall on the s-dimensional tangent space of the torus on which the motion lies. Thus, if we start with $k \leq s$ general deviation vectors, these will remain linearly independent on the s-dimensional tangent space of the torus, since there is no particular reason for them to become linearly dependent. As a consequence GALI$_k$ remains practically constant and different from zero for $k \leq s$. On the other hand, GALI$_k$ tends to zero for $k > s$, since some deviation vectors will eventually have to become linearly dependent. In [93, 316] it was shown that the behavior of GALI$_k$ for such orbits is given by

$$\text{GALI}_k(t) \propto \begin{cases} \text{constant if } 2 \leq k \leq s \\ \frac{1}{t^{k-s}} & \text{if } s < k \leq 2N - s \\ \frac{1}{t^{2(k-N)}} & \text{if } 2N - s < k \leq 2N \end{cases} . \tag{5.58}$$

Note that setting $s = N$ in (5.58) we deduce that GALI$_k$ remains constant for $2 \leq k \leq N$ and decreases to zero as $\sim 1/t^{2(k-N)}$ for $N < k \leq 2N$ in accordance with (5.57).

As a final remark we note that the behavior of the GALIs for stable and unstable periodic orbits, both for Hamiltonian flows and symplectic maps, is studied in [247] (see also Problem 5.1).

5.3.2 Numerical Verification and Applications

5.3.2.1 Low-Dimensional Hamiltonian Systems

In order to apply the GALI method to low-dimensional Hamiltonian systems and verify the theoretically predicted behaviors of Sect. 5.3.1, we shall use two simple examples: the two dof Hénon-Heiles system (5.18)

$$H_2 = \frac{1}{2}(p_x^2 + p_y^2) + \frac{1}{2}(x^2 + y^2) + x^2 y - \frac{1}{3}y^3,$$

and the three dof Hamiltonian

$$H_3 = \sum_{i=1}^{3} \frac{\omega_i}{2}(q_i^2 + p_i^2) + q_1^2 q_2 + q_1^2 q_3, \tag{5.59}$$

studied in [32,104]. We keep the parameters of the two systems fixed at the energies $H_2 = 0.125$ and $H_3 = 0.09$, with $\omega_1 = 1$, $\omega_2 = \sqrt{2}$ and $\omega_3 = \sqrt{3}$. In order to illustrate the behavior of GALI$_k$, for different values of k, we consider some representative cases of chaotic and regular orbits of the two systems. We note here that a MAPLE algorithm for the computation of the GALI for orbits of the Hénon-Heiles system is presented in Appendix B.

Let us start with a chaotic orbit of the two dof Hamiltonian (5.18), with initial conditions $x = 0$, $y = -0.25$, $p_x = 0.42$, $p_y = 0$. In Fig. 5.8a we see the time evolution of K_t^1 (denoted by $L_1(t)$) of this orbit. The computation is carried out until K_t^1 stops having large fluctuations and approaches a positive value (indicating the chaotic nature of the orbit), which could be considered as a good approximation of the maximum LCE, L_1. Actually, for $t \approx 10^5$, we find $L_1 \approx 0.047$.

We recall that two dof Hamiltonian systems have only one positive LCE L_1, since the second largest is $L_2 = 0$. It also holds that $L_3 = -L_2$ and $L_4 = -L_1$ and thus formula (5.39), which describes the time evolution of GALI$_k$ for chaotic orbits, gives

$$\text{GALI}_2(t) \propto e^{-L_1 t}, \quad \text{GALI}_3(t) \propto e^{-2L_1 t}, \quad \text{GALI}_4(t) \propto e^{-4L_1 t}. \tag{5.60}$$

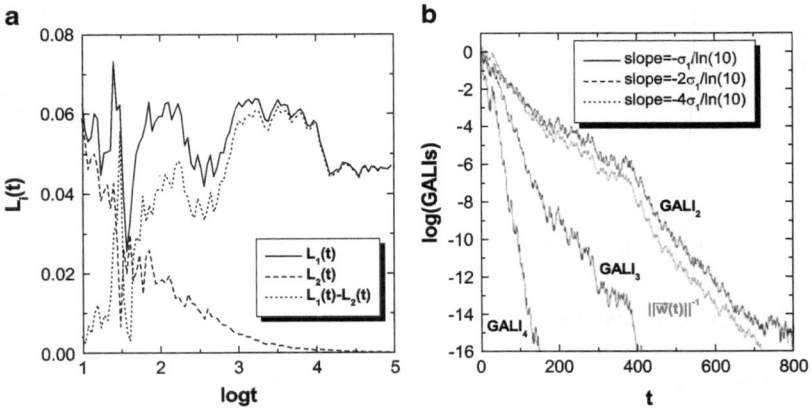

Fig. 5.8 (a) The evolution of the two largest finite time LCEs K_t^1 (*solid curve*), K_t^2 (*dashed curve*) (denoted by $L_1(t)$ and $L_2(t)$ respectively), and $L_1(t) - L_2(t)$ (*dotted curve*) for a chaotic orbit with initial conditions $x = 0$, $y = -0.25$, $p_x \simeq 0.42081$, $p_y = 0$ of system (5.18). (b) The evolution of GALI$_2$, GALI$_3$ and GALI$_4$ of the same orbit. The *plotted lines* correspond to functions proportional to $e^{-L_1 t}$ (*solid line*), $e^{-2L_1 t}$ (*dashed line*) and $e^{-4L_1 t}$ (*dotted line*) for $L_1 = 0.047$. Note that the t-axis is linear. The evolution of the norm of the deviation vector $\mathbf{w}(t)$ (with $\|\mathbf{w}(0)\| = 1$) used for the computation of $L_1(t)$, is also plotted in (b) (*gray curve*) (after [315])

In Fig. 5.8b we plot $GALI_k$, $k = 2, 3, 4$ for the same chaotic orbit as a function of time t. We plot t in linear scale so that, if (5.60) is valid, the slope of $GALI_2$, $GALI_3$ and $GALI_4$ should approximately be $-L_1/\ln 10$, $-2L_1/\ln 10$ and $-4L_1/\ln 10$ respectively. From Fig. 5.8b we see that lines having precisely these slopes, for $L_1 = 0.047$, approximate quite accurately the computed values of the GALIs.

Let us now study the behavior of $GALI_k$ for a regular orbit of the 2D Hamiltonian (5.18) with initial conditions $x = 0$, $y = 0$, $p_x = 0.5$, $p_y = 0$. From (5.57) it follows that in the case of an ordered orbit of a two dof Hamiltonian system the GALIs should evolve as

$$GALI_2(t) \propto \text{constant}, \quad GALI_3(t) \propto \frac{1}{t^2}, \quad GALI_4(t) \propto \frac{1}{t^4}. \quad (5.61)$$

From Fig. 5.9, where we plot the time evolution of SALI, $GALI_2$, $GALI_3$ and $GALI_4$ for this ordered orbit, we see that the indices do evolve according to predictions (5.61).

As we have seen in Sect. 5.2 the different behavior of SALI (or $GALI_2$) for ordered and chaotic orbits can be successfully used for discriminating between regions of order and chaos in various dynamical systems. Figures 5.8 and 5.9 clearly illustrate that $GALI_3$ and $GALI_4$ tend to zero both for ordered and chaotic orbits, but with very different time rates. We may use this difference to distinguish between chaotic and ordered motion following a different approach than SALI or $GALI_2$. Let us illustrate this by considering the computation of $GALI_4$: From (5.60) and (5.61), we expect $GALI_4 \propto e^{-4L_1 t}$ for chaotic orbits and $GALI_4 \propto 1/t^4$ for ordered ones. These time rates imply that, in general, the time needed for the index to become zero is much larger for ordered orbits. Thus, instead of simply registering the value of the index at the end of a given time interval—as we did with SALI ($GALI_2$) in Sect. 5.2—let us record the time, t_{th}, needed for $GALI_4$ to reach a very small threshold, e.g. 10^{-12}, and color each initial according to the value of t_{th}.

Fig. 5.9 Time evolution, in log-log scale, of SALI (*gray curves*), $GALI_2$, $GALI_3$ and $GALI_4$ for the regular orbit of Hamiltonian (5.18) with initial conditions $x = 0$, $y = 0$, $p_x = 0.5$, $p_y = 0$. Note that the curves of SALI and $GALI_2$ are very close to each other and thus cannot be distinguished. *Dashed lines* corresponding to particular power laws are also plotted (after [315])

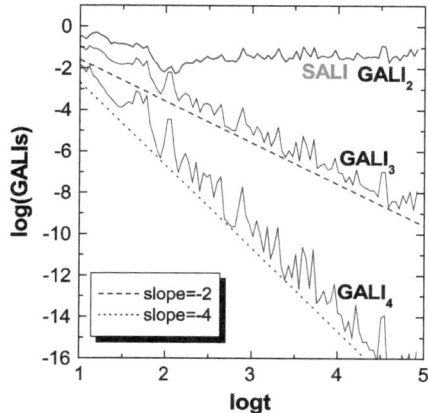

Fig. 5.10 Regions of different values of the time t_{th} needed for GALI$_4$ to become less than 10^{-12} on the PSS defined by $x = 0$ of the two dof Hamiltonian (5.18) (after [315])

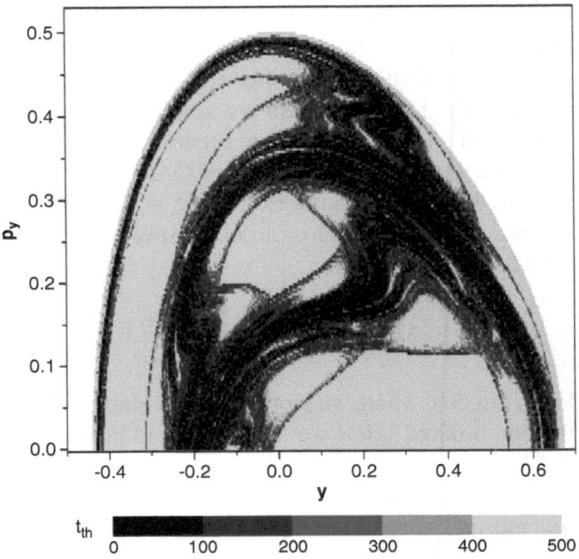

The outcome of this procedure for system (5.18) is presented in Fig. 5.10, which is similar to Fig. 5.5. Each orbit is integrated up to $t = 500$ units and if the value of GALI$_4$, at the end of the integration is larger than 10^{-12} the corresponding initial condition is colored by the light gray color used for $t_{th} \geq 400$. Thus we can clearly distinguish in this figure among various "degrees" of chaotic behavior in regions colored black or dark gray—corresponding to small values of t_{th}—and regions of ordered motion colored light gray, corresponding to large values of t_{th}. At the border between them we find points having intermediate values of t_{th} which belong to the "sticky" chaotic regions.

Let us now study the behavior of the GALIs in the case of the three dof Hamiltonian (5.59). Following [32, 104] the initial conditions of the orbits of this system are defined by assigning arbitrary values to the positions q_1, q_2, q_3, as well as the so-called "harmonic energies" E_1, E_2, E_3 related to the momenta through $p_i = \sqrt{2E_i/\omega_i}$, $i = 1, 2, 3$. Chaotic orbits of three dof Hamiltonian systems generally have two positive LCEs, L_1 and L_2, while $L_3 = 0$. So, for approximating the behavior of GALIs according to (5.39), both L_1 and L_2 are needed. In particular, (5.39) gives

$$\text{GALI}_2(t) \propto e^{-(L_1-L_2)t}, \quad \text{GALI}_3(t) \propto e^{-(2L_1-L_2)t},$$
$$\text{GALI}_4(t) \propto e^{-(3L_1-L_2)t}, \quad \text{GALI}_5(t) \propto e^{-4L_1 t}, \quad \text{GALI}_6(t) \propto e^{-6L_1 t}. \tag{5.62}$$

Consider first the chaotic orbit with initial conditions $q_1 = q_2 = q_3 = 0$, $E_1 = E_2 = E_3 = 0.03$ of system (5.59). Computing L_1, L_2 for this orbit as the long time limits of the finite time LCEs, K_t^1, K_t^2, we present the results in Fig. 5.11a. The computation is carried out until K_t^1 and K_t^2 stop having large fluctuations

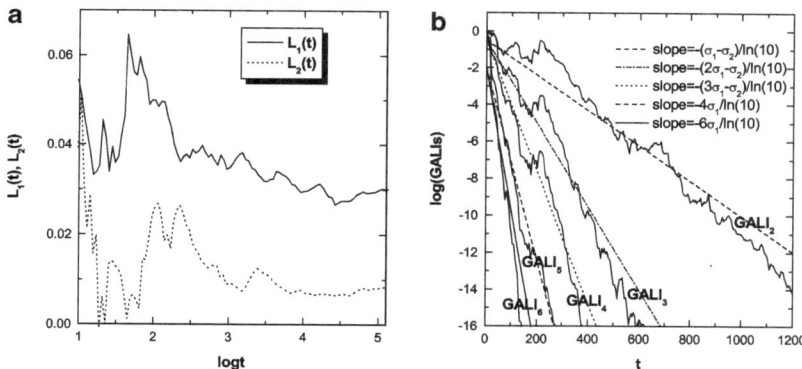

Fig. 5.11 (**a**) The evolution of the two largest finite time LCEs, K_t^1, K_t^2 (denoted by $L_1(t)$, $L_2(t)$), for the chaotic orbit with initial condition $q_1 = q_2 = q_3 = 0$, $E_1 = E_2 = E_3 = 0.03$ of the 3D system (5.59). (**b**) The evolution of GALI$_k$ with $k = 2, \ldots, 6$, of the same orbit. The *plotted lines* correspond to functions proportional to $e^{-(L_1-L_2)t}$, $e^{-(2L_1-L_2)t}$, $e^{-(3L_1-L_2)t}$, e^{-4L_1t} and e^{-6L_1t} for $L_1 = 0.03$, $L_2 = 0.008$. Note that the t-axis is linear (after [315])

and approach some positive values (since the orbit is chaotic), which could be considered as good approximations of their limits L_1, L_2. Actually for $t \approx 10^5$ we have $L_1 \approx 0.03$ and $L_2 \approx 0.008$. Using these values as good approximations of L_1, L_2 we see in Fig. 5.11b that the slopes of all GALIs are well reproduced by (5.62).

Next, we consider the case of ordered orbits of the three dof Hamiltonian system, for which the GALIs should behave as:

$$\text{GALI}_2(t) \propto \text{constant}, \quad \text{GALI}_3(t) \propto \text{constant}, \quad \text{GALI}_4(t) \propto \tfrac{1}{t^2},$$
$$\text{GALI}_5(t) \propto \tfrac{1}{t^4}, \quad \text{GALI}_6(t) \propto \tfrac{1}{t^6}. \tag{5.63}$$

according to (5.57). In order to verify (5.63) we shall follow a specific ordered orbit of system (5.59) with initial conditions $q_1 = q_2 = q_3 = 0$, $E_1 = 0.005$, $E_2 = 0.085$, $E_3 = 0$. The ordered nature of this orbit is revealed by the slow convergence of its K_t^1 to zero, implying that $L_1 = 0$ (Fig. 5.12a). In Fig. 5.12b, we plot the values of all GALIs of this orbit with respect to time t. From these results we see again that the different behaviors of GALIs are very well approximated by (5.57).

From the results of Figs. 5.11 and 5.12, we conclude that in the case of three dof Hamiltonian systems not only GALI$_2$ (or SALI), but also GALI$_3$ has different behavior for ordered and chaotic orbits. Hence, the natural question arises whether GALI$_3$ can be used instead of SALI for the faster detection of chaotic and ordered motion in three dof Hamiltonians and, by extension, whether GALI$_k$, with $k > 3$, should be preferred for systems with $N > 3$. The obvious computational drawback, of course, is that the evaluation of GALI$_k$ requires that we numerically follow the evolution of more than two deviation vectors.

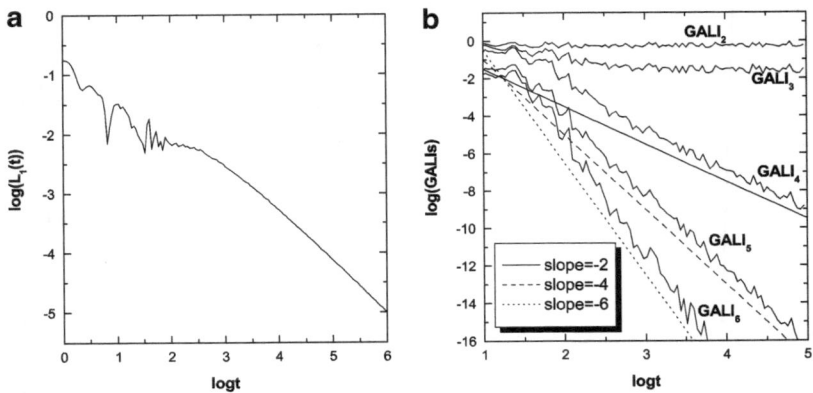

Fig. 5.12 (a) The evolution of K_t^1 (denoted by $L_1(t)$) for the ordered orbit with initial condition $q_1 = q_2 = q_3 = 0$, $E_1 = 0.005$, $E_2 = 0.085$, $E_3 = 0$ of the system (5.59). (**b**) The evolution of $GALI_k$ with $k = 2, \ldots, 6$, of the same orbit. The *plotted lines* correspond to functions proportional to $\frac{1}{t^2}$, $\frac{1}{t^4}$ and $\frac{1}{t^6}$ (after [315])

For example the computation of $GALI_3$ for a given time interval t needs more CPU time than SALI, since we follow the evolution of three deviation vectors instead of two. This is particularly true for ordered orbits as the index does not become zero and its evolution has to be followed for the whole prescribed time interval. In the case of chaotic orbits, however, the situation is different since the index tends exponentially to zero even faster than $GALI_2$ or SALI. Let us consider, for example, the chaotic orbit of Fig. 5.11. The usual technique to characterize an orbit as chaotic is to check, after some time interval, if its SALI has become less than a very small threshold value, e.g. 10^{-8}. For this particular orbit, this threshold value was reached at $t \approx 760$. Adopting the same threshold to characterize an orbit as chaotic, we find that $GALI_3$ becomes less than 10^{-8} after $t \approx 335$, requiring only as much as 65% of the CPU time needed for SALI to reach the same threshold!

So, using $GALI_3$ instead of SALI, we gain considerably in CPU time for chaotic orbits, while we lose for ordered orbits. Thus, the efficiency of using $GALI_3$ for discriminating between chaos and order in a three dof system depends on the percentage of phase space occupied by chaotic orbits (if all orbits are ordered $GALI_3$ requires more CPU time than SALI). More crucially, however, it depends on the choice of the final time, up to which each orbit is integrated. As an example, let us integrate, up to $t = 1,000$ time units, all orbits whose initial conditions lie on a dense grid in the subspace $q_3 = p_3 = 0$, $p_2 \geq 0$ of a four-dimensional surface of section (with $q_1 = 0$) of system (5.59), attributing to each grid point a color according to the value of $GALI_3$ at the end of the integration. If $GALI_3$ of an orbit becomes less than 10^{-8} for $t < 1,000$ the evolution of the orbit is stopped, its $GALI_3$ value is registered and the orbit is characterized as chaotic. The outcome of this experiment is presented in Fig. 5.13.

Fig. 5.13 Regions of different values of the GALI$_3$ on the subspace $q_3 = p_3 = 0$, $p_2 \geq 0$ of the four-dimensional $q_1 = 0$ surface of section of the three dof system (5.59) at $t = 1,000$ (after [315])

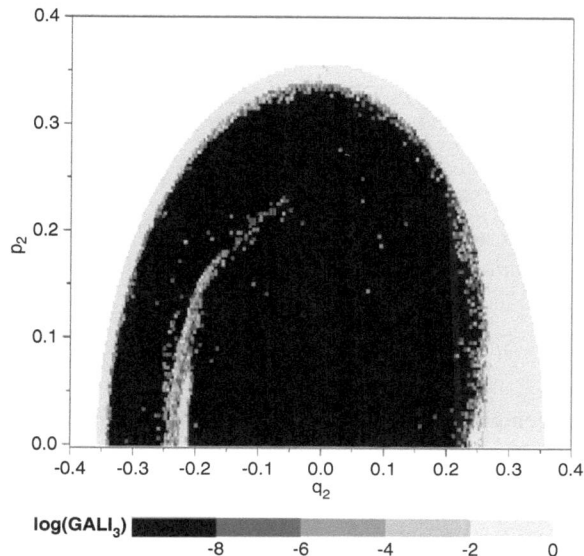

We find that 77% of the orbits of Fig. 5.13 are characterized as chaotic, having GALI$_3$ < 10^{-8}. In order to have the same percentage of orbits identified as chaotic using SALI (i.e. having SALI < 10^{-8}) the same experiment has to be carried out for $t = 2,000$ units, requiring 53% more CPU time. Due to the high percentage of chaotic orbits, in this case, even when the SALI is computed for $t = 1,000$ the corresponding CPU time is 12% higher than the one needed for the computation of Fig. 5.13, while only 55% of the orbits are identified as chaotic. Thus it becomes evident that a carefully designed application of GALI$_3$—or GALI$_k$ for that matter—can significantly diminish the computational time needed for a reliable discrimination between regions of order and chaos in Hamiltonian systems with $N > 2$ dof.

5.3.2.2 High-Dimensional Hamiltonian Systems

Let us now apply the GALI method to study chaotic, ordered and diffusive motion in multi-dimensional Hamiltonian systems. In particular, we consider the FPU$-\beta$ model of N particles with Hamiltonian (3.15)

$$H = \sum_{i=1}^{N} \frac{p_i^2}{2} + \sum_{i=0}^{N} \left[\frac{(q_{i+1} - q_i)^2}{2} + \frac{\beta(q_{i+1} - q_i)^4}{4} \right], \quad (5.64)$$

with $q_1, \ldots, q_N$ being the displacements of the particles with respect to their equilibrium positions, and $p_1, \ldots, p_N$ the corresponding momenta. Defining normal mode variables by (see (4.3))

$$Q_k = \sqrt{\frac{2}{N+1}} \sum_{i=1}^{N} q_i \sin\left(\frac{ki\pi}{N+1}\right) \ , \ P_k = \sqrt{\frac{2}{N+1}} \sum_{i=1}^{N} p_i \sin\left(\frac{ki\pi}{N+1}\right),$$

$$k = 1, \dots, N, \qquad (5.65)$$

the unperturbed Hamiltonian (5.64) with $\beta = 0$ is written as the sum of the so-called *harmonic energies* E_i having the form

$$E_i = \frac{1}{2}\left(P_i^2 + \omega_i^2 Q_i^2\right) \ , \ \omega_i = 2\sin\left(\frac{i\pi}{2(N+1)}\right) i = 1, \dots, N, \qquad (5.66)$$

ω_i being the corresponding *harmonic frequencies*, as we have seen in Sect. 4.1. In our study, we fix the number of particles to $N = 8$ and the system's parameter to $\beta = 1.5$ and impose fixed boundary conditions $q_0(t) = q_{N+1}(t) = p_0(t) = p_{N+1}(t) = 0, \forall t$.

We consider first a chaotic orbit of systems (5.64), having seven positive LCEs, which we compute as the limits for $t \to \infty$ of some appropriate quantities K_t^i, $i = 1, \dots, 7$, to be $L_1 \approx 0.170$, $L_2 \approx 0.141$, $L_3 \approx 0.114$, $L_4 \approx 0.089$, $L_5 \approx 0.064$, $L_6 \approx 0.042$, $L_7 \approx 0.020$ (Fig. 5.14a). Of course, in the case of the $N = 8$ particle FPU$-\beta$ model (5.64) we have $L_i = -L_{17-i}$ for $i = 1, \dots, 8$ with $L_8 = L_9 = 0$. Using the above computed values as good approximations of the real LCEs, we see in Fig. 5.14b, c that the slopes of all GALI$_k$ indices are well reproduced by (5.39).

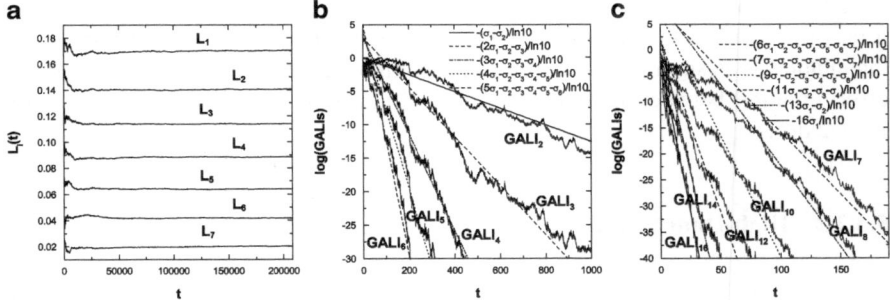

Fig. 5.14 (**a**) The time evolution of quantities K_t^i (denoted by L_i), $i = 1, \dots, 7$, having as limits for $t \to \infty$ the seven positive LCEs L_i, $i = 1, \dots, 7$, for a chaotic orbit with initial conditions $Q_1 = Q_4 = 2$, $Q_2 = Q_5 = 1$, $Q_3 = Q_6 = 0.5$, $Q_7 = Q_8 = 0.1$, $P_i = 0, i = 1, \dots, 8$ of the $N = 8$ particle FPU$-\beta$ lattice (5.64). The time evolution of the corresponding GALI$_k$ is plotted in (**b**) for $k = 2, \dots, 6$ and in (**c**) for $k = 7, 8, 10, 12, 14, 16$. The *plotted lines* in (**b**) and (**c**) correspond to exponentials that follow the asymptotic laws (5.39) for $L_1 = 0.170$, $L_2 = 0.141$, $L_3 = 0.114$, $L_4 = 0.089$, $L_5 = 0.064$, $L_6 = 0.042$, $L_7 = 0.020$. Note that t–axis is linear and that the slope of each line is written explicitly in (**b**) and (**c**) (after [316])

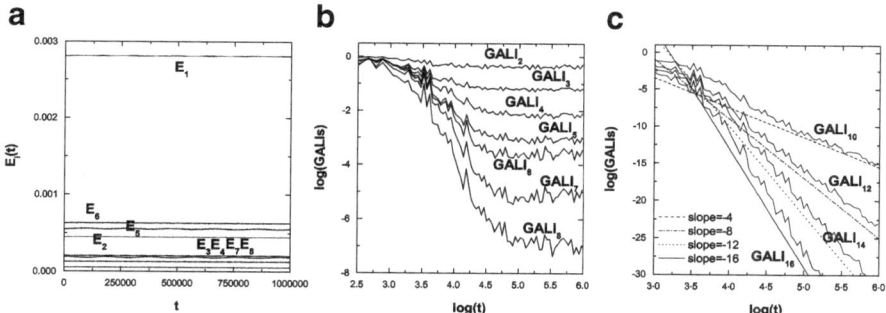

Fig. 5.15 (**a**) The time evolution of harmonic energies E_i, $i = 1, \ldots, 8$, for an ordered orbit with initial conditions $q_1 = q_2 = q_3 = q_8 = 0.05$, $q_4 = q_5 = q_6 = q_7 = 0.1$, $p_i = 0$, $i = 1, \ldots, 8$, of the $N = 8$ particle FPU$-\beta$ lattice (5.64). The time evolution of the corresponding GALI$_k$ is plotted in (**b**) for $k = 2, \ldots, 8$, and in (**c**) for $k = 10, 12, 14, 16$. The *plotted lines* in (**c**) correspond to functions proportional to t^{-4}, t^{-8}, t^{-12} and t^{-16}, as predicted in (5.57) (after [316])

Turning now to the case of an ordered orbit of system (5.64), we plot in Fig. 5.15a the evolution of its harmonic energies E_i, $i = 1, \ldots, 8$. The harmonic energies remain practically constant, exhibiting some feeble oscillations, implying the regular nature of the orbit. In Fig. 5.15b and c we plot the GALIs of this orbit and verify that their behavior is well approximated by the asymptotic formula (5.57) for $N = 8$. Note that the GALI$_k$ for $k = 2, \ldots, 8$ (Fig. 5.15b) remain different from zero, implying that the orbit is indeed quasiperiodic and lies on an eight-dimensional torus. In particular, after some initial transient time, they start oscillating around non-zero values whose magnitude decreases with increasing k. On the other hand, the GALI$_k$ with $8 < k \leq 16$ (Fig. 5.15c) tend to zero following power law decays in accordance with (5.57).

From the results of Figs. 5.14 and 5.15, we conclude that the different behavior of GALI$_k$ for chaotic (exponential decay) and ordered orbits (non-zero values or power law decay) allows for a fast and clear discrimination between the two cases. Consider for example GALI$_8$ which tends exponentially to zero for chaotic orbits (Fig. 5.14c), while it remains small but different from zero in the case of regular orbits (Fig. 5.15b). At $t \approx 150$ GALI$_8$ has values that differ almost 35 orders of magnitude being GALI$_8 \approx 10^{-36}$ for the chaotic orbit, while GALI$_8 \approx 10^{-1}$ for the regular one. This huge difference in the values of GALI$_8$ clearly discriminates between the two cases.

What happens, however, if we choose an orbit that starts near a torus but slowly drifts away from it, presumably through a thin chaotic layer of higher order resonances? This phenomenon is recognized by the GALIs, which provide early predictions that may be quite relevant for applications. To see this let us choose again initial conditions for our 8-particle FPU$-\beta$ model, such that the motion appears quasiperiodic, with its energy recurring between the modes E_1, E_3, E_5 and E_7 (Fig. 5.16a). In particular, we consider an orbit with initial conditions $Q_1 = 2$,

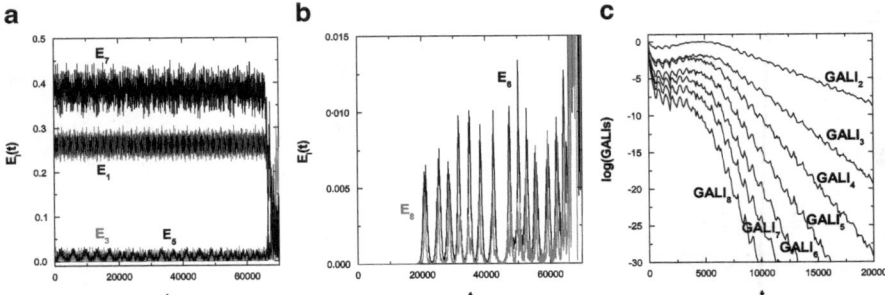

Fig. 5.16 The time evolution of harmonic energies (**a**) E_1, E_3, E_5, E_7 and (**b**) E_6, E_8, for a slowly diffusing orbit of the $N = 8$ particle FPU$-\beta$ lattice (5.64). (**c**) The time evolution of the corresponding GALI$_k$ for $k = 2, \ldots, 8$, clearly exhibits exponential decay already at $t \approx 10,000$ (after [316])

$Q_7 = 0.44$, $Q_3 = Q_4 = Q_5 = Q_6 = Q_8 = 0$, $P_i = 0$, $i = 1, \ldots, 8$, having total energy $H = 0.71$.

This orbit, however, is *not* quasiperiodic, as it drifts away from the initial four-dimensional torus, exciting new frequencies and sharing its energy with more modes, after about $t = 20,000$ time units. This becomes evident in Fig. 5.16b where we plot the evolution of E_6 and E_8. We see that these harmonic energies, which were initially zero, start having non-zero values at $t \approx 20,000$ and exhibit from then on small oscillations (note the different scales of ordinate axis of Fig. 5.16a, b), which look very regular until $t \approx 66,000$. At that time the values of all harmonic energies change dramatically, clearly indicating the chaotic nature of the orbit. As we see in Fig. 5.16c, this type of diffusion is predicted by the *exponential* decay of *all* GALI$_k$, shown already at about $t = 10,000$.

5.3.2.3 Symplectic Maps

We can also demonstrate the validity of the theoretical predictions of Sect. 5.3.1 in the case of symplectic maps, by considering for example the 6D map

$$
\begin{aligned}
x_1' &= x_1 + x_2' \\
x_2' &= x_2 + \frac{K}{2\pi}\sin(2\pi x_1) - \frac{\beta}{2\pi}\{\sin[2\pi(x_5 - x_1)] + \sin[2\pi(x_3 - x_1)]\} \\
x_3' &= x_3 + x_4' \\
x_4' &= x_4 + \frac{K}{2\pi}\sin(2\pi x_3) - \frac{\beta}{2\pi}\{\sin[2\pi(x_1 - x_3)] + \sin[2\pi(x_5 - x_3)]\} \\
x_5' &= x_5 + x_6' \\
x_6' &= x_6 + \frac{K}{2\pi}\sin(2\pi x_5) - \frac{\beta}{2\pi}\{\sin[2\pi(x_1 - x_5)] + \sin[2\pi(x_3 - x_5)]\}
\end{aligned}
\tag{5.67}
$$

which consists of three coupled standard maps [180] and is a typical nonlinear system, in which regions of chaotic and quasi-periodic dynamics are found to

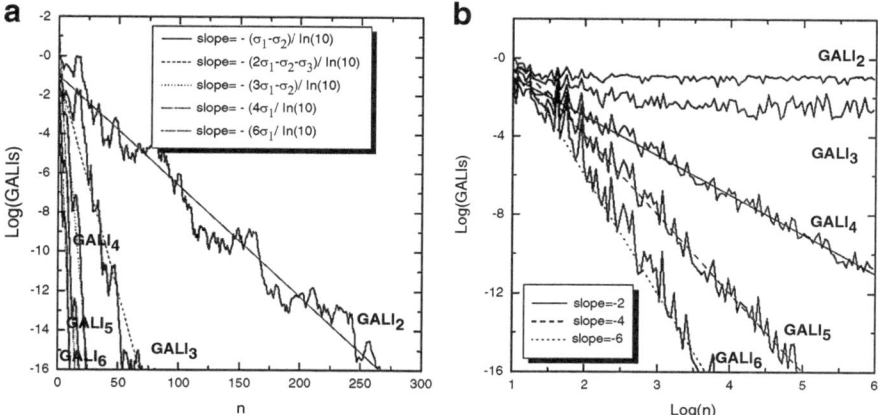

Fig. 5.17 The evolution of $GALI_k$, $k = 2, \ldots, 6$, with respect to the number of iteration n for (**a**) the chaotic orbit C1 and (**b**) the ordered orbit R1 of the 6D map (5.67). The *plotted lines* correspond to functions proportional to $e^{-(L_1-L_2)n}$, $e^{-(2L_1-L_2-L_3)n}$, $e^{-(3L_1-L_2)n}$, e^{-4L_1n}, e^{-6L_1n} for $L_1 = 0.70$, $L_2 = 0.57$, $L_3 = 0.32$ in (**a**) and proportional to n^{-2}, n^{-4}, n^{-6} in (**b**) (after [244])

coexist. Note that each coordinate is given modulo 1 and that in our study we fix the parameters of the map (5.67) to $K = 3$ and $\beta = 0.1$.

In order to verify numerically the validity of (5.39) and (5.57), we study two typical orbits of map (5.67), a chaotic one with initial condition $x_1 = x_3 = x_5 = 0.8$, $x_2 = 0.05$, $x_4 = 0.21$, $x_6 = 0.01$ (orbit C1) and an ordered one with initial condition $x_1 = x_3 = x_5 = 0.55$, $x_2 = 0.05$, $x_4 = 0.01$, $x_6 = 0$ (orbit R1) lying on a three-dimensional torus. In Fig. 5.17 we see the evolution of $GALI_k$, $k = 2, \ldots, 6$, for these two orbits.

For a chaotic orbit of the 6D map (5.67) we have $L_1 = -L_6$, $L_2 = -L_5$, $L_3 = -L_4$ with $L_1 \geq L_2 \geq L_3 \geq 0$, since the LCEs are ordered in pairs of opposite signs. So for the evolution of $GALI_k$ (5.39) gives

$$GALI_2(n) \propto e^{-(L_1-L_2)n}, \quad GALI_3(n) \propto e^{-(2L_1-L_2-L_3)n},$$
$$GALI_4(n) \propto e^{-(3L_1-L_2)n}, \quad GALI_5(n) \propto e^{-4L_1n}, \quad GALI_6(n) \propto e^{-6L_1n}. \tag{5.68}$$

The positive LCEs of the chaotic orbit C1 were found to be $L_1 \approx 0.70$, $L_2 \approx 0.57$, $L_3 \approx 0.32$. From the results of Fig. 5.17a we see that the exponentials in (5.68) for $L_1 = 0.70$, $L_2 = 0.57$, $L_3 = 0.32$ approximate quite accurately the computed values of GALIs.

According to (5.57), the behavior of $GALI_k$, $k = 2, \ldots, 6$ for the regular orbit R1 is given by

$$GALI_2(n) \propto \text{constant}, \quad GALI_3(n) \propto \text{constant}, \quad GALI_4(n) \propto \frac{1}{n^2},$$
$$GALI_5(n) \propto \frac{1}{n^4}, \quad GALI_6(n) \propto \frac{1}{n^6}. \tag{5.69}$$

From the results of Fig. 5.17b we see that these approximations also describe very well the evolution of GALIs.

5.3.2.4   Motion on Low-Dimensional Tori

Let us now turn our attention to an important feature of the GALI method: its ability to detect ordered motion on low-dimensional tori. In order to illustrate this feature of the index we consider again the FPU$-\beta$ lattice (5.64) as a toy model, although the capability of the GALI method to identify motion on low-dimensional tori was also observed for multi-dimensional symplectic maps [63]. Selecting initial conditions such that only a small number of the harmonic energies E_i (5.66) of the FPU$-\beta$ model are initially excited, we observe at small enough energies that the system exhibits the famous FPU recurrences, whereby energy is exchanged quasiperiodically only between the excited E_i (see Chap. 6).

In particular, choosing an orbit with initial conditions $q_i = 0.1$, $p_i = 0$, $i = 1$, ..., 8, and total energy $H = 0.01$ we distribute the energy among 4 modes, E_i, $i = 1, 3, 5, 7$ in the 8-particle FPU$-\beta$ lattice with $\beta = 1.5$. In Fig. 5.18a we observe that actually only 4 modes are excited, while in Fig. 5.18b we see that only the GALI$_k$ for $k = 2, 3, 4$ remain constant, implying that the motion lies on a four-dimensional torus. All the higher order GALIs in Fig. 5.18b, c decay by power laws, whose exponents are obtained from (5.58) for $N = 8$ and $s = 4$.

Thus, for an ordered orbit (5.58) implies that the largest order k of its GALIs that eventually remains constant determines the dimension of the torus on which the motion occurs. Instead of taking particular initial conditions which lie on low-dimensional tori, let us perform a more global investigation of the FPU$-\beta$ model, aiming to track more "globally" the location of such tori. For this purpose, we

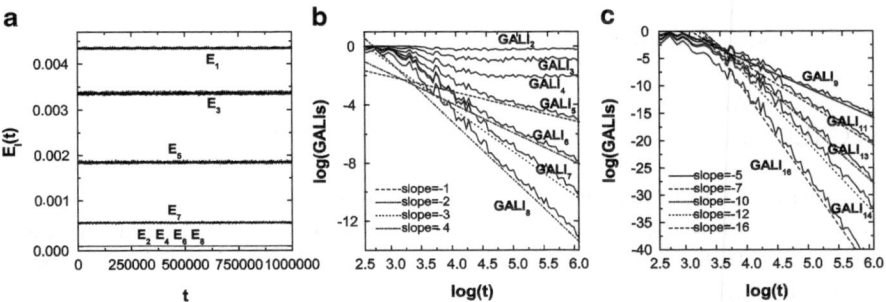

Fig. 5.18 (**a**) The time evolution of harmonic energies E_i for an ordered orbit lying on a four-dimensional torus of the $N = 8$ particle FPU$-\beta$ lattice (5.64). Recurrences occur between E_1, E_3, E_5 and E_7, while all other harmonic energies remain practically zero. The time evolution of the corresponding GALI$_k$ is plotted in (**b**) for $k = 2, \dots, 8$ and in (**c**) for $k = 9, 11, 13, 14, 16$. The *plotted lines* in (**b**) and (**c**) correspond to the precise power laws predicted by (5.58) (after [316])

consider the Hamiltonian system (5.64) with $N = 4$, for which ordered motion can occur on an s-dimensional torus with $s = 2, 3, 4$. According to (5.58) the corresponding GALIs of order $k \leq s$ will be constant, while the remaining ones will tend to zero following particular power laws. Thus, following [147], in order to locate low-dimensional tori we compute the GALI_k, $k = 2, 3, 4$ in the subspace (q_3, q_4) of the system's phase space, considering orbits with initial conditions $q_1 = q_2 = 0.1$, $p_1 = p_2 = p_3 = 0$, while p_4 is computed to keep the total energy H constant at $H = 0.010075$.

Since the constant final values of GALI_k, $k = 2, \ldots, s$, decrease with increasing order k (see for example the GALIs with $2 \leq k \leq 8$ in Fig. 5.15), we chose to "normalize" the values of GALI_k, $k = 2, 3, 4$ of each individual initial condition, by dividing them by the largest GALI_k value, $\max(\text{GALI}_k)$, obtained from all studied orbits. Thus, in Fig. 5.19 we color each initial condition according to its "*normalized* GALI_k" value

$$g_k = \frac{\text{GALI}_k}{\max(\text{GALI}_k)}. \tag{5.70}$$

In each panel of Fig. 5.19, large g_k values (colored in yellow or in light red) correspond to initial conditions whose GALI_k eventually stabilizes to constant, non-zero values. On the other hand, darker regions correspond to small g_k values, which correspond to power law decays of GALIs.

Consequently, motion on two-dimensional tori, which corresponds to large final GALI_2 values and small final GALI_3 and GALI_4 values, should be located in areas of the phase space colored yellow or light red in Fig. 5.19a, and in black in Fig. 5.19b, c. A region of the phase space with these characteristics is for example located on the upper border of the colored areas of Fig. 5.19. A particular initial condition with $q_3 = 0.106$, $q_4 = 0.0996$ in this region is denoted by a triangle in all

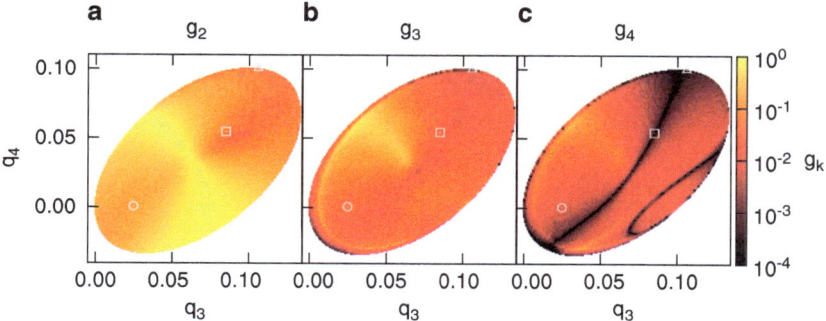

Fig. 5.19 Regions of different g_k (5.70) values, $k = 2, 3, 4$, on the (q_3, q_4) plane of the Hamiltonian system (5.64) with $N = 4$. Each initial condition is integrated up to $t = 10^6$, and colored according to its final (**a**) g_2, (**b**) g_3, and (**c**) g_4 value, while white regions correspond to forbidden initial conditions. Three particular initial conditions of ordered orbits on 2-, 3- and four-dimensional tori are marked by a *triangle*, a *square* and a *circle* respectively (after [147])

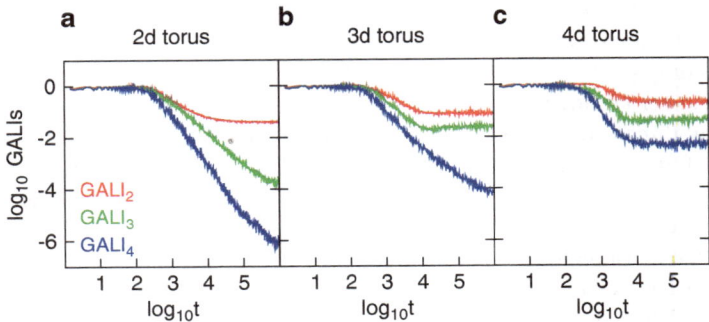

Fig. 5.20 The time evolution of GALIs for ordered orbits lying on a (**a**) two-dimensional torus, (**b**) three-dimensional torus, and (**c**) four-dimensional torus of the Hamiltonian system (5.64) with $N = 4$. The initial conditions of these orbits are marked respectively by a *triangle*, a *square* and a *circle* in Fig. 5.19 (after [147])

panels of Fig. 5.19. This orbit indeed lies on a two-dimensional torus, as we see from the evolution of its GALIs shown in Fig. 5.20a. In a similar way, an ordered orbit on a three-dimensional torus should be located in regions colored in black only in the g_4 plot of Fig. 5.19c. An orbit of this type is the one with $q_3 = 0.085109$, $q_4 = 0.054$ denoted by a square in Fig. 5.19, which actually evolves on a three-dimensional torus, as only its GALI$_2$ and GALI$_3$ remain constant (Fig. 5.20b). Finally, the orbit with $q_3 = 0.025$, $q_4 = 0$ (denoted by a circle in Fig. 5.19) inside a region of the phase space colored in yellow or light red in all panels of Fig. 5.19, is an ordered orbit on a four-dimensional torus and its GALI$_2$, GALI$_3$ and GALI$_4$ remain constant (Fig. 5.20c). It is worth noting, that chaotic motion would lead to very small GALI$_k$ and g_k values, since all GALIs would tend to zero exponentially, and consequently would correspond to regions colored in black in *all* panels of Fig. 5.19. Thus, chaotic motion can be easily distinguished from regular motion on low-dimensional tori.

From the results of Figs. 5.19 and 5.20 it becomes evident that the comparison of color plots of "normalized GALI$_k$" values can facilitate the search for low-dimensional tori in multi-dimensional phase spaces.

Appendix A: Wedge Product

We present here an introduction to the theory of wedge products following [310,315] and textbooks such as [159, 321]. Let us consider an M-dimensional vector space V over the field of real numbers $\mathbb{R}$. The exterior algebra of V is denoted by $\Lambda(V)$ and its multiplication, known as the wedge product (or exterior product), is written as $\wedge$. The wedge product is associative:

$$(\mathbf{u} \wedge \mathbf{v}) \wedge \mathbf{w} = \mathbf{u} \wedge (\mathbf{v} \wedge \mathbf{w}). \tag{5.71}$$

for $\mathbf{u}, \mathbf{v}, \mathbf{w} \in V$ and bilinear

$$(c_1\mathbf{u} + c_2\mathbf{v}) \wedge \mathbf{w} = c_1(\mathbf{u} \wedge \mathbf{w}) + c_2(\mathbf{v} \wedge \mathbf{w}),$$

$$\mathbf{w} \wedge (c_1\mathbf{u} + c_2\mathbf{v}) = c_1(\mathbf{w} \wedge \mathbf{u}) + c_2(\mathbf{w} \wedge \mathbf{v}), \tag{5.72}$$

for $\mathbf{u}, \mathbf{v}, \mathbf{w} \in V$ and $c_1, c_2 \in \mathbb{R}$. The wedge product on V also satisfies

$$\mathbf{u} \wedge \mathbf{u} = \mathbf{0}, \tag{5.73}$$

for all vectors $\mathbf{u} \in V$. Thus we have that

$$\mathbf{u} \wedge \mathbf{v} = -\mathbf{v} \wedge \mathbf{u}, \tag{5.74}$$

for all vectors $\mathbf{u}, \mathbf{v} \in V$ and

$$\mathbf{u}_1 \wedge \mathbf{u}_2 \wedge \cdots \wedge \mathbf{u}_k = \mathbf{0}, \tag{5.75}$$

whenever $\mathbf{u}_1, \mathbf{u}_2, \ldots, \mathbf{u}_k \in V$ are linearly dependent. Elements of the form $\mathbf{u}_1 \wedge \mathbf{u}_2 \wedge \cdots \wedge \mathbf{u}_k$ with $\mathbf{u}_1, \mathbf{u}_2, \ldots, \mathbf{u}_k \in V$ are called k-vectors.

Let $\{\hat{e}_1, \hat{e}_2, \ldots, \hat{e}_M\}$ be an orthonormal basis of V, i.e. $\hat{e}_i$, $i = 1, 2, \ldots, M$, are linearly independent vectors of unit magnitude and

$$\hat{e}_i \cdot \hat{e}_j = \delta_{ij} \tag{5.76}$$

where $(\cdot)$ denotes the inner product in V and

$$\delta_{ij} = \begin{cases} 1 \text{ for } i = j \\ 0 \text{ for } i \neq j \end{cases}. \tag{5.77}$$

It can be easily seen that the set

$$\{\hat{e}_{i_1} \wedge \hat{e}_{i_2} \wedge \cdots \wedge \hat{e}_{i_k} \mid 1 \leq i_1 < i_2 < \cdots < i_k \leq M\} \tag{5.78}$$

is a basis of $\Lambda^k(V)$ since any wedge product of the form $\mathbf{u}_1 \wedge \mathbf{u}_2 \wedge \cdots \wedge \mathbf{u}_k$ can be written as a linear combination of the k-vectors of (5.78).

The coefficients of a k-vector $\mathbf{u}_1 \wedge \mathbf{u}_2 \wedge \cdots \wedge \mathbf{u}_k$ are the minors of the matrix C having as rows the coordinates of vectors $\mathbf{u}_i$, $i = 1, 2, \ldots, k$, with respect to the orthonormal basis $\hat{e}_i$, $i = 1, 2, \ldots, M$. In matrix form we have

$$\begin{bmatrix} \mathbf{u}_1 \\ \mathbf{u}_2 \\ \vdots \\ \mathbf{u}_k \end{bmatrix} = \begin{bmatrix} u_{11} & u_{12} & \cdots & u_{1M} \\ u_{21} & u_{22} & \cdots & u_{2M} \\ \vdots & \vdots & & \vdots \\ u_{k1} & u_{k2} & \cdots & u_{kM} \end{bmatrix} \cdot \begin{bmatrix} \hat{e}_1 \\ \hat{e}_2 \\ \vdots \\ \hat{e}_M \end{bmatrix} = C \cdot \begin{bmatrix} \hat{e}_1 \\ \hat{e}_2 \\ \vdots \\ \hat{e}_M \end{bmatrix}, \tag{5.79}$$

where u_{ij}, $i = 1, 2, \ldots, k$, $j = 1, 2, \ldots, M$ are real numbers. Then the wedge product $\mathbf{u}_1 \wedge \mathbf{u}_2 \wedge \cdots \wedge \mathbf{u}_k$ is written as

$$\mathbf{u}_1 \wedge \mathbf{u}_2 \wedge \cdots \wedge \mathbf{u}_k = \sum_{1 \leq i_1 < i_2 < \cdots < i_k \leq M} \begin{vmatrix} u_{1i_1} & u_{1i_2} & \cdots & u_{1i_k} \\ u_{2i_1} & u_{2i_2} & \cdots & u_{2i_k} \\ \vdots & \vdots & & \vdots \\ u_{ki_1} & u_{ki_2} & \cdots & u_{ki_k} \end{vmatrix} \hat{e}_{i_1} \wedge \hat{e}_{i_2} \wedge \cdots \wedge \hat{e}_{i_k} \ , \quad (5.80)$$

where the sum is performed over all possible combinations of k indices out of the M total indices, and $|\cdot|$ denotes the determinant. The norm of the k-vector $\mathbf{u}_1 \wedge \mathbf{u}_2 \wedge \cdots \wedge \mathbf{u}_k$ is given by

$$\|\mathbf{u}_1 \wedge \mathbf{u}_2 \wedge \cdots \wedge \mathbf{u}_k\| = \left\{ \sum_{1 \leq i_1 < i_2 < \cdots < i_k \leq M} \begin{vmatrix} u_{1i_1} & u_{1i_2} & \cdots & u_{1i_k} \\ u_{2i_1} & u_{2i_2} & \cdots & u_{2i_k} \\ \vdots & \vdots & & \vdots \\ u_{ki_1} & u_{ki_2} & \cdots & u_{ki_k} \end{vmatrix}^2 \right\}^{1/2} . \quad (5.81)$$

Appendix B: Example Algorithms for the Computation of the SALI and GALI Chaos Indicators

We present here MAPLE algorithms for the computation of the SALI and GALIs. As a test model we consider the Hénon-Heiles system (5.18) and compute the indices for the chaotic orbit of Figs. 5.3 and 5.8. We have included several comments in the algorithms in order to facilitate their modification for other systems.

A MAPLE Algorithm for the Computation of the SALI

The following algorithm uses (5.13) for the computation of the SALI.

```
>    restart:
>    with(LinearAlgebra):
>    Digits:=20:
```

Initialization:
The Hénon-Heiles Hamiltonian (5.18)

```
>    H := 1/2*(px^2 + py^2) + 1/2*(x^2 + y^2) + x^2*y - 1/3*y^3;
```

Parameters:
Degrees of freedom

```
>    dof := 2;
```

Variables for the state of the system and the variations

```
>   q := Vector(dof, [x, y]);
>   p := Vector(dof, [px, py]);
>   deltaq := Matrix(dof, 2, [[dx1, dx2], [dy1, dy2]]);
>   deltap := Matrix(dof, 2, [[dpx1, dpx2], [dpy1, dpy2]]);
```

Time range of the integration and time increment

```
>   T0   := 0.0;
>   Tend := 1.0e3;
>   Tinc := 1.0;
```

Initial conditions and initial orthonormal deviation vectors

```
>   ics := x(0)=0.0, y(0)=-0.25, px(0)=0.42081, py(0)=0.0,
>          dx1(0)=1.0, dy1(0)=0.0, dpx1(0)=0.0, dpy1(0)=0.0,
>          dx2(0)=0.0, dy2(0)=1.0, dpx2(0)=0.0, dpy2(0)=0.0;
```

Matrix used for the computation of the SALI

```
>   N := floor((Tend-T0)/Tinc);
>   SALI := Matrix(N, 2, datatype=float[8], []);
```

Equations of motion

```
>   for i from 1 to 2 do
>       deq[i]   := diff(q[i](t), t) =  diff(H, p[i]);
>       deq[2+i] := diff(p[i](t), t) = -diff(H, q[i]):
>   end do:
```

Variational equations

```
>   for nc from 1 to 2 do
>       k := 5 + 4*(nc-1):
>       for nr from 1 to 2 do
>         deq[k]   := diff(deltap[nr,nc](t),t) =
>                       -add(diff(H,q[nr],q[j])*deltaq[j,nc],j=1..dof);
>         deq[k+2] := diff(deltaq[nr,nc](t),t) =
>                       add(diff(H,p[nr],p[j])*deltap[j,nc],j=1..dof);
>         k:=k+1;
>       end do:
>   end do:
```

Make all equations explicit functions of time

```
>   eqs := seq(lhs(deq[i]) = subs(
>       { seq(q[i]=q[i](t),i=1..dof),
>         seq(p[i]=p[i](t),i=1..dof),
>         seq(seq(deltaq[i,j]=deltaq[i,j](t),i=1..dof),j=1..2),
>         seq(seq(deltap[i,j]=deltap[i,j](t),i=1..dof),j=1..2)},
>       rhs(deq[i])), i = 1..12);
```

The integration

```
>   for i from 1 to N do
>       T   := evalf(T0+i*Tinc):
>   # integrate the system of ODEs numerically
>       dsn := dsolve({eqs,ics},numeric,range = T-Tinc..T);
>   # the 2 variational vectors
>       v1 := Vector(4,[
>               subs(dsn(T),dx1(t)),
```

```
>                    subs(dsn(T),dy1(t)),
>                    subs(dsn(T),dpx1(t)),
>                    subs(dsn(T),dpy1(t))]);
>      v2 := Vector(4,[
>                    subs(dsn(T),dx2(t)),
>                    subs(dsn(T),dy2(t)),
>                    subs(dsn(T),dpx2(t)),
>                    subs(dsn(T),dpy2(t))]);
>   # renormalize these vectors
>      v1 := v1/sqrt(v1.v1):
>      v2 := v2/sqrt(v2.v2):
>      tplus   := sqrt((v1+v2).(v1+v2)):
>      tminus  := sqrt((v1-v2).(v1-v2)):
>   # compute the SALI  using Eq. (5.13)
>      SALI[i,1]  := T:
>      SALI[i,2]  := min(tplus,tminus);
>   # use the result as new initial conditions for the next step
>      ics:= x(T)=subs(dsn(T),x(t)), y(T)=subs(dsn(T),y(t)),
>           px(T)=subs(dsn(T),px(t)), py(T)=subs(dsn(T),py(t)),
>           dx1(T)=v1[1], dy1(T)=v1[2], dpx1(T)=v1[3], dpy1(T)=v1[4],
>           dx2(T)=v2[1], dy2(T)=v2[2], dpx2(T)=v2[3], dpy2(T)=v2[4];
>   end do:
```

A MAPLE Algorithm for the Computation of the GALI

The following algorithm computes the whole spectrum of the GALIs using the
Singular Value Decomposition procedure of an appropriate matrix described in
Sect. 5.3. In particular it implements (5.29) for the evaluation of GALIs.

```
>   restart:
>   with(LinearAlgebra):
>   Digits := 20:
```

Initialization:
The Hénon-Heiles Hamiltonian (5.18)

```
>   H := 1/2*(px^2 + py^2) + 1/2*(x^2 + y^2) + x^2*y - 1/3*y^3;
```

Parameters:
Degrees of freedom

```
>   dof := 2;
```

Variables for the state of the system and the variations

```
>   q := Vector(dof, [x, y]);
>   p := Vector(dof, [px, py]);
>   deltaq := Matrix(dof, 4, [[dx1,dx2,dx3,dx4],
>            [dy1,dy2 dy3,dy4]]);
>   deltap := Matrix(dof, 4, [[dpx1,dpx2,dpx3,dpx4],
>            [dpy1,dpy2,dpy3,dpy4]]);
```

Time range of the integration and time increment

```
>   T0 := 0.0;
```

```
>   Tend := 1.0e3;
>   Tinc := 1.0;
```

Initial conditions and initial orthonormal deviation vectors

```
>   ics := x(0)=0, y(0)=-0.25, px(0)=0.42081, py(0)=0.0,
>          dx1(0)=1.0, dy1(0)=0.0, dpx1(0)=0.0, dpy1(0)=0.0,
>          dx2(0)=0.0, dy2(0)=1.0, dpx2(0)=0.0, dpy2(0)=0.0,
>          dx3(0)=0.0, dy3(0)=0.0, dpx3(0)=1.0, dpy3(0)=0.0,
>          dx4(0)=0.0, dy4(0)=0.0, dpx4(0)=0.0, dpy4(0)=1.0;
```

Matrix used for the computation of the GALI

```
>   N := floor((Tend-T0)/Tinc);
>   GALI := Matrix(N, 4, datatype=float[8], []);
```

Equations of motion

```
>   for i from 1 to 2 do
>     deq[i]   := diff(q[i](t), t) =  diff(H, p[i]);
>     deq[2+i] := diff(p[i](t), t) = -diff(H, q[i]):
>   end do:
```

Variational equations

```
>   for nc from 1 to 4 do
>     k := 5 + 4*(nc-1):
>     for nr from 1 to 2 do
>       deq[k]   := diff(deltap[nr,nc](t),t) =
>                     -add(diff(H,q[nr],q[j])*deltaq[j,nc],j=1..dof);
>       deq[k+2] := diff(deltaq[nr,nc](t),t) =
>                     add(diff(H,p[nr],p[j])*deltap[j,nc],j=1..dof);
>       k:=k+1;
>     end do:
>   end do:
```

Make all equations explicit functions of time

```
>   eqs := seq(lhs(deq[i]) = subs(
>       { seq(q[i]=q[i](t),i=1..dof),
>         seq(p[i]=p[i](t),i=1..dof),
>         seq(seq(deltaq[i,j]=deltaq[i,j](t),i=1..dof),j=1..4),
>         seq(seq(deltap[i,j]=deltap[i,j](t),i=1..dof),j=1..4)},
>       rhs(deq[i])), i = 1..20);
```

The integration

```
>   for i from 1 to N do
>     T    := evalf(T0+i*Tinc):
>   # integrate the system of ODEs numerically
>     dsn := dsolve({eqs,ics},numeric,range=T-Tinc..T);
>   # the 4 deviation vectors
>     v1 := Vector(4,[
>             subs(dsn(T),dx1(t)), subs(dsn(T),dy1(t)),
>             subs(dsn(T),dpx1(t)), subs(dsn(T),dpy1(t))]);
>     v2 := Vector(4,[
>             subs(dsn(T),dx2(t)), subs(dsn(T),dy2(t)),
>             subs(dsn(T),dpx2(t)), subs(dsn(T),dpy2(t))]);
>     v3 := Vector(4,[
>             subs(dsn(T),dx3(t)), subs(dsn(T),dy3(t)),
>             subs(dsn(T),dpx3(t)), subs(dsn(T),dpy3(t))]);
```

```
>     v4 := Vector(4,[
>              subs(dsn(T),dx4(t)), subs(dsn(T),dy4(t)),
>              subs(dsn(T),dpx4(t)), subs(dsn(T),dpy4(t))]);
>     # renormalize these vectors
>     v1 := v1/sqrt(v1.v1):
>     v2 := v2/sqrt(v2.v2):
>     v3 := v3/sqrt(v3.v3):
>     v4 := v4/sqrt(v4.v4):
>     # fill the columns of matrix A with these vectors
>     A := Matrix(4,4,[]):
>     A[1..4,1] := v1:
>     A[1..4,2] := v2:
>     A[1..4,3] := v3:
>     A[1..4,4] := v4:
>     # compute GALI_k, k=2,3,4, using Eq. (5.29)
>     for k from 2 to 4 do
>        sv:=SingularValues(A[1..4,1..k],output='list'):
>        GALI[i,k] := mul(sv[i],i=1..k):
>     end do:
>     GALI[i,1] := T:
>     # use the result as new initial conditions for the next step
>     ics:= x(T)=subs(dsn(T),x(t)), y(T)=subs(dsn(T),y(t)),
>           px(T)=subs(dsn(T),px(t)), py(T)=subs(dsn(T),py(t)),
>           dx1(T)=v1[1], dy1(T)=v1[2], dpx1(T)=v1[3], dpy1(T)=v1[4],
>           dx2(T)=v2[1], dy2(T)=v2[2], dpx2(T)=v2[3], dpy2(T)=v2[4],
>           dx3(T)=v3[1], dy3(T)=v3[2], dpx3(T)=v3[3], dpy3(T)=v3[4],
>           dx4(T)=v4[1], dy4(T)=v4[2], dpx4(T)=v4[3], dpy4(T)=v4[4];
>     end do:
```

Exercises

Exercise 5.1. Consider a $2N$-dimensional vector space over the field of real numbers $\mathbb{R}$, which has the usual Euclidean norm and is spanned by the usual orthonormal basis $\{\hat{e}_1, \hat{e}_2, \ldots, \hat{e}_{2N}\}$ with $\hat{e}_1 = (1, 0, 0, \ldots, 0)$, $\hat{e}_2 = (0, 1, 0, \ldots, 0)$, $\ldots, \hat{e}_{2N} = (0, 0, 0, \ldots, 1)$. Consider also two unit vectors $\hat{w}_1$, $\hat{w}_2$ in this space so that

$$\hat{w}_1 = \sum_{i=1}^{2N} w_{1i}\hat{e}_i \quad , \quad \hat{w}_2 = \sum_{i=1}^{2N} w_{2i}\hat{e}_i, \tag{5.82}$$

and $\sum_{i=1}^{2N} w_{1i}^2 = 1$, $\sum_{i=1}^{2N} w_{2i}^2 = 1$. Then, for the 2-vector $\hat{w}_1 \wedge \hat{w}_2$ of (5.80) and its norm from (5.81), show the validity of (5.23)

$$\|\hat{w}_1 \wedge \hat{w}_2\| = \frac{1}{2}\|\hat{w}_1 - \hat{w}_2\| \cdot \|\hat{w}_1 + \hat{w}_2\|.$$

Hint: Consult Appendix B of [315].

Exercise 5.2. Consider the $k \times 2N$ matrix W, (5.27), having as rows the coordinates of k unit deviation vectors $\hat{\mathbf{w}}_i$, $i = 1, 2, \ldots, k$. According to (5.24) GALI_k is equal to the "volume" $\mathrm{vol}(P_k)$ of the parallelepiped P_k having as edges these vectors. This volume is also given by $\mathrm{vol}(P_k) = \sqrt{|W \cdot W^\mathrm{T}|}$, where $|\cdot|$ denotes the determinant of a matrix and $(^\mathrm{T})$ its transpose (see for instance [174, Chap. 5]). Following for example [277, Sect. 2.6], perform the Singular Value Decomposition of W^T and write this matrix as a product of a $2N \times k$ column-orthogonal matrix U, a $k \times k$ diagonal matrix Z with positive or zero elements z_i, $i = 1, \ldots, k$ (the so-called *singular values*), and the transpose of a $k \times k$ orthogonal matrix V. Then show that the GALI_k is equal to the product of these singular values (5.29). Hint: Consult results presented in [316].

Exercise 5.3. (a) Show that a deviation vector which is initially located in the tangent space of an N-dimensional torus will remain there forever.

(b) Consider an ordered orbit of an N dof Hamiltonian system lying on a N-dimensional torus, and k linearly independent deviation vectors such that m ($m \le N$ and $m \le k$) of them are initially located in the tangent space of the torus. Show that for this choice of deviation vectors the evolution of GALI_k is

$$\mathrm{GALI}_k(t) \propto \begin{cases} \text{constant} & \text{if } 2 \le k \le N \\ \frac{1}{t^{2(k-N)-m}} & \text{if } N < k \le 2N \text{ and } 0 \le m < k - N \\ \frac{1}{t^{k-N}} & \text{if } N < k \le 2N \text{ and } m \ge k - N \end{cases} .$$

Hint: Consult results presented in [315].

Exercise 5.4. Following an analysis similar to the one presented in Sect. 5.3.1.2 show that the time evolution of GALI_k for ordered motion on s-dimensional tori with $2 \le s \le N$ for N dof Hamiltonian flows, and $1 \le s \le N$ for $2N$D maps is given by (5.58)

$$\mathrm{GALI}_k(t) \propto \begin{cases} \text{constant if } 2 \le k \le s \\ \frac{1}{t^{k-s}} & \text{if } s < k \le 2N - s \\ \frac{1}{t^{2(k-N)}} & \text{if } 2N - s < k \le 2N \end{cases} .$$

Hint: Consult results presented in [93] and [316].

Problems

Problem 5.1. *The Behavior of GALI for Periodic Orbits.*

(a) Unstable periodic orbits have at least one positive LCE ($L_1 > 0$), which implies that nearby orbits diverge exponentially from them. Applying the analysis presented in Sect. 5.3.1.1 for chaotic orbits which also have $L_1 > 0$, show that GALI_k of unstable periodic orbits tends to zero exponentially following the law

$$\text{GALI}_k(t) \propto e^{-[(L_1-L_2)+(L_1-L_3)+\cdots+(L_1-L_k)]t}. \qquad (5.83)$$

(b) Following an analysis similar to the one presented in Sect. 5.3.1.2 show that the time evolution of GALI_k for stable periodic orbits of N dof Hamiltonian systems is given by

$$\text{GALI}_k \propto \begin{cases} \frac{1}{t^{k-1}} & \text{if } 2 \leq k \leq 2N-1 \\ \frac{1}{t^{2N}} & \text{if } k = 2N \end{cases}. \qquad (5.84)$$

(c) Stable periodic orbits of symplectic maps correspond to stable fixed points of the map, which are located inside islands of stability. Observing that any deviation vector from the stable periodic orbit performs a rotation around the fixed point show that

$$\text{GALI}_k \propto \text{constant}, \quad 2 \leq k \leq 2N. \qquad (5.85)$$

for stable periodic orbits of $2N$D symplectic maps.
Hint: Consult results presented in [247].

Problem 5.2. *Connection Between the Dynamics of Hamiltonian Flows and Symplectic Maps.* N dof Hamiltonian systems can be considered as dynamically equivalent to $2(N-1)$D maps, since the latter can be interpreted as appropriate surfaces of section of the former. Despite this relation the GALIs behave differently for flows and maps. According to (5.85) GALIs remain constant for stable periodic orbits of maps, while they decrease to zero for flows according to (5.84).

In order to understand this difference, consider as an example the Hénon-Heiles system (5.18). Compute for a stable periodic orbit of this system, GALI_2 and GALI_3, using not the whole deviation vectors but *their components orthogonal to the flow*. For a two dof Hamiltonian system it is not necessary to compute GALI_4 with these vectors since $\text{GALI}_4 = 0$ (why?). Check whether the computed GALIs follow predictions (5.84) or (5.85). Hint: Consult results presented in [247].

Chapter 6
FPU Recurrences and the Transition from Weak to Strong Chaos

Abstract The present chapter starts with a historical introduction to the FPU one-dimensional lattice as it was first integrated numerically by Fermi Pasta and Ulam in the 1950s and describes the famous paradox of the FPU recurrences. First we review some of the more recent attempts to explain it based on the nonlinear normal modes (NNMs) representing continuations of the lowest $q = 1, 2, 3, \ldots$ modes of the linear problem, termed q-breathers by Flach and co-workers, due to their exponential localization in Fourier space. We then present an extension of this approach focusing on certain low-dimensional, so-called q-tori, in the neighborhood of these NNMs, which are also exponentially localized and can be constructed by Poncaré-Linstedt series. We demonstrate how q-tori reconcile q-breathers with the "natural packets" approach of Berchialla and co-workers. Finally, we use the GALI method to determine the destabilization threshold of q-tori and study the slow diffusion of chaotic orbits associated with the breakdown of the FPU recurrences.

6.1 The Fermi Pasta Ulam Problem

6.1.1 Historical Remarks

In the history of Hamiltonian dynamics few discoveries have inspired so many researchers and motivated so many theoretical and numerical studies as the one announced in 1955, by E. Fermi, J. Pasta and S. Ulam (FPU) [71, 121, 133]. They had the brilliant idea to use the computers available at that time at the Los Alamos National Laboratory to integrate the ODEs of a chain of 31 identical harmonic oscillators, coupled to each other by cubic nearest neighbor interactions, to investigate how energy was shared by all particles, as soon as the nonlinear forces were turned on. To this end, they imposed fixed boundary conditions and solved numerically the equations of motion

T. Bountis and H. Skokos, *Complex Hamiltonian Dynamics*,
Springer Series in Synergetics, DOI 10.1007/978-3-642-27305-6_6,
© Springer-Verlag Berlin Heidelberg 2012

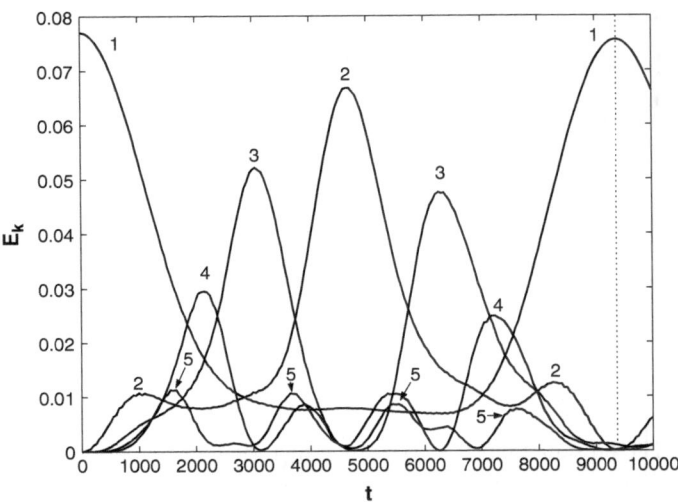

Fig. 6.1 The FPU experiment: Time-integration of the FPU system until just after the first approximate recurrence of the initial state. Shown here is how the energy $E_q = \frac{1}{2}\left(\dot{a}_q^2 + 2a_q^2 \sin^2\left(\frac{q\pi}{2(N+1)}\right)\right)$ is divided over the first five modes $a_q = \sum_{k=1}^{N} x_k \sin\left(\frac{kq\pi}{N+1}\right)$. The numbers in the figure indicate the wave-number q (after [53]) (The figure is a reproduction of Fig. 1 of [121])

$$\ddot{x}_k = (x_{k+1} - 2x_k + x_{k-1}) + \alpha\left[(x_{k+1} - x_k)^2 - (x_k - x_{k-1})^2\right], k = 1, \ldots, 31,$$
(6.1)

with $x_0 = x_{32} = 0$, $\dot{x}_0 = \dot{x}_{32} = 0$, where $x_k = x_k(t)$ represents each particle's displacement from equilibrium and dots denote differentiation with respect to time t. To their great surprise, starting with the initial condition $x_k = \sin\left(\frac{\pi k}{32}\right)$, corresponding to the first ($q = 1$) linear normal mode, they observed for small energies ($\alpha = 0.25$) a remarkable *near-recurrence* of the solutions of (6.1) to this initial condition after a relatively short time period (see Fig. 6.1 below).

This was amazing! Equilibrium statistical mechanics predicted that all higher modes of oscillation would also be excited and eventually equally share the energy of the system, achieving a state of thermodynamic equilibrium. Instead, FPU observed that a very small set of modes participated in the dynamics and the system practically returned to its initial state after relatively short time periods. It is, therefore, no wonder that this so-called "FPU paradox" created a great excitement within the scientific community and generated a number of questions which remained open for many years: Could it be, for example, that low energy recurrences are due to the fact that (6.1) is close to an integrable system? Is it perhaps because this type of nonlinearity is not sufficient for energy equipartition? How can this "energy localization" be explained? A new era of complex phenomena in Hamiltonian dynamics was in the making.

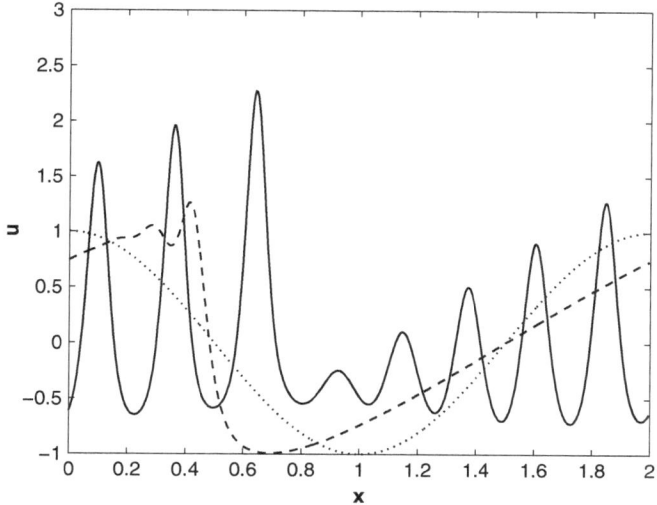

Fig. 6.2 The evolution of the initial state $u = \cos(\pi x)$ in the Korteweg-de Vries equation, at the times $t = 0$ (*dotted line*) $t = 1/\pi$ (*dashed line*). At an intermediate stage ($t = 3.6/\pi$, *solid line*), the solitons are maximally separated, while at $t \approx 30.4/\pi$ the initial state is nearly recovered (after [53]) (The figure is a reproduction of Fig. 1 of [352])

In 1965, attempting to explain the results of FPU, Zabusky and Kruskal [352] derived the Korteweg-de Vries (KdV) partial differential equation of shallow water waves in the long wavelength and small amplitude approximation,

$$u_t + uu_x + \delta^2 u_{xxx} = 0 \quad \delta^2 \ll 1, \tag{6.2}$$

as a continuum limit of the FPU system (6.1). They observed that solitary traveling wave solutions, now known as *solitons*, exist, whose interaction properties result in analogous recurrences of initial states (Fig. 6.2). These results, of course, led to the birth of a new field in applied mathematics dealing with the integrability of evolution equations and their solvability by the theory of Inverse Scattering Transforms [2, 3].

In taking the continuum limit, however, Zabusky and Kruskal overlooked an important aspect of the FPU problem, which is the *discreteness* of the particle chain. What happens often in discrete systems is that resonances between neighboring oscillators can be avoided if their vibration frequencies lie outside the so-called *phonon band* representing the frequency spectrum of the linear problem. This leads to the occurrence of localized oscillations called *discrete breathers*, which certainly prevent equipartition of energy among all particles of the chain. Thus, one might conjecture that FPU recurrences are also the result of limited energy transport caused by discreteness and non-resonance effects. This, however, is an oversimplified way to view the FPU paradox.

The reason is that discrete breathers, of whom we shall have a lot more to say in Chap. 7, are localized in configuration space and hence involve the oscillations of few particles, while the initial state of the FPU experiment involves all particles forming a single Fourier mode of the corresponding linear chain. Thus, we should be looking instead at localization in Fourier q-modal space, as the FPU recurrences were observed after all when all the energy was placed in the $q = 1$ mode!

6.1.2   The Concept of q-breathers

This crucial fact was emphasized by Flach and his co-workers [128, 129], who introduced the concept of *q-breathers* as exact periodic solutions of the problem and pointed out the role of their (linear) stability on the dynamics of FPU systems. In fact, they considered besides the FPU$-\alpha$ version of (6.1) the FPU$-\beta$ version as well, described by the Hamiltonian

$$H = \frac{1}{2} \sum_{k=1}^{N} y_k^2 + \frac{1}{2} \sum_{k=0}^{N} (x_{k+1} - x_k)^2 + \frac{\beta}{4} \sum_{k=0}^{N} (x_{k+1} - x_k)^4 = E \qquad (6.3)$$

where x_k again is the kth particle's displacement from equilibrium, y_k is its canonically conjugate momentum and fixed boundary conditions are imposed by setting $x_0 = x_{N+1} = 0$. A q-breather is defined as the continuation, for $\beta \neq 0$ in (6.3) (or $\alpha \neq 0$ in (6.1)) of the simple harmonic motion exhibited by the qth mode in the uncoupled case ($\alpha = \beta = 0$) and refers to the oscillation of all particles with a single frequency nearly equal to the frequency of the qth linear mode. It is, therefore, exactly what we have called NNM in Chap. 4. For a recent and comprehensive review of these developments see [125].

Clearly, therefore, an important clue to the FPU paradox for small coupling parameters lies in the observation that, if we excite a single low q-breather mode ($q = 1, 2, 3, \ldots$), the total energy remains localized in practice only within a few of these modes, at least as long as the initially excited mode is stable. Furthermore, if we follow numerically FPU trajectories with initial conditions close to such a NNM, we find that they exhibit nearly the same localization profile as the NNM, at least for very long time intervals. Thus, NNMs surface again in our studies as q-breathers offering new insight in understanding the problem of deviations from energy equipartition among all modes, due to FPU recurrences.

Analytical estimates of various scaling laws concerning q-breathers were also obtained via the perturbation method of Poincaré-Linstedt series in [129], which we shall describe in detail later in this chapter. Using this method, the authors of [129] expanded the solutions for all canonical variables up to the lowest non-trivial order with respect to the small parameters and derived two important results: (a) An asymptotic formula for the energy threshold of the first destabilization of the low q-breathers of the FPU$-\beta$ chain (6.3)

$$E_c \approx \frac{\pi^2}{6\beta(N+1)}, \tag{6.4}$$

valid for large N, and (b) an exponentially decaying function for the average harmonic energy of the qth mode, $E(q) \propto \exp(-bq)$, where b depends on the system's parameters, N and the total energy E. Remarkably enough, these q-breather energy profiles $E(q)$ are very similar to those of "FPU trajectories", i.e. solutions generated by exciting initially one or a few low-q modes (as in the original FPU experiments), whose exponential localization in q-space had already been noted earlier by other authors [110, 139, 140, 228].

Based on this similarity, we may therefore conjecture that there is a close connection between q-breathers and the energy localization properties of FPU trajectories. This sounds as if we are on the right track, yet two intriguing questions arise: What is the precise nature of this connection and why do FPU recurrences persist at parameter values for which the corresponding q-breather solution becomes *unstable*? In this chapter, we shall try to answer these questions.

Before doing this, however, we need to make one more crucial remark: In 2004, Berchialla et al. [37] explored in detail the phenomenon of energy equipartition in FPU experiments, exciting initially a fixed (low-frequency) fraction of the spectrum, which is detached from zero i.e. $[N/64, 5N/64]$. The direct implication of their numerical study was that the flow of energy across the high-frequency parts of the spectrum takes place exponentially slowly, by a law of the form $T \propto \exp(\varepsilon^{-1/4})$, where T is the time needed for the energy to be nearly equally partitioned, and $\varepsilon = E/N$ is the specific energy of the system. In fact, this dependence appears as a piecewise power law, with different "best-fit" slopes over different ranges of values of ε.

Recently, a new study on equipartition times [30] offers new insight to this question. These authors show numerically that the *full* FPU system, including both α and β terms (see (6.1), (6.3)), does *not* exhibit exponentially long times to equipartition in the thermodynamic limit, while the FPU-β model itself does so only for initial conditions of the type chosen in [37]. Still, it is important to emphasize that all results on equipartition times in the thermodynamic limit depend on the specific energy of the system.

For example, a power-law behavior of the type $T \propto \varepsilon^{-3}$ was found in a subinterval of ε values in [37], which fits nicely previous results reported in [112] on the scaling of the equipartition time with ε beyond a critical "weak chaos" threshold. Nevertheless, the question of which scaling laws correctly characterize the approach to equipartition remains open, since no rigorous results have so far been provided, while numerical results are available over a limited range of values of E and N. For a recent review of various semi-analytical and numerical approaches to this issue see [227].

In this context, an important observation made in [38] turns out to be quite relevant: Even if one starts by exciting one low q-mode with large enough energy, one observes, long before equipartition, the formation of *metastable states* coined "natural packets", in which the energy undergoes first a kind of equipartition among a subset of (low-frequency) modes, as if only a fraction of the spectrum was initially excited. These packets appear quite clearly in the FPU$-\alpha$ model, where the set of participating modes exhibits a "plateau" in the energy spectrum. In the FPU$-\beta$ model on the other hand, although no clear plateaus are observed, the low-frequency part of the spectrum contains most of the energy given initially to the system, in a way that allows us to define an effective packet width. In fact, one such packet consisting of five modes is exactly what FPU observed in their original experiment (see Fig. 6.1). We may thus conclude that FPU trajectories are characterized by a kind of *metastability* resulting from all types of initial excitations of the low frequency part of the spectrum [34].

Empirical scaling laws were established [38] concerning the dependence of the "width" of a packet (i.e. the effective number of participating modes), on the specific energy ε. These are consistent with the laws of energy localization obtained via either a continuous Hamiltonian model which interpolates the FPU dynamics in the space of Fourier modes [24, 275], or the q-breather model [126]. A difference, however, between the two models is that in the continuous model the constant b in $E(q) \propto \exp(-bq)$ turns out to be independent of E.

6.1.3   The Concept of q-tori

Let us now think of q-breathers (or NNMs) as tori of dimension one, forming a "backbone" in phase space that describes the motion of the normal mode variables. The following question naturally arises: Since we are dealing with N dof Hamiltonian systems, why not consider also the relevance of tori of any dimension lower than N? In particular, do such tori exist, and if so, do they exhibit similar localization properties as the q-breathers? As we explain in this chapter, a more complete interpretation of the "paradox" of FPU recurrences can be put forth, by introducing the concept of *q-tori*, which reconciles q-breathers and their energy localization properties, with the occurrence of metastable packets of low-frequency modes. To achieve this deeper level of understanding of lack of equipartition, we need to generalize the results obtained in [37, 112] for FPU trajectories, observing that the packet of modes excited in these experiments corresponds to spectral numbers satisfying the condition $(N + 1)/64 \leq q \leq 5(N + 1)/64$, with N equal to a power of two minus one.

We shall thus consider classes of special solutions lying not only on one-dimensional tori (as is the case with q-breathers), but also on tori of *any low dimension $s \ll N$*, i.e. solutions with s independent frequencies, representing the continuation of motions resulting from exciting s modes of the uncoupled case, where s is allowed to vary proportionally to N. In what follows, we shall focus on exploring the properties of such *q-tori* solutions, both analytically and numerically. In particular, we shall establish energy localization laws analogous to those for q-breathers, using the Poincaré-Linstedt perturbative method. We will then demonstrate by numerical experiments that these laws describe accurately the properties not only of exact q-tori solutions, but also of FPU trajectories with nearby initial conditions.

The q-tori approach has been described in detail in a recent publication [94], so we will only briefly sketch it here. Note that our aim in studying q-tori is to establish the existence of exact invariant manifolds of lower dimension in the FPU problem. A similar approach can be found in the work of Giorgilli and Muraro [150], who also explored FPU trajectories confined in lower-dimensional manifolds of the $2N$-dimensional FPU phase space. They proved that this confinement persists for times exponentially long in the inverse of ε, in the spirit of the famous Nekhoroshev theory [33,256]. In fact, the theory of Nekhoroshev offers a natural framework for studying analytically the metastability scenario.

In [94], it was shown that use of the Poincaré-Linstedt method leads to accurate scaling laws for the energy profile $E(q)$ of a trajectory lying exactly on a q-torus, while the consistency of the Poincaré-Linstedt construction on a Cantor set of perturbed frequencies (or amplitudes) was explicitly demonstrated. Numerically, one finds that energy localization persists for long times also for trajectories near a q-torus, even beyond the energy threshold where the q-torus itself becomes unstable. Of course, the determination of linear stability for an s-dimensional torus ($s > 1$) is a subtle question, since no straightforward application of Floquet theory is available, as in the $s = 1$ case. Nevertheless, as shown in [94], one can employ the method of GALI (see Chap. 5) to approximately determine critical parameter values at which a low-dimensional torus turns unstable, in the sense that orbits in its immediate vicinity begin to display chaotic behavior.

Thus, in the next sections we briefly describe analytical and numerical results on the existence, stability and scaling laws of q-tori, using the FPU$-\beta$ model as an example. We focus on the subset of q-tori corresponding to zeroth order excitations of a set of adjacent modes $q_{0,i}$, $i = 1,\ldots,s$, whose energy profile $E(q)$ can be evaluated analytically and examine the energy localization of FPU trajectories when the linear stability of a nearby q-torus is lost. Finally, we address the question of the breakdown of FPU recurrences and the onset of equipartition in connection with the slow diffusion of orbits starting in the vicinity of unstable q-tori.

6.2 Existence and Stability of q-tori

Let us begin our study of the FPU$-\beta$ system (6.3) introducing normal mode canonical variables Q_q, P_q through the (linear) set of canonical transformations (see our discussion in Sects. 4.1 and 4.2)

$$x_k = \sqrt{\frac{2}{N+1}} \sum_{q=1}^{N} Q_q \sin\left(\frac{qk\pi}{N+1}\right), \quad y_k = \sqrt{\frac{2}{N+1}} \sum_{q=1}^{N} P_q \sin\left(\frac{qk\pi}{N+1}\right).$$

$$(6.5)$$

Direct substitution of (6.5) into (6.3) allows us to write the Hamiltonian in the form $H = H_2 + H_4$, in which the quadratic part represents N uncoupled harmonic oscillators

$$H_2 = \frac{1}{2} \sum_{q=1}^{N} \left(P_q^2 + \Omega_q^2 Q_q^2 \right) \tag{6.6}$$

with (linear) normal mode frequencies

$$\Omega_q = 2 \sin \left(\frac{q\pi}{2(N+1)} \right), \quad 1 \le q \le N, \tag{6.7}$$

and the quartic part is as given in (4.9). Note that in this chapter, in contrast with Chap. 4, we shall denote harmonic frequencies by the symbol Ω and use ω for the perturbed frequencies in the context of the Poincaré-Linstedt approach described below. Thus, we write our equations of motion in the new canonical variables as

$$\ddot{Q}_q + \Omega_q^2 Q_q = -\frac{\beta}{2(N+1)} \sum_{l,m,n=1}^{N} C_{q,l,m,n} \Omega_q \Omega_l \Omega_m \Omega_n Q_l Q_m Q_n \tag{6.8}$$

(see (4.11)), using for the non-zero values of the coefficients $C_{q,l,m,n}$ the expressions given in (4.10). As we have already explained in Sect. 4.2, the individual harmonic energies $E_q = (P_q^2 + \Omega_q^2 Q_q^2)/2$, which are constants of the motion for $\beta = 0$, become functions of time for $\beta \ne 0$ and only the total energy E of the system is conserved. It is, therefore, useful to define a specific energy as $\varepsilon = E/N$ and an average harmonic energy of each mode over a time interval $0 \le t \le T$ by $\bar{E}_q(T) = \frac{1}{T} \int_0^T E_q(t) dt$.

In classical FPU experiments, one starts with the total energy shared only by a small subset of modes. Thus, for short time intervals T, we have $E_q(T) \approx 0$ for all q corresponding to non-excited modes. Equipartition, therefore, means that (due to nonlinearity) the energy will eventually be shared equally by all modes, i.e.

$$\lim_{T \to \infty} \bar{E}_q(T) = \varepsilon , \quad q = 1, \ldots, N. \tag{6.9}$$

Classical ergodic theory predicts that (6.9) will be violated only for orbits resulting from a zero measure set of initial conditions. The FPU "paradox", on the other hand, owes its name to the crucial observation that large deviations from the approximate equality $\bar{E}_q(T) \approx \varepsilon$ occur for many other orbits as well. Depending on the initial conditions, these deviations, termed FPU recurrences, are seen to persist even when T becomes very large.

6.2.1 Construction of q-tori by Poincaré-Linstedt Series

We now introduce the main elements of the method of q-torus construction presented in [94], using an explicitly solved example for $N = 8$, whose solutions lie on a two-dimensional torus representing the continuation, for $\beta \ne 0$, of the quasi-periodic solution of the uncoupled ($\beta = 0$) system $Q_1(t) = A_1 \cos \Omega_1 t$, $Q_2 = A_2 \cos \Omega_2 t$, for a suitable choice of A_1 and A_2.

To this end, we follow the Poincaré-Linstedt method and look for solutions $Q_q(t)$, $q = 1, \ldots, 8$, expanded as series of powers of the (small) parameter $\sigma = \beta/2(N+1)$, namely

$$Q_q(t) = Q_q^{(0)}(t) + \sigma Q_q^{(1)}(t) + \sigma^2 Q_q^{(2)}(t) + \ldots, \quad q = 1, \ldots, 8. \tag{6.10}$$

For the motion to be quasiperiodic on a 2-torus, the functions $Q_q^{(r)}(t)$ must, at any order r, be trigonometric polynomials involving only two frequencies (and their multiples). Furthermore, in a continuation of the unperturbed solutions Q_1 and Q_2, the new frequencies ω_1 and ω_2 of the perturbed system must represent small corrections of the linear normal mode frequencies Ω_1, Ω_2. According to the Poincaré-Linstedt method, these new frequencies are also given by series in powers of σ, as

$$\omega_q = \Omega_q + \sigma \omega_q^{(1)} + \sigma^2 \omega_q^{(2)} + \ldots \quad q = 1, 2. \tag{6.11}$$

As is well-known, the above corrections are determined by the requirement that all terms in the differential equations of motion, giving rise to secular terms (of the form $t \sin \omega_q t$ etc.) in the solutions $Q_q(t)$, be eliminated.

Let us consider the equation of motion for the first mode, whose first few terms on the right side are

$$\ddot{Q}_1 + \Omega_1^2 Q_1 = -\sigma(3\Omega_1^4 Q_1^3 + 6\Omega_1^2 \Omega_2^2 Q_1 Q_2^2 + 3\Omega_1^3 \Omega_3 Q_1^2 Q_3 + \ldots). \tag{6.12}$$

Proceeding with the Poincaré-Linstedt series, the frequency Ω_1 is substituted on the left side of (6.12) by its equivalent expression obtained by squaring (6.11) and solving for Ω_1^2. Up to first order in σ this gives

$$\Omega_1^2 = \omega_1^2 - 2\sigma \Omega_1 \omega_1^{(1)} + \ldots. \tag{6.13}$$

Substituting the expansions (6.10) into (6.12), as well as the frequency expansion (6.13) into the left hand side (lhs) of (6.12), and grouping together terms of like orders, we find at zeroth order $\ddot{Q}_1^{(0)} + \omega_1^2 Q_1^{(0)} = 0$, while at first order

$$\ddot{Q}_1^{(1)} + \omega_1^2 Q_1^{(1)} = 2\Omega_1 \omega_1^{(1)} Q_1^{(0)} - 3\Omega_1^4 (Q_1^{(0)})^3 - 6\Omega_1^2 \Omega_2^2 Q_1^{(0)} (Q_2^{(0)})^2 -$$
$$-3\Omega_1^3 \Omega_3 (Q_1^{(0)})^2 Q_3^{(0)} + \ldots. \tag{6.14}$$

Repeating the above process for modes 2 and 3, we find that their zeroth order equations also take the harmonic oscillator form

$$\ddot{Q}_2^{(0)} + \omega_2^2 Q_2^{(0)} = 0, \quad \ddot{Q}_3^{(0)} + \Omega_3^2 Q_3^{(0)} = 0. \tag{6.15}$$

Note that the corrected frequencies ω_1, ω_2 appear in the zeroth order equations for the modes 1 and 2, while the uncorrected frequency Ω_3 appears in the zeroth order equation of mode 3 (similarly, $\Omega_4, \ldots, \Omega_8$, appear in the zeroth order equations of modes 4 to 8). Thus to construct a solution which lies on a 2-torus, we start from particular solutions of (6.15) (with zero velocities at $t = 0$) which read

$$Q_1^{(0)}(t) = A_1 \cos \omega_1 t, \quad Q_2^{(0)}(t) = A_2 \cos \omega_2 t, \quad Q_3^{(0)}(t) = A_3 \cos \Omega_3 t,$$

where the amplitudes A_1, A_2, A_3 are arbitrary. If the solution is to lie on a 2-torus with frequencies ω_1, ω_2, we must set $A_3 = 0$, so that no third frequency is introduced in the solutions. In the same way, the zeroth order equations $\ddot{Q}_q^{(0)} + \Omega_q^2 Q_q^{(0)} = 0$ for the remaining modes $q = 4, \ldots, 8$, yield solutions $Q_q^{(0)}(t) = A_q \cos \Omega_q t$ for which we require $A_4 = A_5 = \ldots = A_8 = 0$. We are, therefore, left with only two non-zero free amplitudes A_1, A_2.

Now, consider (6.14) for the first order term $Q_1^{(1)}(t)$. Only zeroth order terms $Q_q^{(0)}(t)$ appear on its right hand side (rhs), allowing for the solution to be found recursively. The crucial remark here is that the choice $A_3 = \ldots = A_8 = 0$, also yields $Q_3^{(0)}(t) = \ldots = Q_8^{(0)}(t) = 0$, whence only a small subset of the terms appearing in the original equations of motion survive on the right side of (6.14), namely those in which none of the functions $Q_3^{(0)}(t), \ldots, Q_8^{(0)}(t)$, appears. As a result, (6.14) is dramatically simplified and upon substitution of $Q_1^{(0)}(t) = A_1 \cos \omega_1 t$, $Q_2^{(0)}(t) = A_2 \cos \omega_2 t$, reduces to

$$\ddot{Q}_1^{(1)} + \omega_1^2 Q_1^{(1)} = 2\Omega_1 \omega_1^{(1)} A_1 \cos \omega_1 t - 3\Omega_1^4 A_1^3 \cos^3 \omega_1 t -$$
$$- 6\Omega_1^2 \Omega_2^2 A_1 A_2^2 \cos \omega_1 t \cos^2 \omega_2 t. \tag{6.16}$$

This can now be used to fix $\omega_1^{(1)}$ so that no secular terms appear in the solution, yielding

$$\omega_1^{(1)} = \frac{9}{8} A_1^2 \Omega_1^3 + \frac{3}{2} A_2^2 \Omega_1 \Omega_2^2,$$

while, after some simple operations, we find

$$Q_1^{(1)}(t) = \frac{3A_1^3 \Omega_1^4 \cos 3\omega_1 t}{32\omega_1^2} + \frac{3A_1 A_2^2 \Omega_1^2 \Omega_2^2 \cos(\omega_1 + 2\omega_2)t}{2[(\omega_1 + 2\omega_2)^2 - \omega_1^2]} +$$
$$+ \frac{3A_1 A_2^2 \Omega_1^2 \Omega_2^2 \cos(\omega_1 - 2\omega_2)t}{2[(\omega_1 - 2\omega_2)^2 - \omega_1^2]}. \tag{6.17}$$

By the same analysis, we fix the frequency correction of the second mode to be

$$\omega_2^{(1)} = \frac{9}{8} A_2^2 \Omega_2^3 + \frac{3}{2} A_1^2 \Omega_1^2 \Omega_2,$$

and obtain the solution

$$Q_2^{(1)}(t) = \frac{3A_2^3\Omega_2^4\cos 3\omega_2 t}{32\omega_2^2} + \frac{3A_1^2A_2\Omega_1^2\Omega_2^2\cos(2\omega_1+\omega_2)t}{2[(2\omega_1+\omega_2)^2-\omega_2^2]} +$$
$$+\frac{3A_1^2A_2\Omega_1^2\Omega_2^2\cos(2\omega_1-\omega_2)t}{2[(2\omega_1-\omega_2)^2-\omega_2^2]}, \tag{6.18}$$

which has a similar structure as the first order solution of the first mode. For the third order term there is no frequency correction, and we find

$$Q_3^{(1)}(t) = \frac{A_1^3\Omega_1^3\Omega_3}{4}\left(\frac{3\cos\omega_1 t}{\omega_1^2-\Omega_3^2} + \frac{\cos 3\omega_1 t}{9\omega_1^2-\Omega_3^2}\right) + \tag{6.19}$$
$$+\frac{3A_1A_2^2\Omega_1\Omega_2^2\Omega_3}{4}\left(\frac{\cos(\omega_1-2\omega_2)t}{(\omega_1-2\omega_2)^2-\Omega_3^2} + \frac{\cos(\omega_1+2\omega_2)t}{(\omega_1+2\omega_2)^2-\Omega_3^2} + \frac{2\cos\omega_1 t}{\omega_1^2-\Omega_3^2}\right).$$

We may thus proceed to the sixth mode to find solutions in which all the functions $Q_3^{(0)}, \ldots, Q_6^{(0)}$, are equal to zero, while the functions $Q_3^{(1)}, \ldots, Q_6^{(1)}$, are non zero. However, a new situation appears when we arrive at the seventh and eighth modes. A careful inspection of the equation for the term $Q_7^{(1)}(t)$, i.e.

$$\ddot{Q}_7^{(1)} + \Omega_7^2 Q_7^{(1)} = -\sum_{l,m,n=1}^{8} C_{7,l,m,n}\Omega_7\Omega_l\Omega_m\Omega_n Q_l^{(0)} Q_m^{(0)} Q_n^{(0)}, \tag{6.20}$$

shows that there can be no term on the rhs which does not involve some of the functions $Q_3^{(0)}, \ldots, Q_8^{(0)}$. Again, this follows from the selection rules for the coefficients in (4.10). Since all these functions are equal to zero, the rhs of (6.20) is equal to zero. Taking this into account, we set $Q_7^{(1)}(t) = 0$, so as not to introduce a third frequency Ω_7 in the solutions, whence the series expansion (6.10) for $Q_7(t)$ necessarily starts with terms of order at least $\mathcal{O}(\sigma^2)$. The same holds true for the equation determining $Q_8^{(1)}(t)$.

Let us now make a number of important remarks regarding the above construction:

1. *Consistency.* The solutions (6.17)–(6.19) (and those of subsequent orders) are meaningful only when the frequencies appearing in the denominators are *incommensurable*, so that denominators do not vanish. Note that the spectrum of uncorrected frequencies Ω_q, given by (6.7), is fully incommensurable if: (1) N is a prime number minus one, or (2) a power of two minus one (see Problem 6.1 and [163]).

   In other cases, there are commensurabilities among the unperturbed frequencies, which do not affect the consistency of the construction of the Poincaré-Linstedt series, because it can be shown that no divisors of the form $\sum_{q=1}^{N} n_q\Omega_q$, appear in the series at any order and any integer vector $\mathbf{n} \equiv (n_1, n_2, \ldots, n_q)$. The proof of this statement follows from the fact that differential equations like (6.16) determining all terms $Q_q^{(k)}$, $q = 1, \ldots, 8$, at order k read

$$\ddot{Q}_q^{(k)}(t) + \omega_q^2 Q_q^{(k)}(t) = B_q^{(k)} \omega_q^{(k)} \cos \omega_q t + \sum_{\substack{n_1, n_2 \in Z \\ |n_1| + |n_2| \neq 0}}^{k-1} A_{q, n_1, n_2}^{(k)} \cos[(n_1 \omega_1 + n_2 \omega_2)t],$$

(6.21)

if $q = 1, 2$, while for $q = 3, \ldots, 8$, the same equation holds with ω_q^2 on the left side replaced by Ω_q^2. The coefficients $B_q^{(k)}$ and $A_{q, n_1, n_2}^{(k)}$ in (6.21) are determined at previous steps of the construction. It follows, therefore, that all new divisors appearing at successive orders, belong to one of the following sets

$$(n_1 \pm 1)\omega_1 + n_2 \omega_2, \quad n_1 \omega_1 + (n_2 \pm 1)\omega_2, \quad \Omega_q \pm (n_1 \omega_1 + n_2 \omega_2), \quad q = 3, \ldots, 8.$$

Thus, we deduce that no commensurabilities between the unperturbed frequencies Ω_q can ever show up in the above series, as a zero divisor. For a more detailed analysis of the consistency of the Poincaré-Linstedt method as applied to the FPU β-model (6.3) we refer the reader to [94]. In fact, the proof of consistency can be generalized to construct s-dimensional q-tori, according to the nonlinear continuation of the set of linear modes q_i, $i = 1, \ldots, s$, permitted by Lyapunov's theorem [232]. For example, if the first s modes are excited with amplitudes A_k, $k = 1, \ldots, s$, the formula for the perturbed frequencies reads

$$\omega_q = \Omega_q + \frac{3\sigma}{2} \Omega_q \sum_{k=1}^{s} A_k^2 \Omega_k^2 - \frac{3}{8} A_q^2 \Omega_q^3 + \mathcal{O}(\sigma^2; A_1, \ldots, A_s), \quad (6.22)$$

for $q = 1, \ldots, s$. Fixing the frequencies ω_q in advance implies that (6.22) should be regarded as yielding the (unknown) amplitudes A_k for which the quasi-periodic solution exhibits the chosen set of frequencies. The case $s = 1$, of course, implies that the same property holds for the Poincaré-Linstedt representation of q-breathers described in [129].

2. *Convergence.* Even after establishing consistency, one still has to prove convergence of the series. However, as demonstrated in [120, 142, 143], the question of convergence of the Poincaré-Linstedt series is notoriously difficult even for simple Hamiltonian systems. This is because the above series for orbits on invariant tori are convergent, but *not absolutely*, as explained in [149]. Indeed, the problem of a suitable Kolmogorov construction of lower-dimensional tori yielding absolutely convergent series is still open (see e.g. [148]).

Thus, in [94] the construction of q-tori is justified by numerical simulations, taking initial conditions from the analytical expressions (6.10) at $t = 0$ and using the GALI method of Chap. 5 to demonstrate that the computed orbits lie on two-dimensional tori. The numerical evaluation of the modes $Q_1(t)$, $Q_3(t)$ and

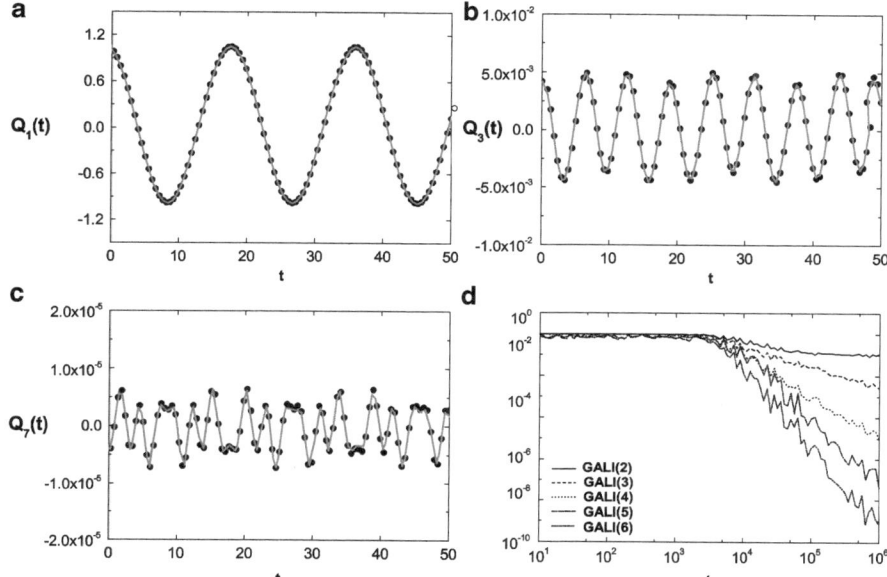

Fig. 6.3 Comparison of numerical (*points*) versus analytical (*solid grey curve*) solutions, using the Poincaré-Linstedt series up to order $\mathcal{O}(\sigma^2)$, for the time evolution of the modes (**a**) $q = 1$, (**b**) $q = 3$, (**c**) $q = 7$, when $A_1 = 1$, $A_2 = 0.5$, and $N = 8$, $\beta = 0.1$. (**d**) The indices $GALI_2$ to $GALI_6$ up to a time $t = 10^6$ clearly indicate that the motion lies on a two-dimensional torus (after [94])

$Q_7(t)$ follows remarkably well the analytical solution provided by the Poincaré-Linstedt series truncated to second order in σ, for $N = 8$, $\beta = 0.1$, and $A_1 = 1$, $A_2 = 0.5$, as shown in Fig. 6.3a–c. The size of the error is as expected by the truncation order in (6.10), while the GALI method shows, in Fig. 6.3d, that the numerical orbit lies on a 2-torus.

Let us recall from Chap. 5 that the indicators GALI$_k$, $k = 2, 3, \ldots$, decay exponentially for chaotic orbits, due to the attraction of all deviation vectors by the most unstable direction corresponding to the MLE. On the other hand, if an orbit lies on a stable s-dimensional torus, the indices GALI$_2$, ..., GALI$_s$ oscillate about a *non-zero* value, while the GALI$_{s+j}$, $j = 1, 2, \ldots$ decay asymptotically by a power law. This is precisely what we observe in Fig. 6.3d. Namely, after an initial transient interval, the GALI$_2$ index stabilizes at a constant value GALI$_2 \approx$ 0.1, while all higher indices, starting with GALI$_3$, decay following power laws expected by the theory. Thus, we conclude that the motion lies on a 2-torus, exactly as predicted by the Poincaré-Linstedt construction, despite the fact that some excitation was provided initially to all modes.

3. *Presence and accumulation of small divisors.* Even though zero denominators do not occur, we cannot avoid the notorious problem of *small divisors* appearing in terms of all orders beyond the zeroth. Since the low-mode frequencies satisfy

$\omega_q \sim \pi q/N$, divisors like ω_1^2, or $(\omega_1 - 2\omega_2)^2 \pm \omega_1^2$ (appearing e.g. in (6.17)) are small and care must be taken regarding their effect on the growth of terms of the series at successive orders. For example, one of the most important such *nearly resonant* divisors is $9\omega_1^2 - \Omega_3^2$ in (6.19). However, since the first order corrections of the frequencies ω_1 and ω_2 are $\mathcal{O}(\beta A_j^2/N^4)$, for A_j, β sufficiently small, one may still use for all these frequencies the approximation given by the first two terms in the sinus expansion of (6.7) namely

$$\omega_q \simeq \frac{\pi q}{(N+1)} - \frac{\pi^3 q^3}{24(N+1)^3},\tag{6.23}$$

the error being $\mathcal{O}(A_j^2 \beta/N^4)$ for ω_1, ω_2, and $\mathcal{O}((q/N)^5)$ for all other frequencies. This implies that a divisor like $9\omega_1^2 - \Omega_3^2$ can be approximated by the relation

$$|q^2\omega_1^2 - \Omega_q^2| = |(q\omega_1 + \Omega_q)(q\omega_1 - \Omega_q)| \simeq \frac{2q\pi}{(N+1)}\frac{\pi^3(q^3-q)}{24(N+1)^3}$$

$$\simeq \frac{\pi^4 q^4}{12N^4} + O\left(\frac{q^6}{N^6}\right).$$

Defining the quantity

$$a_q = \frac{\pi^4 q^4}{12N^4},\tag{6.24}$$

we conclude that a divisor of order a_3 appears in (6.19), which is the solution of an equation of the first order in the recursive scheme. In general, the terms produced at consecutive orders involve products of the form $a_{q_1} a_{q_2} \cdots a_{q_r}$ in the denominators of the terms produced at the rth order of the recursive scheme. Consequently, this type of accumulation of small divisors does not lead to a divergent series, but can be exploited instead to obtain estimates of the profile of energy localization for the q-tori solutions, as we shall explain in the next section.

4. *Sequence of mode excitations.* The profile of energy localization along a q-torus solution can be determined by the sequence of mode excitations as the recursive scheme proceeds to higher orders. The term "excitation" here means that the solutions of the Poincaré-Linstedt method are non-zero for the first time at the order where it is claimed that the excitation arises. For example, as already explained in the construction of a 2-torus solution, the modes 1 and 2 are excited at zeroth order, the modes 3, 4, 5, and 6 at first order, and the modes 7 and 8 at second order.

Figure 6.4 presents one more example showing the comparison between analytical and numerical results for the modes $Q_1(t)$, $Q_5(t)$, and $Q_{13}(t)$, along a 4-torus solution constructed precisely as described above, with $N = 16$, $\beta = 0.1$, by exciting modes 1–4 at zeroth order, via the amplitudes $A_1 = 1$, $A_2 = 0.5$, $A_3 = 1/3$, $A_4 = 0.25$. In this case, we find that the modes excited at the first

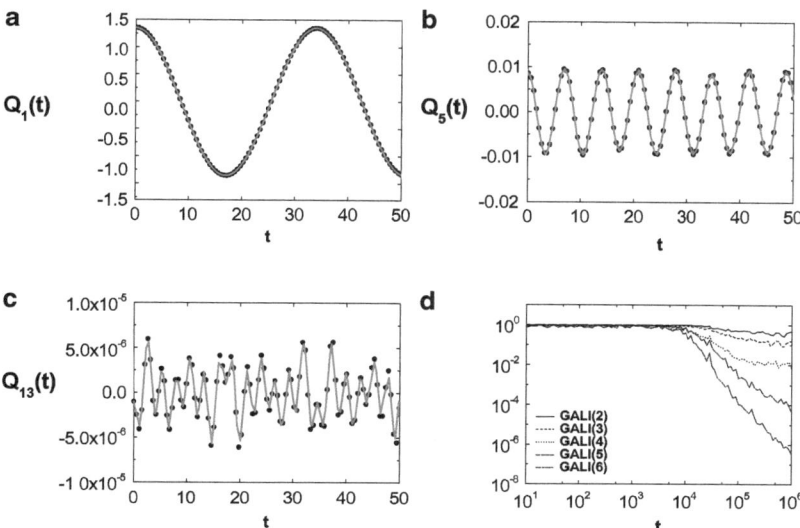

Fig. 6.4 Same as in Fig. 6.3, showing the existence of a 4-torus in the system with $N = 16$, $\beta = 0.1$, when $A_1 = 1$, $A_2 = 0.5$, $A_3 = 0.333...$, $A_4 = 0.25$. The time evolution $Q_q(t)$ is shown for the modes (**a**) $q = 1$, (**b**) $q = 5$, (**c**) $q = 13$. (**d**) The $GALI_k$, $k = 2, 3, 4$ indices are again seen to stabilize after $t \approx 10^4$, while for $k \geq 5$ they continue to decrease by power laws (after [94])

order of the recursive scheme are $q = 5$ to $q = 12$, while the modes excited at second order are $q = 13$ to $q = 16$. Thus, according to the GALI method the motion lies on a 4-torus, since GALI$_5$ is the first index to fall asymptotically as t^{-1} (see Fig. 6.4d).

The sequence of excitation of different modes, whose amplitudes at the rth order have a pre-factor $\sigma^r = (\beta/2(N+1))^r$, plays a crucial role in estimating the profile of energy localization. However, the contribution of small divisors to such a pre-factor must also be examined. This is the subject of the next subsection, in which a Proposition is provided for the sequence of mode excitations, in the generic case of arbitrary N and arbitrary dimension s of the torus (with the restriction $s < N$). The consequences of this proposition are then examined on the localization profile of q-tori and nearby FPU trajectories.

6.2.2 Profile of the Energy Localization

Let us see what happens in the case of the solutions shown in Figs. 6.3 and 6.4. Figure 6.5 shows the average harmonic energy of each mode over a time span $T = 10^6$ in the cases of the q-torus of Figs. 6.3 and 6.4, shown in Fig. 6.5a, b respectively. The numerical result (open circles) compares excellently with the

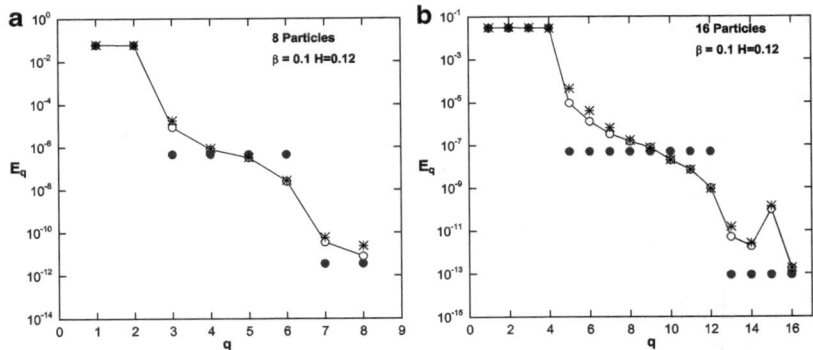

Fig. 6.5 The average harmonic energy E_q of the qth mode as a function of q, after a time $t = 10^6$ for (**a**) the 2-torus and (**b**) 4-torus solutions (*open circles*) corresponding to the initial conditions used in Figs. 6.3 and 6.4 respectively. The stars are E_q values calculated via the analytical representation of the solutions $Q_q(t)$ by the Poincaré-Linstedt series. The *filled circles* show a theoretical estimate based on the average energy of suitably defined groups of modes (see (6.34) and relevant discussion in the text) (after [94])

analytical result (stars) obtained via the Poincaré-Linstedt method. The filled circles in each plot represent "piecewise" estimates of the localization profile in groups of modes excited at consecutive orders of the recursive scheme.

The derivation and exact meaning of these theoretical estimates will be analyzed in detail below. Here we point out their main feature showing a clear-cut separation of the modes into groups following essentially the sequence of excitations predicted by the Poincaré-Linstedt series construction. Namely, in Fig. 6.5a we see clearly that the decrease of the average energy $\bar{E}_q$ along the profile occurs by abrupt steps, with three groups formed by nearby energies, namely the group of modes 1, 2, then 3 to 6, and then 7, 8. The same phenomenon is also seen in Fig. 6.5b, where the grouping of the different modes follows precisely the sequence of excitations 1 to 4, 5 to 12, 13 to 16 as predicted by the theory.

Let us now extend this approach to a more general consideration of the structure of solutions lying on low-dimensional tori. In particular, let us consider the case where only a subset of modes $1 \leq q_1 < q_2 < \ldots < q_s < N$ are excited at the zeroth order of the perturbation theory, assuming that $Q_q^{(0)} \neq 0$ iff $q \in \{q_1, q_2, \ldots, q_s\}$. The modes $q_1, \ldots, q_s$, need not be consecutive. We wish to see how this type of zeroth order excitation propagates at subsequent orders. More specifically, we wish to determine for which values of q one has $Q_q^{(k'<k)} = 0$, $Q_q^{(k)} \neq 0$, i.e., the modes q are first excited at the kth order. The answer is provided by the following Proposition proved in [94]:

Proposition 6.1. *Let the starting terms of a Poincaré-Linstedt series solution with s frequencies be set as*

$$Q_{q_i}^{(0)} = A_{q_i} \cos(\omega_{q_i} t + \phi_{q_i}), \quad for \; i = 1, 2, \ldots, s, \;\; 1 \le q_1 \le q_2 \le \ldots \le q_s \le N$$

$$Q_q^{(0)} = 0 \;\; for \; all \; q \ne q_i.$$

Then, besides the terms $Q_{q_i}^k(t)$, the Poincaré-Linstedt series terms $Q_q^{(k)}(t)$ which are permitted to be non-zero, at the kth order of the series expansion, are given by the values of $q = q^{(k)}$ satisfying

$$q^{(k)} = |\; 2\lambda(N+1) - m_k \;|, \quad (m_k \; mod \; (N+1)) \ne 0, \tag{6.25}$$

where m_k can take any of the values $|\; \pm q_{i_1} \pm q_{i_2} \pm \ldots \pm q_{i_{2k+1}} \;|$, with $i_1, i_2, \ldots, i_{2k+1} \in \{1, \ldots, s\}$ for any possible combination of the $\pm$ signs, and λ the integer part of the fraction $(m_k + N)/2(N+1)$.

It is instructive to clarify the use of this Proposition on some simple examples:

q-breathers: If we excite only one mode q_1 at zeroth order, new modes are excited one by one at subsequent orders of perturbation theory and one has $m_k = (2k+1)q_1$. As long as $m_k \le N$, the newly excited modes are $q^{(k)} = m_k = (2k+1)q_1$, i.e., exactly as predicted in [128].

two-dimensional q-tori: Assume we excite the modes $q_1 = 1$, $q_2 = 2$ at zeroth order. At first order ($k = 1$) we have again $q = |q_{i_1} \pm q_{i_2} \pm q_{i_3}|$ with $i_1, i_2, i_3 \in \{1, 2\}$. We readily find that the newly excited modes are $q = 1 + 1 + 1 = 3$, $q = 1 + 1 + 2 = 4$, $q = 1 + 2 + 2 = 5$, and $q = 2 + 2 + 2 = 6$. At order $k = 2$, the first newly excited mode is $q = 1 + 1 + 1 + 2 + 2 = 7$, while the last newly excited mode is $q = 2 + 2 + 2 + 2 + 2 = 10$. In general, at order $k \ge 1$ the newly excited modes are $2(2k - 1) + 1 \le q \le 2(2k + 1)$.

s-dimensional q-tori: Now, assume we excite the modes $q_1 = 1$, $q_2 = 2$, ..., $q_s = s$ at the zeroth order. In the same way as above we find that the newly excited modes at order $k \ge 1$ are $s(2k - 1) + 1 \le q \le s(2k + 1)$.

In order to study energy localization phenomena associated with FPU trajectories, we need to construct estimates for all $Q_q(t)$ terms participating in a particular q-torus solution in which the s first modes $(q_1, q_2, \ldots, q_s) = (1, 2, \ldots, s)$ are excited at zeroth order of the theory, with amplitudes A_1, A_2, ..., A_s, respectively. Denoting by $q^{(k)}$ the indices of all modes that are newly excited at kth order, according to the above Proposition, we have

$$q^{(k)} \in \mathcal{M}_{s,k} \equiv \{\max(1, (2k-1)s+1), \ldots, (2k+1)s\}, \tag{6.26}$$

whence the following useful estimates are obtained

$$\forall q^{(k)} \in \mathcal{M}_{s,k}, k \ge 1 \quad q^{(k)} \lesssim (2k+1)s, \quad \omega_{q^{(k)}} \lesssim \frac{(2k+1)\pi s}{N+1}. \tag{6.27}$$

Let us now define, in the space of trigonometric polynomials f containing (for simplicity) only cosine terms

$$f = \sum_{\mathbf{k}} A_{\mathbf{k}} \cos(\mathbf{k} \cdot \phi), \tag{6.28}$$

the "majorant" norm

$$\|f\| = \sum_{\mathbf{k}} |A_{\mathbf{k}}|. \tag{6.29}$$

Since the terms of mode $Q_{q^{(k)}}(t)$ satisfy $Q_{q^{(k)}}^{(n)} = 0, \forall n < k,\ Q_{q^{(k)}}^{(k)} \neq 0$, we deduce that the equation determining $Q_{q^{(k)}}^{(k)}$ reads

$$\ddot{Q}_{q^{(k)}}^{(k)} + \omega_{q^{(k)}}^2 Q_{q^{(k)}}^{(k)} = \tag{6.30}$$

$$= -\Omega_{q^{(k)}} \sum_{\substack{n_{1,2,3}=0 \\ n_1+n_2+n_3=k-1}}^{k-1} \left(\sum_{(q^{(n_1)},q^{(n_2)},q^{(n_3)}) \in \mathscr{D}_{q^{(k)}}} \Omega_{q^{(n_1)}} \Omega_{q^{(n_2)}} \Omega_{q^{(n_3)}} Q_{q^{(n_1)}}^{(n_1)} Q_{q^{(n_2)}}^{(n_2)} Q_{q^{(n_3)}}^{(n_3)} \right),$$

where

$$\mathscr{D}_q = \{(q_{j_1}, q_{j_2}, q_{j_3}) : q \pm q_{j_1} \pm q_{j_2} \pm q_{j_3} = 0\}.$$

Assuming that the Poincaré-Linstedt series are convergent, we may omit higher order terms $Q_{q^{(k)}}^{(k+1)}$, $Q_{q^{(k)}}^{(k+2)}$, ..., as long as we seek an estimate of the size of the oscillation amplitude of the mode $q^{(k)}$. The average size of the oscillation amplitudes of each mode on an s-dimensional q-torus now follows from estimates on the norms of the various terms appearing in (6.30). If we denote by $A^{(k)}$ the mean value of all the norms $\|Q_{q^{(k)}}\|$, we obtain the following estimate:

$$A^{(k)} = \frac{(C s)^k A_0^{2k+1}}{2k+1}, \tag{6.31}$$

where $A_0 \equiv A^{(0)}$ is the mean amplitude of the oscillations of all modes excited at the zeroth order of the perturbation theory, and C is a constant of order $\mathcal{O}(1)$ [94]. In fact, (6.31) is a straightforward generalization of the estimate given in [129] for q-breathers, while the two estimates become identical (except for the value of C) if one sets $s = 1$ in (6.31), and $q_0 = 1$ in Flach's q-breathers formulae.

The above analysis suggests that, starting with initial conditions near q-breathers, a "backbone" is formed in phase space by a hierarchical set of solutions which are, precisely, the solutions lying on low-dimensional q-tori (of dimension $s = 1, 2, \ldots,\ s \ll N$). This also means that all FPU trajectories with initial conditions within this set exhibit a profile of energy localization characterized by a "stepwise" exponential decay, with step size equal to $2s$, as implied by (6.26). Thus, all the modes $q^{(k)}$ in this group share a nearly equal mean amplitude of oscillations given by the estimate of (6.31).

Using (6.31), it is also possible to derive 'piecewise' estimates of the energy of each group using a formula for the average harmonic energies $E^{(k)}$ of the modes $q^{(k)}$. To achieve this, note that the total energy E given to the system can be estimated as the sum of the energies of the modes $1, \ldots, s$ (the remaining modes yield only small corrections to the total energy), i.e.

$$E \sim s\omega_{q^{(0)}}^2 A_0^2 \sim \frac{\pi^2 s^3 A_0^2}{(N+1)^2}.$$

On the other hand, the energy of each mode $q^{(k)}$ can be estimated from

$$E^{(k)} \sim \frac{1}{2}\Omega_{q^{(k)}}^2 \left(\frac{\beta}{2(N+1)}\right)^{2k} (A^{(k)})^2 \sim \frac{\pi^2 s^2 (C s \beta)^{2k} A_0^{4k+2}}{2^{2k+1}(N+1)^{2k+2}},$$

which, in terms of the total energy E, yields

$$E^{(k)} \sim \frac{E}{s}\left(\frac{C^2\beta^2(N+1)^2 E^2}{\pi^4 s^4}\right)^k. \tag{6.32}$$

Once again, the similarity of (6.32) with the corresponding equation for q-breathers [129]

$$E_{(2k+1)q_0} \sim E_{q_0}\left(\frac{9\beta^2(N+1)^2 E^2}{64\pi^4 q_0^4}\right)^k \tag{6.33}$$

is evident, where q_0 is the unique mode excited at zeroth order of the perturbation theory. Note, in particular, that the integer s in (6.32) plays a role similar to that of q_0 in (6.33). This means that the energy profile of a q-breather with $q_0 = s$ obeys the same exponential law as the energy profile of the s-dimensional q-torus. But the most important feature of the latter solutions is that their profile remains unaltered as N increases, provided that: (1) a constant fraction $M = s/N$ of the spectrum is initially excited, (i.e. that s increases proportionally to N), and (2) the specific energy $\varepsilon = E/N$ remains constant. Indeed, in terms of the specific energy ε, (6.32) takes the form

$$E^{(k)} \sim \frac{\varepsilon}{M}\left(\frac{C^2\beta^2\varepsilon^2}{\pi^4 M^4}\right)^k, \tag{6.34}$$

i.e. the profile becomes *independent* of N. A similar behavior is observed with q-breather solutions provided that the "seed" mode q_0 varies linearly with N, as was shown explicitly in [130, 179].

The same q-torus construction can be performed for the FPU$-\alpha$ model as well. Here we only discuss the main results regarding the exponential localization of q-torus solutions, whose profile is quite similar to the one given in (6.34).

6.3   A Numerical Study of FPU Trajectories

Examples of the "stepwise" profiles predicted by (6.32) in the case of exact q-tori
solutions are shown by filled circles in Fig. 6.5, corresponding to the solutions
shown in Figs. 6.3 and 6.4 (in all fittings we set $C = 1$ for simplicity). From
these we see that the theoretical "piecewise" profiles yield nearly the same average
exponential slope as the profiles obtained numerically, or analytically by the
construction of the solutions via the Poincaré-Linstedt method . Thus, the estimates
(6.32), or (6.34), appear quite satisfactory for characterizing the localization profiles
of exact q-tori solutions.

The key question now, regarding the relevance of the q-tori solutions for the
interpretation of the FPU paradox, is whether (6.32) or (6.34) retain their predictive
power in the case of generic FPU trajectories which, by definition, are started close
to but not exactly on a q-torus.

To answer this question, let us examine the results plotted in Figs. 6.6 and 6.7.
Figure 6.6 shows the energy localization profile in numerical experiments on the

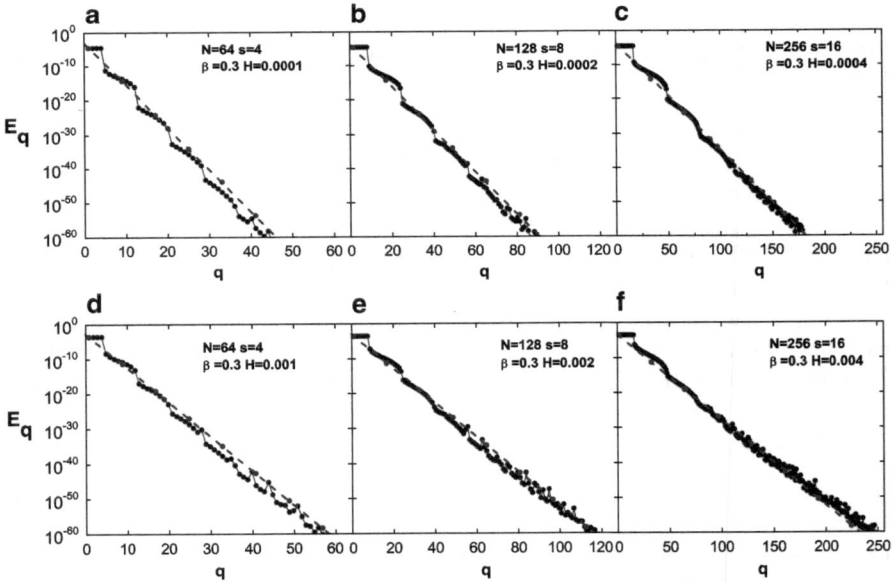

Fig. 6.6 The average harmonic energy E_q of the qth mode over a time span $t = 10^6$ as a function
of q in various examples of FPU trajectories, for $\beta = 0.3$, in which the $s(= N/16)$ first modes
are only excited initially via $Q_q(0) = A_q$, $\dot{Q}_q(0) = 0$, $q = 1, \ldots, s$, with the A_q selected so that
the total energy is equal to the value $E = H$ indicated in each panel. We thus have (**a**) $N = 64$,
$E = 10^{-4}$, (**b**) $N = 128$, $E = 2 \times 10^{-4}$, (**c**) $N = 256$, $E = 4 \times 10^{-4}$, (**d**) $N = 64$, $E = 10^{-3}$,
(**e**) $N = 128$, $E = 2 \times 10^{-3}$, (**f**) $N = 256$, $E = 4 \times 10^{-3}$. The specific energy is constant
in each of the two rows, i.e. $\varepsilon = 1.5625 \times 10^{-6}$ in the *top row* and $\varepsilon = 1.5625 \times 10^{-5}$ in the
bottom row. The *dashed lines* represent the average exponential profile E_q obtained theoretically
by the hypothesis that the depicted FPU trajectories lie close to q-tori governed by the profile (6.34)
(after [94])

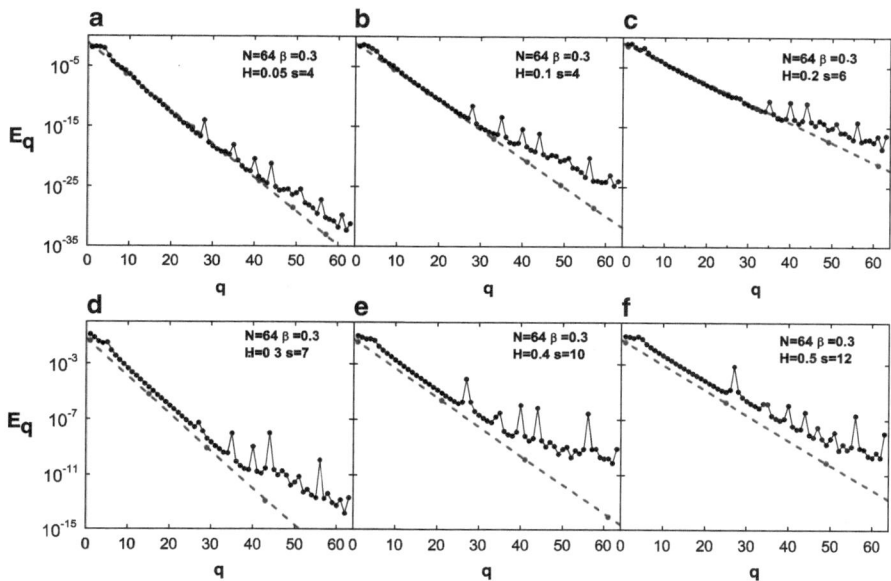

Fig. 6.7 Same as in Fig. 6.6a but for larger energies, namely (**a**) $E = 0.05$, (**b**) $E = 0.1$, (**c**) $E = 0.2$, (**d**) $E = 0.3$, (**e**) $E = 0.4$, (**f**) $E = 0.5$. Beyond the threshold $E \approx 0.05$, theoretical profiles of the form (6.32) yield the correct exponential slope if s is gradually increased from $s = 4$ in (**a**) and (**b**) to $s = 6$ in (**c**), $s = 7$ in (**d**), $s = 10$ in (**e**), and $s = 12$ in (**f**) (after [94])

FPU$-\beta$ model in which β is kept fixed ($\beta = 0.3$), while N takes the values $N = 64$, $N = 128$ and $N = 256$. Although the derivation of explicit q-tori solutions is practically feasible by computer algebra only up to a rather small value of N ($N = 16$), in the present section we numerically follow trajectories for much larger values of N.

In all six panels of Fig. 6.6 the FPU trajectories are computed starting with initial conditions in which only the $s = 4$ (for $N = 64$), $s = 8$ (for $N = 128$) and $s = 16$ (for $N = 256$) modes are excited at $t = 0$, with the excitation amplitudes corresponding to the total energy E indicated in each panel, and constant specific energy $\varepsilon = 1.5625 \times 10^{-6}$ in the top row and $\varepsilon = 1.5625 \times 10^{-5}$ in the bottom row of Fig. 6.6.

The resulting trajectories differ from q-tori solutions in the following sense: For the q-tori *all* modes have an initial excitation whose size was estimated in (6.31) or (6.32), while in the case of the FPU trajectories only the s first modes are excited initially, and one has $Q_q(0) = 0$ for all modes $q > s$. Furthermore, since in the q-tori solutions one also has $\|Q_{q>s}(t)\| \ll \|Q_{q \le s}(t)\|$ for all t, the FPU trajectories can be considered as lying in the neighborhood of a q-torus, at least initially. The numerical results suggest that if E is small, FPU trajectories remain close to the q-tori even after relatively long times, e.g. $t = 10^6$.

This is depicted in Fig. 6.6, in which one sees that the average energy profiles of the FPU trajectories (at $t = 10^6$) exhibit the same behavior as predicted by (6.32), for an exact q-torus solution with the same total energy as the FPU trajectory in each panel. For example, based on the values of their average harmonic energy, the modes in Fig. 6.6a (in which $s = 4$) are clearly separated in groups, 1 to 4, 5 to 12, 13 to 20, etc., as foreseen by (6.26) for an exact 4-torus solution. The energies of the modes in each group have a sigmoid variation around a level value characteristic of the group, which is nearly the value predicted by (6.32). The grouping of the modes is distinguishable in all panels of Fig. 6.6, a careful inspection of which verifies that the grouping follows the laws found for q-tori. Also, if we superpose the numerical data of the three top (or bottom) panels we find that the average exponential slope is nearly identical in all panels of each row, a fact consistent with (6.34), according to which, for a given fraction M of initially excited modes, this slope depends only on the specific energy, i.e. it is independent of N for constant ε.

If we now increase the energy, the FPU trajectories resulting from s initially excited modes start deviating from their associated exact q-tori solutions. As a consequence, the energy profiles of the FPU trajectories also start to deviate from the energy profiles of the exact s-tori. This is evidenced by the fact that the profiles of the FPU trajectories become smoother, and the groups of modes less distinct, while retaining the average exponential slope as predicted by (6.34). This "smoothing" of the profiles is discernible in Fig. 6.7a, in which the energy is increased by a factor of 50 with respect to Fig. 6.6d, for the same values of N and β. Also, in Fig. 6.7a we observe the formation of the so-called "tail", i.e. an overall rise of the localization profile at the high-frequency part of the spectrum, accompanied by spikes at particular modes. This is a precursor of the breakdown of FPU recurrences and the evolution of the system towards energy equipartition, which occur earlier in time as the energy becomes larger.

It is important to note that exponential localization of FPU trajectories persists and is still characterized by laws like (6.32) even when the energy is substantially increased. Furthermore, at energies beyond a threshold value, an interesting phenomenon is observed: For fixed N (see e.g. Fig. 6.7, where $N = 64$), as the energy increases, a progressively higher value of s needs to be used in (6.32), so that the theoretical profile yields an exponential slope that agrees with the numerical data. When $\beta = 0.3$, $N = 64$, this threshold value, $E \approx 0.1$, splits the system in two distinct regimes: One for $E < 0.1$, where the numerical data are well fitted by a constant choice of $s = 4$ in (6.32) (indicating that the FPU trajectories are indeed close to a 4-torus), and another for $E > 0.1$, where "best-fit" models of (6.32) occur for values of s increasing with the energy, e.g. $s = 6$ for $E = 0.2$, rising to $s = 12$ for $E = 0.5$. This indicates that the respective FPU trajectories are close to q-tori with a progressively higher value of $s > 4$, despite the fact that only the four first modes are excited by the initial conditions.

This behavior is analogous to the "natural packet" scenario described by Berchialla et al. [38], who observed that the localization profile of their metastable states begins to stabilize as the energy increases beyond a certain threshold. Indeed, according to (6.32), such a stabilization implies that in the second regime the width

s depends asymptotically on E as $s \propto E^{1/2}$, or, from (6.34) $M \propto \varepsilon^{1/2}$. This agrees well with estimates on the width of natural packets in the FPU$-\beta$ model as described in [227].

6.3.1   Long Time Stability near q-tori

It follows from the above results that as the exponential localization of the energy in the low q-modes relaxes, the FPU recurrences start to breakdown sooner and equipartition will eventually take place beyond some energy threshold. The question remains, however: What does all this have to do with the stability of the q-tori? How is the dynamics in their vicinity related to the properties of the motion more globally as the energy increases?

The linear stability of q-breathers, of course, just as any other periodic solution, can be studied by the implementation of Floquet theory (see [129]), which demonstrates that a q-breather is linearly stable as long as

$$R = \frac{6\beta E_{q_0}(N+1)}{\pi^2} < 1 + O(1/N^2).   \qquad (6.35)$$

This result is obtained by analyzing the eigenvalues of the monodromy matrix of the linearized equations about a q-breather solution constructed by the Poincaré-Linstedt series.

In the case of q-tori, however, the above techniques are no longer available. Nevertheless, a reliable numerical criterion for the stability of q-tori is provided by the GALI method. According to this approach, if a q-torus becomes unstable beyond a critical energy threshold, the deviation vectors of trajectories started on (or nearby) the torus follow its unstable manifold, whence all GALIs fall to zero exponentially fast. Most trajectories in that neighborhood are weakly chaotic and begin to diffuse slowly away to larger chaotic seas, where no FPU recurrences occur and equipartition reigns. Since all $GALI_k$ in that vicinity decay exponentially after a transient time, this weak chaos transition at low energies can be directly inferred from the time evolution of the lowest index, i.e. $GALI_2$.

Using this criterion, one can study the stability of q-tori and determine approximately the value of the critical energy E_c at which a q-torus turns from stable to unstable. Note first that, due to the obvious scaling of the FPU Hamiltonian by β, E_c is well fitted by a power law $E_c = A\beta^{-1}$ for all N considered, as shown in Fig. 6.8a for $N = 64$. Here, the quantity A is obtained by keeping N fixed and varying β, until a critical energy is determined, beyond which the $GALI_2$ index loses its (nearly) constant behavior. All calculations in Fig. 6.8 refer to a maximum time $t_{max} = 10^7$ up to which the exponential fall-of of $GALI_2$ is observed. In general, E_c decreases as t_{max} increases, but appears to tend to a limiting value as $t_{max} \to \infty$.

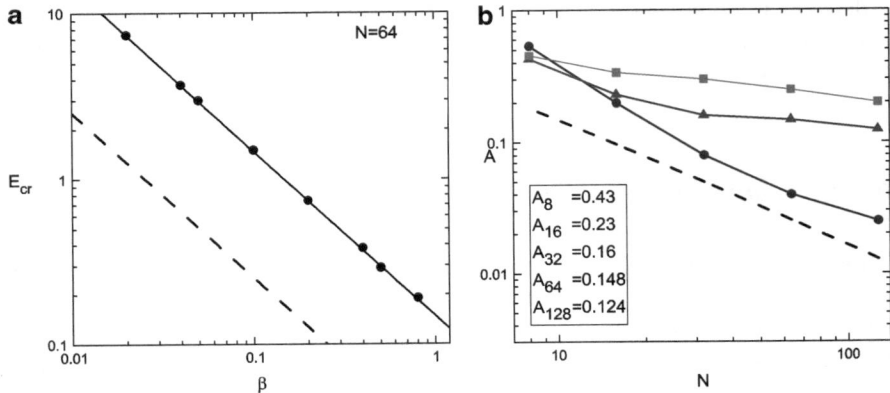

Fig. 6.8 (a) The critical energy E_c, at which the GALI$_2$ index shows that a q-torus destabilizes, is fitted by the *solid line* $E_c = A\beta^{-1}$, with $A = 0.164$, for fixed $N = 64$ (*the dashed line* shows the -1 slope separately for clarity). (b) Plotting A as a function of N for an FPU trajectory started by exciting initially the $q = 1, 2, 3$ modes (*squares*), or the $q = 1, 2$ modes (*triangles*), for fixed $\beta = 0.3$ yields curves that lie well above the threshold (*filled circles*) at which FPU trajectories near a $q = 1$ breather solution destabilize (the *dashed line* depicts the $A \sim N^{-1}$ law (6.35)). The total integration time for both panels is $t_{max} = 10^7$ (after [92])

Thus, the values of E_c found in this way provide an upper estimate for the transition energy at which the exact q-torus turns unstable [92].

The fitting constant A, however, does depend on N, as seen in Fig. 6.8b, where FPU trajectories are studied exciting the $q = 1, 2, 3$ modes, the $q = 1, 2$ modes and a $q = 1$ breather. The data marked by squares (3-torus) and triangles (2-torus) in Fig. 6.8b clearly show that the dependence of A on N is considerably *weaker* than the N^{-1} law, (6.35), derived for FPU trajectories started near the $q = 1$ breather. In that case, the data shown by filled circles in Fig. 6.8b have a slope much closer to that predicted by (6.35), although the numerical curve is shifted upwards with respect to the dashed line. These results evidently demonstrate that the GALI$_2$ method yields critical energies E_c which are quite *higher* than the values at which the q-breather becomes unstable.

In summary, the upper two curves of Fig. 6.8b strongly suggest that the q-tori solutions are *more robust* than the q-breathers, regarding their stability under small perturbations. Furthermore, they imply that the destabilization of the simple periodic orbit represented by a q-breather does *not* imply that the tori surrounding the q-breather are also unstable.

At any rate, it is important to remark that the exponential localization of FPU trajectories persists even after the associated q-breathers or q-tori have been identified as unstable by the GALI criterion. This is exemplified in Fig. 6.9, where panels (a) and (c) show the time evolution of the GALI$_2$ index for two FPU trajectories started in the vicinity of 2-tori of the $N = 32$ and $N = 128$ systems, when $\beta = 0.1$ and $t_{max} = 10^7$. In both cases, the energy satisfies $E > E_c$,

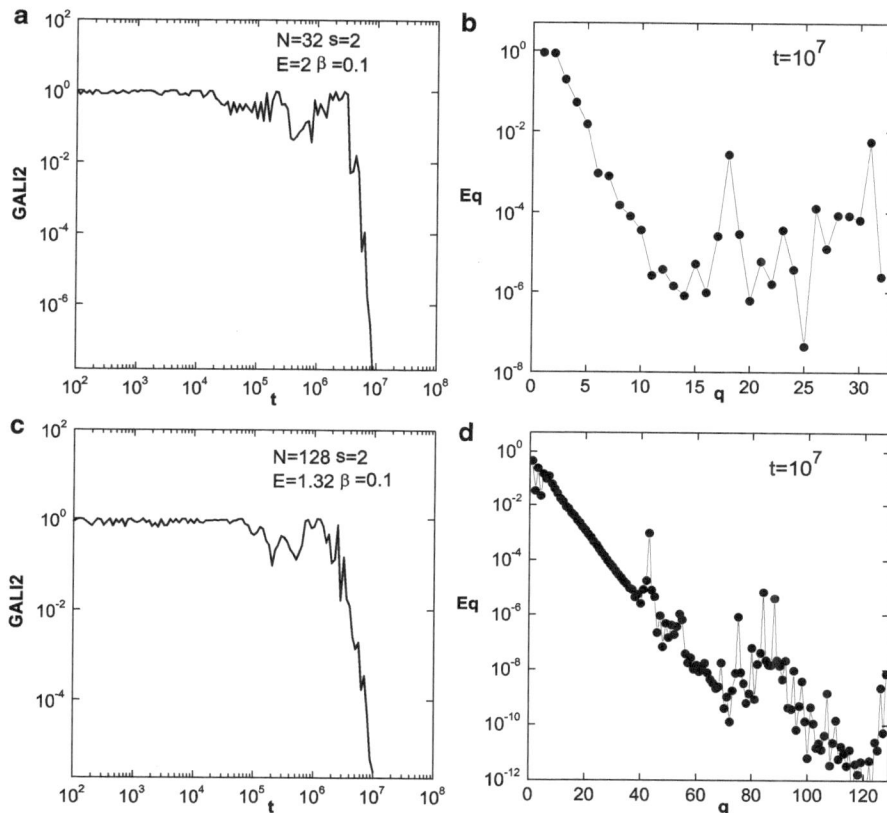

Fig. 6.9 (a) Time evolution of GALI$_2$ up to $t_{max} = 10^7$ for an FPU trajectory started by exciting the $q = 1$ and $q = 2$ modes in the system $N = 32$, $\beta = 0.1$, with total energy $E = 2$, i.e. higher than $E_c = 1.6$. (b) *Instantaneous* localization profile of the FPU trajectory of (a) at $t = 10^7$. (c) Same as in (a) but for $N = 64$, $E = 1.323$ (in this case the critical energy is $E_c = 1.24$). (d) Same as in (b) but for the trajectory of (c) (after [94])

as the exponential fall-off of the GALI$_2$ index is already observed at $t = 10^7$. Nevertheless, a simple inspection of Figs. 6.9b, d clearly shows that the exponential localization of the energy persists in the Fourier space of both systems.

6.4 Diffusion and the Breakdown of FPU Recurrences

Let us now study the physically important phenomenon of diffusion in an FPU chain, when initial conditions are given near the first normal mode of the system. In particular, we shall consider here the FPU-α model and show that the evolution of its dynamics breaks down into two stages during which the qualitative behavior of the motion is considerably different.

In direct analogy with the FPU-β model, the FPU–α system describes a one-dimensional array of N particles governed by the Hamiltonian

$$H = \frac{1}{2}\sum_{k=1}^{N} y_k^2 + \frac{1}{2}\sum_{k=0}^{N}(x_{k+1} - x_k)^2 + \frac{\alpha}{3}\sum_{k=0}^{N}(x_{k+1} - x_k)^3, \qquad (6.36)$$

where we have imposed again fixed boundary conditions $x(0) = x(N + 1) = 0$.

In the harmonic limit ($\alpha = 0$), the Hamiltonian is identical to its quadratic part H_2 (see (6.6)). Thus, (6.36) can be expressed, through the canonical transformation (6.5), in P_q, Q_q variables as

$$H = \frac{1}{2}\sum_{q=1}^{N}(P_q^2 + \Omega_q^2 Q_q^2) + \frac{\alpha}{3\sqrt{2(N+1)}}\sum_{q,l,m=1}^{N} B_{q,l,m}\Omega_q\Omega_l\Omega_m Q_q Q_l Q_m, \qquad (6.37)$$

with Ω_q denoting the linear normal mode frequencies (6.7) and $B_{q,l,m}$ the coupling coefficient between the modes given by the formula

$$B_{q,l,m} = \sum_{\pm}(\delta_{q\pm l\pm m,0} - \delta_{q\pm l\pm m,2(N+1)}), \qquad (6.38)$$

while the equations of motion of the system are given by

$$\ddot{Q}_q + \Omega_q^2 Q_q = -\frac{\alpha}{\sqrt{2(N+1)}}\sum_{q,l,m=1}^{N} B_{q,l,m}\Omega_q\Omega_l\Omega_m Q_l Q_m. \qquad (6.39)$$

6.4.1 Two Stages of the Diffusion Process

The dynamics of the FPU-α chain is studied systematically in [276], where two stages are found to occur: (I) The energy is suddenly transferred to a small number of modes, (II) a metastable state arises and a slow diffusive process is observed leading the system finally to equipartition. Let us use as initial condition the first normal mode $q = 1$ of the FPU-α chain given by

$$x_k = A \sin\frac{\pi k}{N+1}, \quad \dot{x}_k = 0, \qquad (6.40)$$

where $k = 1, 2, \ldots, N$ and A is arbitrary. If we denote by $e_q = E_q/E$ the normalized instantaneous energies of the normal modes, one finds for small times (typically of the order of $T = 10^3$) that they are governed by an expression of the form

$$e_q \sim t^{2(q-1)}, \quad \forall q \in \{1, 2, \ldots, N\}. \tag{6.41}$$

This formula, in fact, represents Stage I of the diffusion process, as we discuss below (for more details see [276]).

Stage I:

In Stage I the energy diffuses from the initially excited normal mode to the remaining ones at a rate that is well described by the power law (6.41), as has also been verified in [92] at energies $E = 0.01$, $E = 0.1$ and $E = 1$, with $N = 31$ and $\alpha = 0.33$. This first type of energy transport occurs over a short time interval, beyond which (see Stage II below) the linear mode energies settle down to an exponential profile in q space. When the energy is small enough the spectral energy profile is quite close to the corresponding q-breather of the FPU-α model.

However, when the energy attains higher values, the system forms a packet of modes that undergoes a type of internal equipartition of its energy. Referring to (6.41), the authors of [276] explain this behavior through near resonances of the form $\omega_q - q\omega_1$, which appear as denominators in the perturbation scheme. For short times, these denominators are inversely proportional to t and hence the energy e_2 is proportional to t^2. Recursively, one can thus find that the remaining modes have normalized energies which increase as $e_q \sim t^{2(q-1)}$, signifying that the resonance that develops over that time period is responsible for the sudden transport of energy from the long wavelength normal modes to those of short wavelength.

Let us denote by T_I the time needed for the FPU-α chain to pass from Stage I to Stage II of the dynamics. Computations show that $T_I \approx 4000$ for $E = 0.01$, $T_I \approx 1000$ for $E = 0.1$ and $T_I \approx 400$ for $E = 1$.

Stage II:

Stage I lasts until the frequencies in the denominators of the solution cease to be in resonance and the harmonic energies settle down to a definite value. Given that we have chosen as initial condition the $q = 1$ normal mode, the first modes to be successively excited are $\alpha^{1/2}\epsilon^{1/4}N$ [24, 37] and share within them almost all the energy of the system, forming a well-defined packet as described in Sect. 6.1.2. The remaining ones, i.e. the tail modes, have exponentially localized energy spectrum. This occurs for $t \geq T_I$ and is called Stage II of the process. The FPU-α chain remains in that state until the resonances between the modes in the packet and tail modes complete the transfer of energy between them and the system moves away from the exponential profile.

Let us examine this stage of the dynamics for the following values of the total energy: $E = 0.01$, $E = 0.1$ and $E = 1$, setting $N = 31$ and $\alpha = 0.33$. For $E = 0.01$ and $E = 0.1$ one observes that the energy spectrum of the FPU-α model exhibits "spikes" at some modes, due to resonances with the tail modes, similar to those seen in Fig. 6.7 for the FPU$-\beta$ model. The exponential slope of the profile, however, is the same, while the linear energy spectrum is invariant for exponentially long times. On the contrary, if the total energy of the FPU-α system exceeds the value 0.1, the transfer of energy from the packet to the tail modes occurs at a much faster rate.

It follows from these results that there is a critical energy value below which FPU trajectories remain close to the q-breather for exponentially long times. This critical value can be calculated if we consider that the length of the mode packet is nearly equal to

$$\frac{q}{N} \simeq \sqrt{\mu} = \alpha^{1/2}\varepsilon^{1/4}, \tag{6.42}$$

whence for $q = 1$ it follows that this critical energy threshold is

$$E_c = \frac{1}{\alpha^2 N^3}. \tag{6.43}$$

We conclude, therefore, that for low energy values the E_q spectrum of the FPU-α chain is localized for exponentially long times, due to the FPU recurrence phenomenon and the existence of q-tori discussed in the previous section. It is interesting that the spectrum remains localized for energy values $E = 0.1$ much higher than the critical value $E_c = 2.8 \times 10^{-4}$. In fact, for $E = 0.1$ ($\alpha = 0.33$, $N = 31$) we observe through (6.42), that we have a packet of length 4, i.e. the energy is equally distributed among the 4 modes $q = 1, 2, 3, 4$. This means that the FPU trajectory converges to a four-dimensional torus, even though initially only one mode was excited.

Stage II concludes with the localized energy of the FPU-α model diffusing from the normal modes of the initial packet to the tail modes leading the system to equipartition. Studying the rate of diffusion from the low to the high frequency modes, the following has been recently discovered in [276]: (a) The FPU-α model is characterized by *sub-diffusion* and (b) by measuring the diffusion rate one can estimate the time it takes for the system to reach equipartition. As is well-known, this is a delicate issue, which has been studied to date by many authors (see e.g. [37, 38, 110, 112, 275, 276]).

The main difficulty lies in the fact that the onset of energy equipartition is practically impossible to observe, based only on the numerical integration of the equations of motion. Consequently the only tool we have to predict this onset with some degree of accuracy is to use the theoretical prediction of the diffusion rate of the motion described below.

6.4.2   Rate of Diffusion of Energy to the Tail Modes

We shall now obtain a law of diffusion using the sum of the last third of the normalized linear energies e_q. The reason for choosing this set with the shortest wavelengths is because they exhibit the largest variations in their harmonic energies

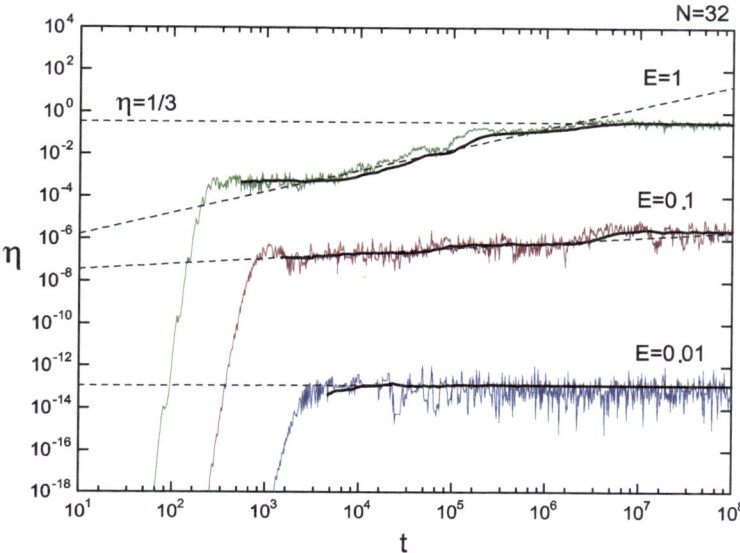

Fig. 6.10 Time evolution of the energy of tail modes for the FPU-α model, with $N = 31$ and $\alpha = 0.33$. The *bottom curve* corresponds to total energy $E = 0.01$, the middle one to $E = 0.1$ and the *top curve* to $E = 1$, for the instantaneous normalized harmonic energies e_q (after [92,276])

and consequently yield a more efficient criterion for predicting of the time required for energy equipartition.

The sum of normalized harmonic energies for the mode interval $q \in [2N/3, N]$ is called *tail mode energy* in [276] and is defined by

$$\eta(t) = \sum_{q=2N/3}^{N} e_q(t),\qquad (6.44)$$

for the spectrum of energies e_q.

In Fig. 6.10 we plot the function $\eta(t)$, for the energy values $E = 0.01$, $E = 0.1$ and $E = 1$, in log-log scale as a function of time and observe a diffusion law of the form

$$\eta \propto Dt^{\gamma}.\qquad (6.45)$$

Let us now determine the dependence of the parameters D and γ on the total energy of the system E, given that $\mu = \alpha^{1/2}\varepsilon^{1/4}$ is the only independent parameter of the problem. To this end, let us examine the FPU-α system with $N = 31$ and $\alpha = 0.33$, for 200 values of the total energy, starting with $E = 0.01$ by 0.01 increments up to the value $E = 2$. In Fig. 6.11 we estimate the value of γ for the energies of the tail modes η as a function of E, using the method of least squares, for suitably long time intervals during which the tail mode energies satisfy the law

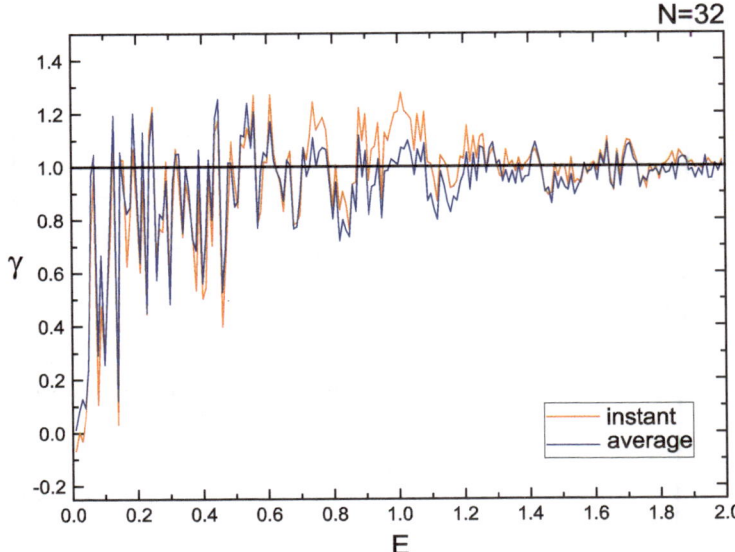

Fig. 6.11 Plot of the value of γ of the energies of the tail modes η of the FPU-α chain, with $N = 31$ and $\alpha = 0.33$ (after [92, 276])

(6.45). Since for every energy value we have a different time interval, during which the tail mode energies grow linearly (in log-log scale), it is practically impossible to repeat the calculation for every value of E. For this reason we split the computation in three energy ranges: (a) $E \in [0, 0.51]$, where the appropriate time interval is $[10^4, 3 \times 10^6]$, (b) $E \in [0.52, 0.66]$, where this interval is $[3 \times 10^3, 10^6]$ and (c) $E \in [0.67, 2]$, for which the time interval is $[10^3, 3 \times 10^5]$.

Thus, in Fig. 6.11, plotting γ in (6.45) as a function of E, we find that the system exhibits *subdiffusion*, since the power of t in (6.45) is generally smaller than unity.

6.4.3   Time Interval for Energy Equipartition

As we deduce from the results shown in Fig. 6.10, the system is characterized by energy equipartition when $\eta = 1/3$. Thus, solving from the expression (6.45) for $t = T^{eq}$, with $\eta = 1/3$, we can estimate the equipartition time of the FPU-α system to be

$$T^{eq} \approx (3D)^{-1/\gamma}. \tag{6.46}$$

Evaluating the parameters D and γ from our knowledge of the total energy E of the FPU-α model with $N = 31$ and $\alpha = 0.33$, we use (6.46) to calculate the equipartition time of the system. In Fig. 6.12 we plot the logarithm of this time as a function of energy of the system, in semi-log plot, for 200 energy values within the intervals: (a) $E \in [0.001, 0.2]$ where the energy increment is 0.001

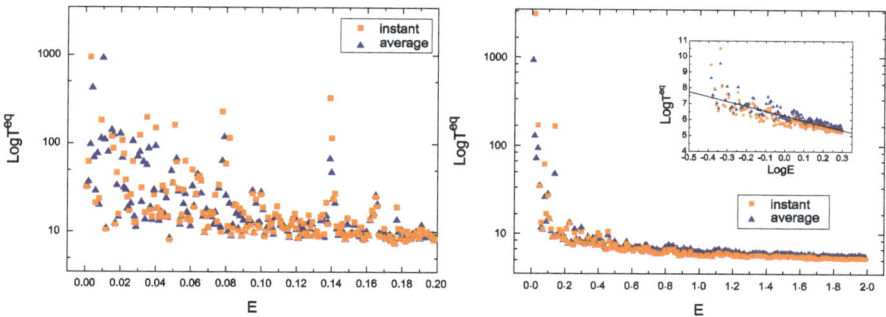

Fig. 6.12 The logarithm of the equipartition time T^{eq} for the FPU-α chain, with $N = 31$ and $\alpha = 0.33$, as a function of the total energy, in semi-log scale, for: (**a**) $E \in [0.001, 0.2]$ and (**b**) $E \in [0.01, 2]$. In the insert of (**b**) we show the logarithm of the equipartition time T^{eq} as a function of the logarithm of E (after [92, 276])

and (b) $E \in [0.01, 2]$, where the energy increases by steps of 0.01. In the small insert of Fig. 6.12b we depict the quantity $\log T^{eq}$ vs. $\log E$, and find that in the energy interval $E \in [0.4, 2]$, the law for the time required for equipartition may be expressed as

$$T^{eq} \approx 10^{6.25} E^{-3}. \tag{6.47}$$

This agrees very well with the formula given in [112] for the same energy interval. Even so, for energy values within $0.1 < E < 0.4$ the system takes exponentially long to reach equilibrium, while for $E \leq 0.1$ equipartition has not yet been observed. The theoretical prediction of the relation (6.46) diverges from the numerical results for $0.1 < E < 0.4$, while for $E \leq 0.1$, it is impossible to check its validity with present day available numerical techniques.

As we see from Fig. 6.10, the slope of the η-curve for $E = 0.01$ is practically zero. The problem is that its value is smaller than the errors of the least square method. Therefore, the conclusion is that the validity of these results is questionable for $E \leq 0.1$. To date there is no theoretical estimate in the literature allowing us to decide whether the system reaches equipartition when E becomes too small. Thus, there are two possibilities: Either the equipartition times T^{eq} are exponentially long, or the system remains trapped in the neighborhood of a q-torus or a q-breather. The formulas described above cannot answer the question whether the system reaches equipartition for $E \leq 0.1$, but do confirm the validity of the law $T^{eq} \sim E^{-3}$ given in [112], for energies $E \in [0.4, 2]$.

Exercises

Exercise 6.1. Show that the spectrum of the unperturbed (harmonic) frequencies Ω_q, given by (6.7), is fully incommensurable if N is either a prime number minus one, or a power of two minus one. Hint: See Exercise 4.1.

Problems

Problem 6.1. (a) Carry out in detail the steps outlined in Sect. 6.2.1 and apply the
Poincaré-Linstedt series method to construct a 2-torus solution with $N = 8$,
and a 4-torus solution with $N = 16$ for the FPU$-\beta$ model (6.3) with $\beta = 0.1$.
(b) Choosing initial conditions on these tori, implement numerically the methods
of SALI and GALI described in Chap. 5 to show that these tori are stable for
small enough values of the total energy E.
(c) Increasing the value of E, determine critical values $E = E_c(2)$ and $E = E_c(4)$,
at which these tori first become unstable.

Problem 6.2. Repeat the calculations of Problem 6.1 for the FPU-α model, with
Hamiltonian (6.36) and $\alpha = 0.1$. Consult the beginning of Sect. 6.4, where the
appropriate transformations to canonical variables Q_q, P_q are described. Compare
the results with the corresponding findings obtained for the FPU-β model in
Problem 6.1.

Chapter 7
Localization and Diffusion in Nonlinear One-Dimensional Lattices

Abstract In this chapter we focus on localization properties of nonlinear lattices in the configuration space of their spatial coordinates. In particular, we begin by discussing the phenomenon of exponentially localized periodic oscillations, called discrete breathers, in homogeneous one-dimensional lattices. Demonstrating that they are directly connected with homoclinic orbits of invertible maps, we describe a precise numerical procedure for constructing them to arbitrarily high accuracy. Furthermore, the homoclinic approach allows for the symbolic classification of all multibreathers (having more than one major amplitudes) symmetric and asymmetric and can be extended to 2D lattices. We also analyze the (linear) stability of breathers and show how their bifurcations can be monitored using methods of feedback control. Chapter 7 ends with a description of important recent results on *delocalization* and diffusion of wavepackets in lattices with random disorder, pointing out the complexity of these phenomena and the open questions that still remain.

7.1 Introduction and Historical Remarks

Localization phenomena in systems of many (often infinite) degrees of freedom have frequently attracted attention in solid state physics, nonlinear optics, superconductivity and quantum mechanics. It is well-known, of course, that localization in lattices can be due to disorder (see e.g. [1]), while, if the disorder is random, one encounters the famous Anderson localization phenomenon [10]. There is, however, another very important type of localization, occurring in homogeneous infinite dimensional Hamiltonian lattices, which is *dynamic*: It refers to localized periodic oscillations arising not because of the presence of some defect, but due to the interaction between nonlinearity and the spectrum of linear normal modes (or phonons) of the system.

We are talking, in particular, about a remarkable phenomenon called *discrete breathers*, representing periodic oscillations of very few sites, in nonlinear lattices consisting, in principle, of infinitely many particles! It is surprisingly common in

T. Bountis and H. Skokos, *Complex Hamiltonian Dynamics*,
Springer Series in Synergetics, DOI 10.1007/978-3-642-27305-6_7,
© Springer-Verlag Berlin Heidelberg 2012

discrete systems, in contrast to continuum systems (like strings, membranes or fluid surfaces), where it only appears in certain very special integrable examples. Since the reader may not be too familiar with these concepts, we shall first present a brief review of the history of discrete breathers and then describe how one can study them using homoclinic orbits of multi-dimensional maps. This approach not only provides a very efficient method for computing discrete breathers, but also leads to a systematic way of classifying them and allows us to accurately follow their existence and stability properties as certain physical parameters of the problem are varied.

At the beginning we will deal exclusively with homogeneous lattices (of identical masses and spring constants) and proceed in subsequent sections to treat the problem of diffusion in disordered lattices starting with an initial excitation of very few sites, which has recently received a great deal of attention in the literature.

7.1.1 Localization in Configuration Space

In 1969, while studying a system of coupled nonlinear oscillators modeling finite-sized molecules, Ovchinnikov [261] showed in a very beautiful way how discreteness in combination with an intrinsic nonlinearity of the system can cause energy localization of the type discovered by FPU. He demonstrated, in particular, that resonances between neighboring oscillators can be avoided when the vibrations of the system lie outside the so-called *phonon band*, or spectrum of linear normal mode frequencies (see Fig. 7.1). Thus, the recurrence phenomena in the FPU experiments can be explained as the result of limited energy transport between Fourier modes, caused by discreteness and nonlinearity effects. Today, of course, our understanding of this phenomenon is considerably enhanced through the analysis of exponential localization in Fourier space described in detail in Chap. 6.

Twenty years later (1988), the subject of localized oscillations in nonlinear lattices was revived in a seminal paper by Sievers and Takeno [306], who used analytical arguments to show that energy localization occurs generically in FPU systems of *infinitely many* particles in one dimension, obeying (6.1).

Combining their perturbative analysis with numerical experiments, they demonstrated that a new class of solutions, known to date as discrete breathers, exist as oscillations which are both time-periodic and spatially localized. In their simplest form, these solutions exhibit significant excitations only of the middle and nearby particles. However, a great many patterns are also possible, the so-called *multibreathers*, in which several particles oscillate with large amplitudes, as shown here in Fig. 7.2. How can one determine all the possible shapes? Which of them are stable under small perturbations? These are the kind of questions we would like to address in the next section.

Besides the FPU system, the existence of these localized oscillations was soon verified numerically by other research groups on a variety of lattices, including the

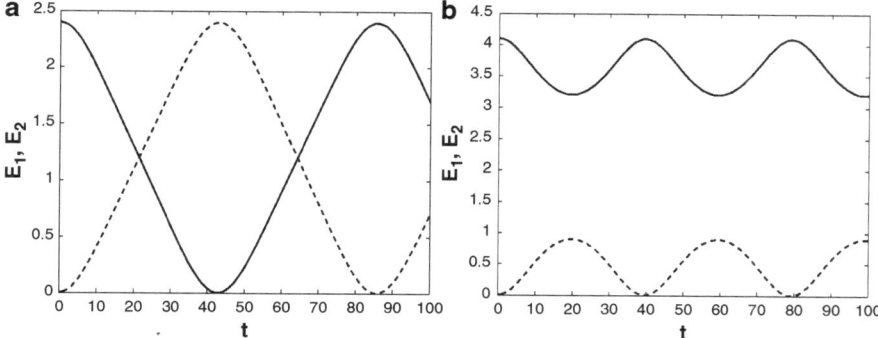

Fig. 7.1 Local energies $E_1(t)$ (*solid line*) and $E_2(t)$ (*dashed line*) of the two coupled nonlinear oscillators considered in [261], showing how energy transfer is impeded by discreteness and nonlinearity. (**a**) Complete energy transfer when the first oscillator has initial amplitude $a = 2.0$, for $E_1(0) = 2.4$) and (**b**) Incomplete energy transfer when the first oscillator has initial amplitude $a = 2.5$ and $E_1(0) \approx 4.1$ (after [53])

Klein-Gordon (KG) system of ODEs written in the form

$$\ddot{u}_n = -V'(u_n) + \alpha(u_{n+1} - 2u_n + u_{n-1}), \quad V(x) = \frac{1}{2}Kx^2 + \frac{1}{4}x^4, \quad -\infty < n < \infty \tag{7.1}$$

where $V(x)$ is the on-site potential, $\alpha > 0$ is a parameter indicating the strength of coupling between nearest neighbors, and $(')$ denotes differentiation with respect to the argument of $V(x)$.

It was not, however, until 1994, that a mathematical proof of the existence of discrete breathers was published by MacKay and Aubry [235] in the case of one-dimensional lattices of the type (7.1). Under the general assumptions of nonlinearity and non-resonance, such chains of interacting oscillators were rigorously shown to possess discrete breather solutions for small enough values of the coupling parameter α, as a continuation of their obvious existence at $\alpha = 0$.

Section 7.2 below is devoted to a study of discrete breathers and multibreathers based on the concept of homoclinic orbits of invertible maps. As has been pointed out in a number of papers [41, 42, 61, 62], homoclinic dynamics offers a very convenient way to construct such solutions and study their stability properties, away from the $\alpha = 0$ limit, through the "geometry" of the homoclinic solutions of *nonlinear recurrence relations*. The results we shall describe have been presented in [39, 40], while for the most recent developments in this field the reader is strongly advised to consult the recent review article [125].

It is also important to remark at this point that, although strictly speaking discrete breathers are proved to exist in infinite lattices, in practice we construct them as solutions of a finite number of ODEs of the form (7.1), taking $-N < n < N$, where N is large enough so that particles near the ends of the lattice are practically motionless.

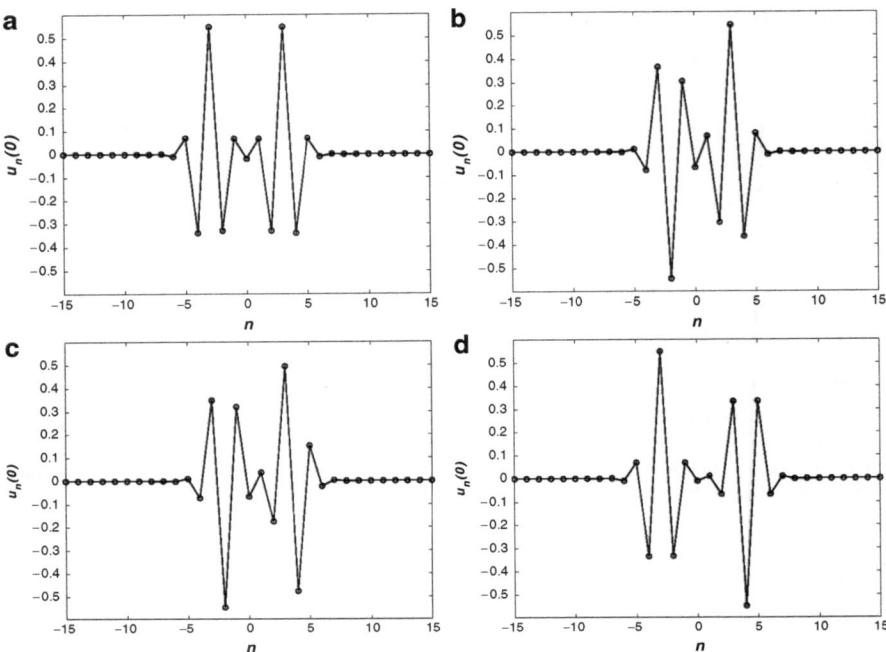

Fig. 7.2 Examples of discrete multibreathers of an FPU lattice of the form (6.1) with cubic rather than quadratic forces (the so-called FPU−β model), obtained by starting a Newton-Raphson search using homoclinic orbits of a map of the form (7.6) as an initial guess. The term 'multibreather' refers to the fact that more than one site exhibit oscillations of comparably large amplitude (after [53])

7.2 Discrete Breathers and Homoclinic Dynamics

Focusing on the property of spatial localization, Flach was the first to show that discrete breathers in simple one-dimensional chains can be accurately represented by homoclinic orbits in the Fourier amplitude space of time-periodic functions [123]. Indeed, inserting a Fourier series

$$u_n(t) = \sum_{k=\infty}^{\infty} A_n(k) \exp(ik\omega_b t) \tag{7.2}$$

into the equations of motion of either the FPU or KG (7.1) lattice and setting the amplitudes of terms with the same frequency equal to zero, leads to the system of equations

$$-k^2 \omega_b^2 A_n(k) = \langle -V'(u_n) + W'(u_{n+1} - u_n) - W'(u_n - u_{n-1}), \exp(\mathbb{I}ik\omega_b t)\rangle,$$

$$\forall k, n \in \mathbb{Z}, \tag{7.3}$$

where W is the potential function, $\mathbb{I}$ an infinite unit matrix and ω_b the frequency of the breather. This is an infinite-dimensional mapping of the Fourier coefficients $A_n(k)$, with the brackets $\langle .,. \rangle$ indicating a normalized inner product. Time-periodicity is ensured by the Fourier basis functions $\exp(\mathbb{I}ik\omega_b t)$. Spatial localization requires that $A_n(k) \to 0$ exponentially as $n \to \pm\infty$. Hence a discrete breather is a homoclinic orbit in the space of Fourier coefficients, i.e. a doubly infinite sequence of points beginning at 0 for $n \to -\infty$ and ending at 0 for $n \to +\infty$.

As a simple example, let us consider an infinite one-dimensional model which plays an important role in many physical systems [188]: It is described by a special form of the discrete nonlinear Schrödinger (DNLS) equation

$$i\dot{u}_n + \gamma|u_n|^2 u_n + \left(1 + |u_n|^2\right)(u_{n-1} + u_{n+1}) = 0, \quad n \in \mathbb{Z}. \tag{7.4}$$

Suppose now that we seek periodic solutions of the form $u_n = x_n \exp(i\omega t)$, with x_n real. Substituting this expression of u_n in (7.4) and eliminating the exponential $\exp(i\omega t)$ leads to the two-dimensional map

$$x_{n+1} + x_{n-1} = \frac{\omega - \gamma x_n^2}{1 + x_n^2} x_n, \quad x_n \in \mathbb{R}. \tag{7.5}$$

This map has three fixed points, in general, in the (x_n, x_{n+1}) plane: One at the origin and two located symmetrically along the diagonal $x_n = x_{n+1}$. The eigenvalues of the linearized equations about these equilibria dictate e.g. that for $\omega = 0.25$ and $\gamma = -3$ the ones on the diagonal are saddles, while $(0,0)$ is a center (see Exercise 7.1).

On the other hand, for $\omega = 4$ and $\gamma = 1$ the fixed point at the origin is a saddle, while the other two fixed points are centers. This is the case shown in Fig. 7.3, where the stable and unstable manifolds of $(0,0)$ have been plotted. Clearly, their intersections form homoclinic orbits, which yield discrete breathers (and multibreathers) of the DNLS equation with frequency $\omega = \omega_b$ as follows: Since, by definition, homoclinic orbits satisfy $x_n \to 0$ as $n \to \pm\infty$, very few of their x_n, x_{n+1} coordinates have appreciable magnitude, corresponding to sites u_n which are oscillating periodically with frequency ω.

Now, if a homoclinic orbit is formed by intersections of the first part of the stable (resp. unstable) manifold with the first "lobes" of the unstable (resp. stable) manifold of the saddle point at $(0,0)$, this yields an orbit that makes one loop around the elliptic point and leads to oscillations with a single major amplitude belonging to a simple breather. If, on the other hand, the homoclinic orbit consists of intersections between "lobes" of both manifolds and makes several loops about an elliptic point, it leads to oscillations with large amplitudes at several sites and hence corresponds to a multibreather.

Of course, if the Fourier expansion of a breather consists of a single term, as in the DNLS case, the associated two-dimensional map represents the exact dynamics.

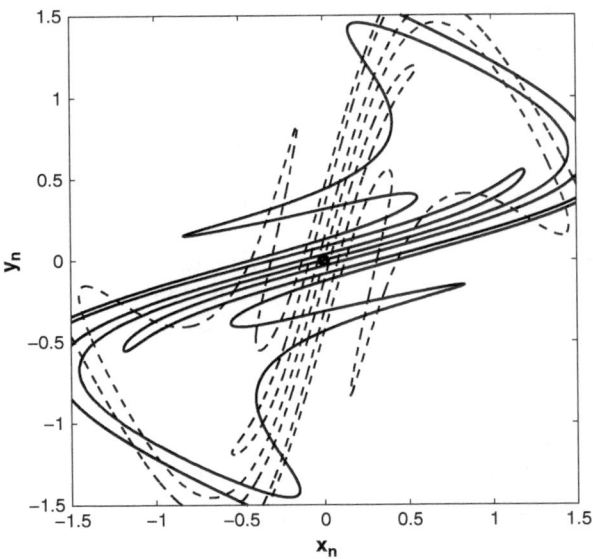

Fig. 7.3 The stable (*dashed curve*) and unstable (*solid curve*) manifolds of the fixed point at $(0, 0)$ of a 2D DNLS map of the form (7.5), with $y_n = x_{n+1}$, or $x_n = a_n$, $y_n = a_{n+1}$ as in (7.6). The manifolds are clearly seen to intersect at infinitely many points, hence a wealth of homoclinic orbits exists (after [53])

What happens, however, in a case like the KG system (7.1), where the substitution of (7.2) leads to an infinite dimensional map like (7.3) for the Fourier coefficients? Not to worry. Owing to the exponential decay of their magnitudes, we can get away keeping only the largest one ($k = 1$), reducing (7.3) to a simple 2D map, for $a_n = A_n(1)$, of the form

$$a_{n+1} = g\left(a_n, a_{n-1}\right). \tag{7.6}$$

The invariant manifolds, emanating out of the saddle point at the origin, could then be plotted, as in Fig. 7.3 for example, by the method described in Exercise 7.1. It is easy to apply this approach to the KG chain, where the above reduction yields the map

$$a_{n+1} = -a_{n-1} - C a_n + \frac{1}{\alpha} a_n^3, \quad C = -2 + \frac{K - \omega^2}{\alpha} \tag{7.7}$$

(see Exercise 7.2 and Problem 7.1). As explained in [61], one can use this map to find approximate analytical expressions for the coefficients a_n that will allow us to obtain numerically not only simple breathers but also multibreathers of the KG chain. Moreover, interesting effects arise when one applies this approach to KG "soft" potentials with a minus sign before the quartic term (see (7.1), [262]).

7.2.1 How to Construct Homoclinic Orbits

It is hopefully clear by now from the above discussion (as well as Exercises 7.1 and 7.2), that the accurate computation of invariant manifolds and their intersections is extremely important for constructing discrete breathers and multibreathers. For this reason, we shall now describe a particularly useful and efficient method that has been developed in [42] to locate homoclinic orbits of invertible maps of arbitrary (but finite) dimension. As it turns out, we need to exploit certain *symmetry properties* of the dynamics and understand the "geometry" of the invariant manifolds near the origin, which we shall henceforth assume to be a saddle point of the maps under study.

To see how this can be done, let us consider a general first order map (or recurrence relation) of the form

$$\mathbf{x}_{n+1} = f(\mathbf{x}_n), \quad \mathbf{x}_n, \ \mathbf{x}_{n+1} \in \mathbb{R}^d, \tag{7.8}$$

where $d \geq 2$ is a positive integer. Recall that homoclinic orbits are solutions which satisfy $\mathbf{x}_n \to 0$ as $n \to \pm\infty$ and concentrate on all orbits satisfying a symmetry condition of the form

$$\mathbf{x}_n = M\mathbf{x}_{-n}, \tag{7.9}$$

where M is a $d \times d$-matrix with constant entries and $\det M \neq 0$. Furthermore, observe that if an orbit obeys such a symmetry and $\mathbf{x}_n \to 0$ as $n \to -\infty$, it is also true that $\mathbf{x}_n \to 0$ as $n \to \infty$. Hence, any orbit which obeys this symmetry and also satisfies $\mathbf{x}_n \to 0$ as $n \to -\infty$ is a homoclinic orbit.

To construct such an orbit, we first need to specify numerically its asymptotic behavior as $n \to -\infty$. In other words, it has to be on the unstable manifold of $\mathbf{x} = 0$. Thus, given that the origin is a saddle fixed point, it is known that its unstable manifold is well approximated in its vicinity by the corresponding Euclidean subspace of the linearized equations, which is tangent to the nonlinear manifold and has the same number of dimensions.

Now, the dimension of the linear unstable subspace equals the number of coordinates necessary to determine a point $\mathbf{x}_{-N}$, $N \gg 1$ uniquely. So, by choosing this point to be on the linear unstable manifold, very close to the origin, it will also lie approximately on the corresponding nonlinear manifold. Thus, when mapped forward $N + 1$ times, we can test whether it satisfies the above symmetry relation (7.9). By this reasoning, locating *symmetric* homoclinic orbits reduces to a search for solutions of the system

$$\begin{cases} \mathbf{x}_1 - M\mathbf{x}_{-1} = 0 \\ \mathbf{x}_0 - M\mathbf{x}_0 = 0 \end{cases}, \tag{7.10}$$

since, given $\mathbf{x}_{-N}$, the values of $\mathbf{x}_1$, $\mathbf{x}_0$ and $\mathbf{x}_{-1}$ are uniquely obtained by direct iteration of the map (7.8).

Now, given that $x_1 = F(x_0)$ and $x_{-1} = F^{-1}(x_0)$, (7.10) can be solved for x_0, searching for solutions of the equation

$$Z(x_0) = F(x_0) - MF^{-1}(x_0) = 0. \qquad (7.11)$$

The number of unknowns in this equation is $2d$, d being the dimension of the unstable manifold. However, a zero of this function automatically implies $x_0 = Mx_0$ (i.e. the second component of x_0 is equal to zero, $w_0 = 0$), reducing the number of unknowns to d.

Suppose now that x_{eq} is a fixed point of the map (7.8) and a state x_{-N} exists, with $N \gg 1$, for which $x_{-N-n} \to x_{eq}$ as $n \to \infty$, i.e. x_{-N} is on the unstable manifold of x_{eq}. Thus, the point x_{-N} can be uniquely identified by a d-dimensional coordinate, say σ. Since $x_0 = F^N(x_{-N})$ this also determines all the unknowns in (7.11), which can thus be written as an equation $Z(\sigma) = 0$. Clearly, every coordinate σ solving this equation defines an orbit of (7.8) having the desired asymptotic behavior. The search for homoclinic and heteroclinic solutions, therefore, reduces to finding a state x_{-N} on the unstable manifold of the equilibrium point x_{eq}, determined by the coordinate σ, for which $Z(\sigma) = 0$. This yields a set of d equations with d unknowns, and hence is, in general, solvable e.g. by Newton iterative methods (see Problem 7.2 for more details).

In the case of an invertible map, we can use this method to also find all *asymmetric* homoclinic orbits, i.e. those which do not obey the symmetry condition (7.10). This can be done by introducing the new "sum" and "difference" variables

$$\begin{cases} v_n = x_n + x_{-n} \\ w_n = x_n - x_{-n} \end{cases},$$

which always possess the symmetry

$$\begin{cases} v_n = v_{-n} \\ w_n = -w_{-n} \end{cases}. \qquad (7.12)$$

In this way, we can apply again the above strategy and look for symmetric homoclinic orbits of a new map (whose dimension is twice that of the original f) described by the equations

$$F : \begin{cases} v_{n+1} = f\left(\frac{v_n+w_n}{2}\right) + f^{-1}\left(\frac{v_n-w_n}{2}\right) \\ w_{n+1} = f\left(\frac{v_n+w_n}{2}\right) - f^{-1}\left(\frac{v_n-w_n}{2}\right) \end{cases}, \qquad (7.13)$$

yielding homoclinic orbits x_n of the original map f (7.8), that are not themselves necessarily symmetric. On the other hand, each homoclinic orbit of f is a symmetric homoclinic orbit of the new map F. Therefore, by determining all symmetric homoclinic orbits of F, we find all homoclinic orbits of f. In fact, it is possible by this approach to classify all multibreathers of an N-particle one-dimensional

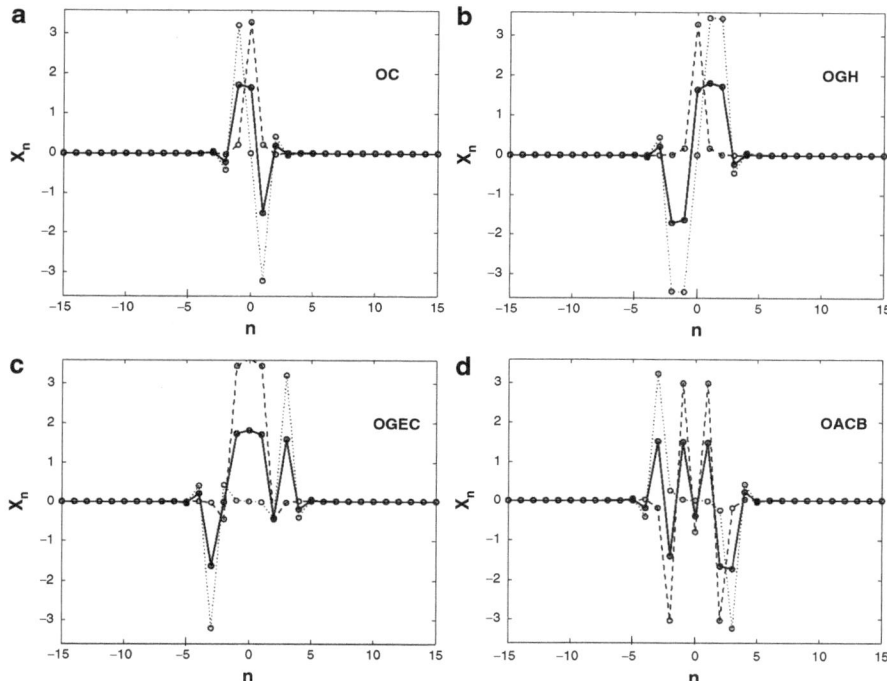

Fig. 7.4 Several homoclinic orbits of the map F (7.13), determined by a zero-search of the system of equations $\mathbf{v}_1 - \mathbf{v}_{-1} = 0$ and $\mathbf{w}_0 = 0$, related by $\mathbf{v}_n = \mathbf{x}_n + \mathbf{x}_{-n}$ and $\mathbf{w}_n = \mathbf{x}_n - \mathbf{x}_{-n}$. Shown here are $\mathbf{v}_n$ (*dashed*), $\mathbf{w}_n$ (*dotted*) and $\mathbf{x}_n$ (*solid*). Also indicated is the symbolic name of the orbit assigned by the procedure outlined in the text and described in detail in [41, 42] (after [53])

lattice by assigning to them symbolic sequences in a systematic way, as follows: If, at a moment when all velocities are zero, a particle has a large positive (resp. negative) displacement, it is assigned a $+$ (resp. $-$) sign, while if its displacement is small it is assigned the symbol 0. We then introduce nine new symbols: $A = -+$, $B = 0+$, $C = ++$, $D = -0$, $O = 00$, $E = +0$, $F = --$, $G = 0-$ and $H = +-$ corresponding to nine regions of the associated 2D map where these points are located. One may thus locate multibreathers of increasing complexity, whose number, of course, grows with increasing N. The interested reader is referred for more details to [41, 42] and Fig. 7.4 for an example of the results presented in these papers.

It is important to mention that, besides the above approach based on homoclinic orbits, there also exist other methods to compute breathers and multibreathers that one may find simpler or more straightforward to implement. Perhaps the most popular of them starts from the so called "anticontinuum limit", where the particles are uncoupled and uses Newton iterative schemes to continue the solutions to the regime of finite coupling [21, 248]. The homoclinic approach, however, also has important advantages: It uses Fourier decomposition to turn the problem to an invertible map whose low-dimensional approximations yield very accurate initial conditions for the homoclinic orbits, provides a clear geometric and symbolic description of all breathers and multibeathers and can be used to locate such solutions in lattices of more than one dimension, as we demonstrate in the section that follows.

7.3 A Method for Constructing Discrete Breathers

Based on the above results, we now describe a method for computing discrete breathers that can be efficiently applied to: (1) lattices of more than one spatial dimension and (2) systems with vector valued variables assigned to each lattice site. The main idea is to write a breather solution as the product of a space-dependent and a time-dependent part and reduce the problem to the computation of homoclinic orbits of a 2D map, under the constraint that the given ODEs possess simple periodic oscillations of a well-defined type and specified period.

As we have already discovered, it is possible to formulate a map in Fourier amplitude space linking discrete breathers to homoclinic orbits. This map can be reduced to a finite-dimensional recurrence relation, by neglecting Fourier components with a wave number k larger than some cutoff value $k_{\max}$. Then, one can use the methods described in [41,42] to approximate discrete breather solutions by finding all homoclinic solutions of the corresponding recurrence relations.

This problem has also been examined from a different perspective: In 2002, inspired by the work of Flach [122] and Kivshar [189], Tsironis [334] suggested a new way to approximately separate amplitude from time-dependence, yielding in some cases ODEs with known solutions (for example elliptic functions) while keeping the dimension of the recurrence relation as low as possible. This led to an improved accuracy of the calculations and provided analytical expressions of discrete breathers for a special class of FPU and KG systems.

In [39], Bergamin extended Tsironis' work by developing a numerical procedure for which the time-dependent functions need not be known analytically. In this way, a much wider class of nonlinear lattices can be treated involving scalar or vector valued variables in one or more spatial dimensions. In particular, the approximation proposed by Tsironis can be formulated as follows

$$\begin{cases} u_{n+1}(t) - u_n(t) \approx (a_{n+1} - a_n) T_n(t) \\ u_{n-1}(t) - u_n(t) \approx (a_{n-1} - a_n) T_n(t) \end{cases}, \tag{7.14}$$

where a_n denotes the time-independent amplitude of $u_n(t)$ and $T_n(t)$ is its time-dependence, defined by $T_n(0) = 1$ and $\dot{T}_n(0) = 0$.

Note now, that since all FPU and KG systems are described by a Hamiltonian of the form

$$H = \frac{1}{2} \sum_{n=-\infty}^{\infty} \dot{u}_n^2 + \sum_{n=-\infty}^{\infty} \{V(u_n) + W(u_{n+1} - u_n)\},$$

we may use the approximation (7.14) to transform the equations of motion

$$\ddot{u}_n = -V'(u_n) + W'(u_{n+1} - u_n) - W'(u_n - u_{n-1})$$

into

$$a_n \ddot{T}_n = -V'(a_n T_n) + W'((a_{n+1} - a_n) T_n) - W'((a_n - a_{n-1}) T_n). \quad (7.15)$$

This is an ODE for $T_n(t)$ which can, in principle, be solved since the initial conditions are known.

Of course, an analytical solution of (7.15) is in general very difficult to obtain. However, since we are primarily interested in the amplitudes a_n, what we ultimately need to do is develop a numerical procedure to find a recurrence relation linking a_n, a_{n+1} and a_{n-1} without having to solve the ODE beforehand.

Let us observe first, that the knowledge of a_n, a_{n+1} and a_{n-1} permits us to solve the above ODE numerically. Under mild conditions, solutions $T_n(t)$ can thus be obtained, which are time-periodic, while for a discrete breather all functions $T_n(t)$ have the same period. Choosing a specific value for this period, allows us to invert this process and determine a_{n+1} as a function of a_n and a_{n-1}, just as we did in (7.6). In the same way, we also determine a_{n-1} as a function of a_n and a_{n+1}. Thus, a two-dimensional invertible map has been constructed for the a_n, ensuring that all oscillators have the same frequency.

As is explicitly shown in [39,40], on a variety of examples, the homoclinic orbits of this map provide highly accurate approximations to the discrete breather solutions with the given period and the initial state $u_n(0) = a_n$, $\dot{u}_n(0) = 0$. In fact, we can now apply this approach to more complicated potentials and higher dimensional lattices, as demonstrated in Fig. 7.5, where we compute a discrete breather solution of a two-dimensional lattice, with indices n, m in the x, y directions and dependent variable $u_{n,m}(t)$.

Having thus discovered new and efficient ways of calculating discrete breathers in a wide class of nonlinear lattices, we may now turn to a study of their stability, continuation and control properties in parameter space. More specifically, we will show that it is possible to use the above methods to extend the domain of existence of breathers to parameter ranges that cannot easily be accessed by other more standard continuation techniques.

7.3.1   Stabilizing Discrete Breathers by a Control Method

So far, we have seen that transforming nonlinear lattice equations to low-dimensional maps and using numerical techniques to compute their homoclinic orbits provides an efficient method for approximating discrete breathers in any (finite) dimension and classifying them in a systematic way. This clears the path for an investigation of important properties of large numbers of discrete breathers of

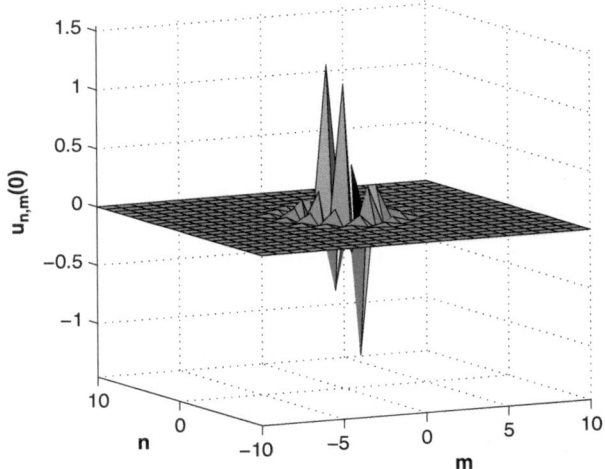

Fig. 7.5 A breather in a two-dimensional lattice is obtained by the method described in the text, for equations of motion where the particles at each lattice site (n, m) are subject to an on-site potential of the KG type and experience harmonic interactions with their four nearest neighbors (after [39])

increasing complexity. One such property, which is relevant to many applications and requires that a solution be known to great accuracy is (linear) stability in time.

In [62], the accurate knowledge of a discrete breather solution was used in a rather uncommon way to study stability properties: As is well known, a familiar task in physics and engineering is to try to influence the behavior of a system, by applying control methods. By control we refer here to the addition of an external force to the system which allows us to influence its dynamics. In particular, our objective is to use this force to change the stability type of a discrete breather solution from what it was in the absence of control.

It is important to emphasize, of course, that during the application of control our system is altered in such a way that the solution itself does not change. In other words, the controlled system possesses solutions which are exactly the same as in the uncontrolled case. Thus, if a solution is unstable in the uncontrolled system, the extra terms added to the equations by the external forcing can cause the solution to become stable and vice versa. This approach is often called feedback control and was first used very successfully to influence the motion of dynamical systems by Pyragas in [278].

The system that was studied in [62], using the above feedback control method, is a KG one-dimensional chain, whose equations of motion are written in the form

$$\ddot{u}_n = -V'(u_n) + \alpha(u_{n+1} - 2u_n + u_{n-1}) + L\frac{d}{dt}(\hat{u}_n - u_n), \qquad (7.16)$$

where $\hat{u}_n = \hat{u}_n(t)$ is the known discrete breather solution of the equations when $L = 0$. The parameter L indicates how strongly the control term influences the

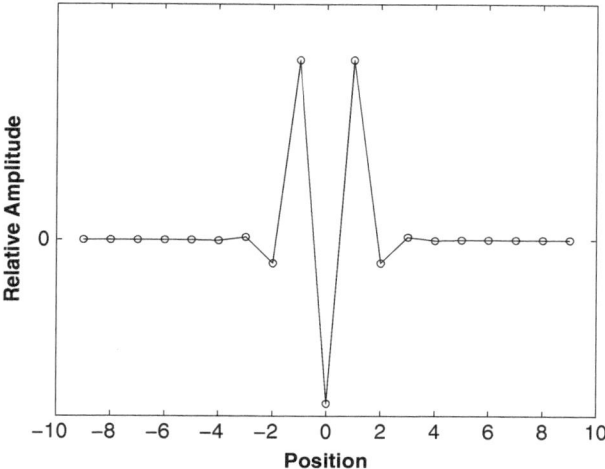

Fig. 7.6 A breather solution of (7.16), $u_n(t) = \hat{u}_n(t)$, which is unstable for $L = 0$ (after [62])

KG system. Thus, for any value of L, $u_n(t) = \hat{u}_n(t)$ is clearly a solution of both the controlled as well as the uncontrolled equations. Clearly, for $L > 0$, $L\frac{d}{dt}u_n$ is a dissipative-like term, while $L\frac{d}{dt}\hat{u}_n$ represents a kind of periodic forcing. It is therefore reasonable to investigate whether, by increasing L, the dissipative part of the process will force the system to converge to a stable solution. If this solution is the $\hat{u}_n(t)$ we started with, the latter is stable. If this is not the case, the original solution is unstable. Indeed, it is not difficult to prove (see Exercise 7.3) the following proposition:

Proposition 7.1. *Let $u_n = \hat{u}_n(t)$ be a periodic discrete breather solution of the lattice equations*

$$\ddot{u}_n = -V'(u_n) + \alpha(u_{n+1} - 2u_n + u_{n-1}), \quad -N < n < N,$$

where N is large enough so that the solution $\hat{u}_n(t)$ exists. Then there exists an $L > 0$ such that $u_n = \hat{u}_n(t)$ is an asymptotically stable solution of the modified (controlled) system

$$\ddot{u}_n = -V'(u_n) + \alpha(u_{n+1} - 2u_n + u_{n-1}) + L\frac{d}{dt}(\hat{u}_n - u_n).$$

This result clearly implies that, by increasing L, it is possible to stabilize the original solution, independent of its stability character in the uncontrolled situation. Let us demonstrate this by taking the breather of Fig. 7.6, which is unstable at $L = 0$, substitute its (known) solution form $\hat{u}_n(t)$ in the above equations and increase the value of L. As we see in Fig. 7.7, it is quite easy to stabilize it at $L = 1.17$, since

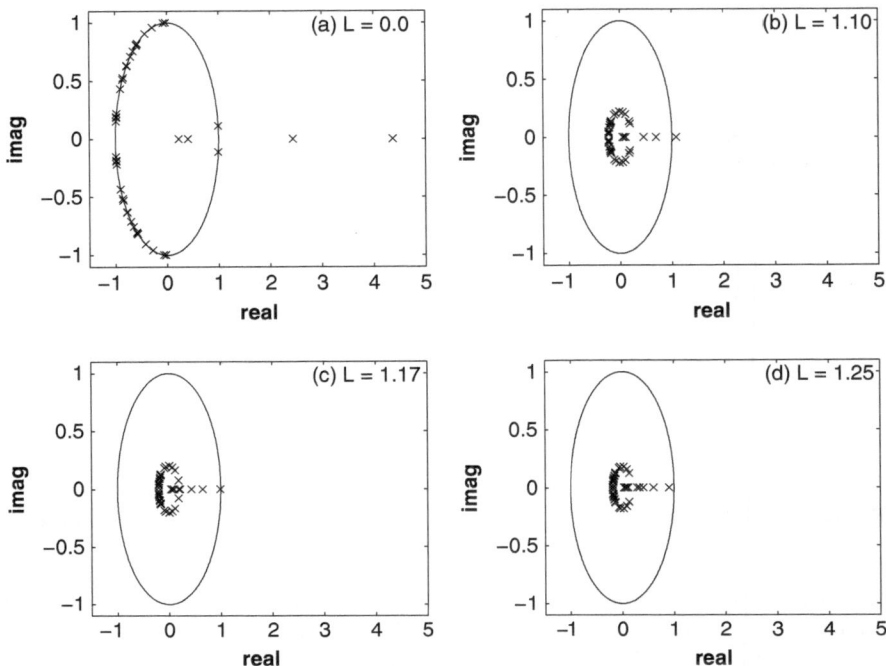

Fig. 7.7 Increasing the control parameter L from $L = 0$ in the system (7.16), using as initial shape the one given in Fig. 7.6 moves the eigenvalues of the monodromy matrix of the orbit inside the unit circle. Initially, some eigenvalues lie outside the unit circle, but eventually, for $L > 1.17$, all eigenvalues attain magnitudes less than one, thus achieving stability and control (after [62])

increasing the value of L gradually brings all eigenvalues of the monodromy matrix of the solution inside the unit circle.

The above Proposition has an additional significant advantage: It gives us the opportunity to address the question of the *existence* of discrete breathers in ranges of the coupling parameter α where other more straightforward continuation techniques do not apply. In order to do this, it is important to recall how this question was originally answered in the existence proof of MacKay and Aubry, using the notion of the so-called anti-continuum limit $\alpha = 0$ [235].

Let us observe that the KG equations of motion

$$\ddot{x}_i = V'(x) + \alpha\,(x_{i+1} - 2x_i + x_{i-1}),$$

describe a system of uncoupled oscillators for $\alpha = 0$. Obviously, in that case, any initial condition, where only a finite number of oscillators have a non-zero amplitude, produces a discrete breather solution. In their celebrated paper of 1994, MacKay and Aubry prove that, under the conditions of nonresonance with the phonon band and nonlinearity of the function $V'(x)$, this solution can be continued to the regime where $\alpha > 0$.

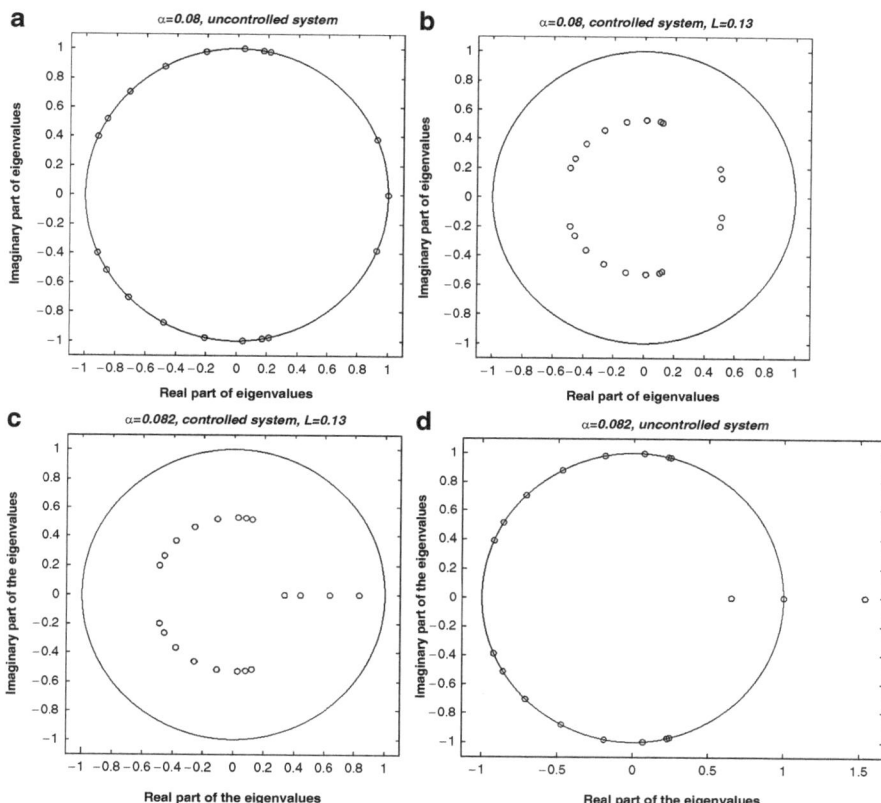

Fig. 7.8 (**a**) As the uncontrolled system approaches a bifurcation point at some critical value $\alpha_{cr} = 0.081$ when $L = 0$, the bifurcation can be avoided as follows: (**b**) Fixing $\alpha = 0.08$ and increasing the control parameter, for example to $L = 0.13$, where the breather solution is asymptotically stable, (**c**) increasing the coupling strength to $\alpha = 0.082$ and (**d**) switching off the control to return to the breather of the uncontrolled system, which is now unstable (after [62])

According to their approach, however, continuation for $\alpha > 0$ is possible, only as long as the eigenvalues of the Floquet matrix of the solution do not cross the value $+1$. This means that the typical occurrence of a bifurcation, through which the breather becomes unstable, prevents such a continuation approach from following the breather beyond that critical α value. This is where the control method proposed in [62] comes to the rescue: As we can see in Fig. 7.8, following a path in $\alpha > 0$ and $L > 0$ space, a discrete breather solution can be continued to a higher value of α, by choosing L in the controlled system such that the solution remains stable!

Therefore, since by the above Proposition one can always find L such that stability is possible for any coupling α, this allows the continuation of any solution of the $\alpha = 0$ case to $\alpha > 0$ values beyond bifurcation, demonstrating the existence of breather solutions in the corresponding parameter regime. The fact that $\hat{u}_n(t)$—which is a stable solution of the controlled system—is by definition also an unstable

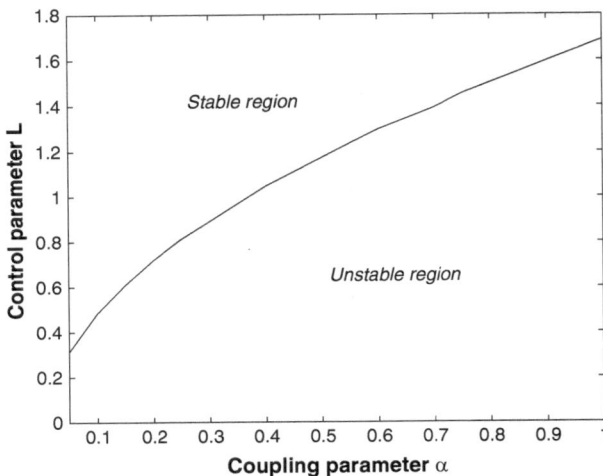

Fig. 7.9 Relation between the control parameter L and the coupling strength α: If α is increased, L has to be larger to achieve stability and successful control. Shown here are the regions for which the breather of Fig. 7.6 is a stable or unstable solution $u_n(t) = \hat{u}_n(t)$ of (7.16) (after [62])

solution of the uncontrolled system, implies that we have succeeded in continuing a discrete breather solution to higher values of α, beyond the bifurcation point.

In Fig. 7.9, we show in (L, α) space the regions of stability of this particular breather. As is well-known by the work of Segur and Kruskal [300], breathers are not expected to exist in these systems in the continuum limit of α going to infinity. At what coupling value though and how do they disappear? Can we use our control aided continuation methods to follow them at arbitrarily high α to answer such questions? These are open research problems that offer interesting applications of the methods we have described in this chapter.

7.4 Disordered Lattices

7.4.1 Anderson Localization in Disordered Linear Media

Let us now turn our attention to the topic of one-dimensional *disordered* lattices that have applications to many physical problems.

One fundamental issue of interest to condensed matter physics was (and still remains) the study of conductivity of electrons in solids. Since in an infinite perfect crystal electrons can propagate ballistically, a natural question to ask is what happens in a more realistic situation when disorder is present in the crystal due to impurities or defects? Will an increase of the degree of disorder lead to a decrease of conductivity, or not? These questions were first answered in a seminal paper by Philip Anderson [10], where it was shown that if disorder is large enough the diffusive motion of the electron will come to a halt. In particular, Anderson studied an unperturbed lattice of uncoupled sites, where the perturbation was considered to be the coupling between them, and randomness was introduced in the on-site energies. For this model he showed that for a large degree of randomness, the transmission of a wave decays exponentially with the length of the lattice.

This absence of wave diffusion in disordered media is nowadays called *Anderson localization*, and is a general wave phenomenon that applies to the transport of different types of classical or quantum waves, like electromagnetic, acoustic and spin waves. Its origin is the wave interference between multiple scattering paths; i.e. the introduction of randomness can drastically disturb the constructive interference, leading to the halting of waves. Anderson localization plays an important role in several physical phenomena. For example, the localization of electrons has dramatic consequences regarding the conductivity of materials, since the medium no longer behaves like a metal, but becomes an insulator when the strength of disorder exceeds a certain threshold. This transition is often referred as the metal-insulator transition.

Although today the significance of Anderson localization is well recognized as indicated for example by the large number of theoretical and numerical papers related to this topic, its physical relevance was not fully realized for many years. The fact that experimental observation of Anderson localization was (and still is) a cumbersome task played an important role in this situation. It is worth mentioning that Anderson himself failed initially to understand the significance of his discovery, as he admits in his Nobel price winning lecture in 1978, where he stated about localization that '... *very few believed it at the time, and even fewer saw its importance, among those who failed to fully understand it at first was certainly its author. It has yet to receive adequate mathematical treatment, and one has to resort to the indignity of numerical simulations to settle even the simplest questions about it.*'

Many theoretical and numerical approaches of localization start with the Anderson model: a standard tight-binding (i.e. nearest-neighbor hopping) system with on-site potential disorder. This can be represented in one dimension by a time-dependent Schrödinger equation

$$i\frac{\partial \psi_l}{\partial t} = \epsilon_l \psi_l - \psi_{l+1} - \psi_{l-1}, \qquad (7.17)$$

where ϵ_l are the random on-site energies, drawn from an uncorrelated uniform distribution in $[-W/2, W/2]$, where W parameterizes the disorder strength and ψ_l is the complex wave function associated with lattice site l. Using the substitution $\psi_l = A_l \exp(-i\lambda t)$ yields a time-independent system of difference equations

$$\lambda A_l = \epsilon_l A_l - A_{l+1} - A_{l-1}, \qquad (7.18)$$

whose solution consists of a set of eigenvectors called the normal mode eigenvectors (NMEs), A_l^ν, (normalized as $\sum_l (A_l^\nu)^2 = 1$), and a set of eigenvalues called the normal frequencies, λ_ν. All eigenvectors are exponentially localized, meaning that their *asymptotic behavior* can be described by an exponential decay $|A_l^\nu| \sim e^{-l/\xi(\lambda_\nu)}$, where $\xi(\lambda_\nu)$ is a characteristic energy-dependent length, called the *localization length*. Naturally, $\xi \to \infty$ corresponds to an extended eigenstate. Several approaches have been developed for the evaluation of ξ, such as the transfer matrix method, schemes based on the transport properties of the lattice, as well as perturbative techniques. For more information on such approaches, the reader is referred to [206] and references therein.

The mathematical description of Anderson localization based on the above formalism is useful for theoretical approaches, but not so much for experimental studies. Clearly, unbounded media do not exist and eigenvalues or eigenstates are rarely measured in real experiments, where mainly measurements of transition and conductivity are performed. So the need for a connection between conductivity and the spectrum of the system becomes apparent. The basic approach towards the fulfilment of this goal was the establishment of a relationship between conductivity and the sensitivity of the eigenvalues of the Hamiltonian of a finite (but very large) system to changes in the boundary conditions [118]. This sensitivity turned out to be conceptually important for the formulation of a scaling theory for localization [5]. The main hypothesis of this single-parameter scaling theory is that, close to the transition between localized and extended states, there should only be one scaling variable which depends on the conductivity for the metallic behavior and the localization length for the insulating behavior. This single parameter turned out to be a dimensionless conductance (often called *Thouless conductance* or *Thouless number*) defined as

$$g(N) = \frac{\delta E}{\Delta E}, \qquad (7.19)$$

where δE is the average energy shift of eigenvalues of a finite system of size N due to the change in the boundary conditions, and ΔE is the average spacing of the eigenvalues. For localized states and large N, δE becomes very small and $g(N)$ vanishes, while in the metallic regime the boundary conditions always influence the energy levels, even in the limiting case of infinite systems. The introduction of the Thouless conductance led to the formulation of a simple criterion for the occurrence of Anderson localization: $g(N) < 1$. In one and two-dimensional random media this criterion can be reached for any degree of disorder by just increasing the size of the medium, while in higher dimensions a critical threshold exists.

The experimental verification of Anderson localization is not easy, for example due to the electron–electron interactions in cases of electron localization, and the difficult discrimination between localization and absorption in experiments of photon localization. Nevertheless, nowadays the observation of Anderson localization has been reported in several experiments, a few of which we quote here: (1) light localization in three-dimensional random media [323, 345], (2) transverse localization of light for two [299] and one [208] dimensional photonic lattices, (3) localization of a Bose-Einstein condensate in an one-dimensional optical potential [45, 284], and (4) elastic waves propagating in a three-dimensional disordered medium [173]. In addition, the observation of the metal-insulator transition in a three-dimensional model with atomic matter waves has been described in [78].

7.4.2   Diffusion in Disordered Nonlinear Chains

As was discussed in the previous subsection, the presence of uncorrelated spatial disorder in one-dimensional linear lattices results in the localization of their NMEs. An interesting question, therefore, that naturally arises is what happens if nonlinearity is introduced to the system. Understanding the effect of nonlinearity on the localization properties of wave packets in disordered systems is a challenging task, which, has attracted the attention of many researchers in recent years [47, 48, 124, 131, 144, 202, 210, 252–255, 271, 311, 317, 338, 339]. Most of these studies consider the evolution of an initially localized wave packet and show that wave packets spread subdiffusively for moderate nonlinearities. On the other hand, for weak enough nonlinearities, wave packets appear to be frozen over the complete available integration time, thereby resembling Anderson localization, at least on finite time scales. Recently, it was conjectured in [178] that these states may be localized for infinite times on Kolmogorov-Arnold-Moser (KAM) torus-like structures in phase space. Whether this is true or not remains open to the present day.

7.4.2.1   Two Basic Models

In order to present the characteristics of subdiffusive spreading, let us consider two one-dimensional lattice models. The first one is a variant of system (7.4). It represents a disordered discrete nonlinear Schrödinger equation (DDNLS) described by the Hamiltonian function

$$H_D = \sum_l \epsilon_l |\psi_l|^2 + \frac{\beta}{2} |\psi_l|^4 - (\psi_{l+1}\psi_l^* + \psi_{l+1}^*\psi_l), \qquad (7.20)$$

in which ψ_l are complex variables, ψ_l^* their complex conjugates, l are the lattice site indices and $\beta \geq 0$ is the nonlinearity strength. The random on-site energies ϵ_l are chosen uniformly from the interval $\left[-\frac{W}{2}, \frac{W}{2}\right]$, with W denoting the disorder strength. The equations of motion are generated by $\dot{\psi}_l = \partial H_D / \partial(i\psi_l^*)$:

$$i\dot{\psi}_l = \epsilon_l \psi_l + \beta |\psi_l|^2 \psi_l - \psi_{l+1} - \psi_{l-1}. \qquad (7.21)$$

This set of equations conserves both the energy of (7.20), and the norm $S = \sum_l |\psi_l|^2$.

The second model we consider is the quartic KG lattice (7.1), described by the Hamiltonian

$$H_K = \sum_l \frac{p_l^2}{2} + \frac{\tilde{\epsilon}_l}{2} u_l^2 + \frac{1}{4} u_l^4 + \frac{1}{2W}(u_{l+1} - u_l)^2, \qquad (7.22)$$

where u_l and p_l respectively are the generalized coordinates and momenta on site l, and $\tilde{\epsilon}_l$ are chosen uniformly from the interval $\left[\frac{1}{2}, \frac{3}{2}\right]$. The equations of motion are $\ddot{u}_l = -\partial H_K / \partial u_l$ and yield

$$\ddot{u}_l = -\tilde{\epsilon}_l u_l - u_l^3 + \frac{1}{W}(u_{l+1} + u_{l-1} - 2u_l). \tag{7.23}$$

This set of equations only conserves the energy H of (7.22). The scalar value $H \geq 0$ serves as a control parameter of nonlinearity, similar to β for the DDNLS case.

For $\beta = 0$ and $\psi_l = A_l \exp(-i\lambda t)$, (7.21) reduces to the linear eigenvalue problem (7.18). The width of the eigenfrequency spectrum λ_ν in (7.18) is $\Delta_D = W + 4$ with $\lambda_\nu \in \left[-2 - \frac{W}{2}, 2 + \frac{W}{2}\right]$. In the limit $H \to 0$ (in practice by neglecting the nonlinear term $u_l^4/4$) the KG model of (7.22) is reduced (with $u_l = A_l \exp(i\omega t)$) to the same linear eigenvalue problem of (7.18), under the substitutions $\lambda = W\omega^2 - W - 2$ and $\epsilon_l = W(\tilde{\epsilon}_l - 1)$. The width of the squared frequency ω_ν^2 spectrum is $\Delta_K = 1 + \frac{4}{W}$ with $\omega_\nu^2 \in \left[\frac{1}{2}, \frac{3}{2} + \frac{4}{W}\right]$. As in the case of DDNLS, W determines the disorder strength.

In the case of weak disorder, $W \to 0$, the localization length of NMEs is approximated by $\xi(\lambda_\nu) \leq \xi(0) \approx 100/W^2$ [206, 207]. On average the NME localization volume (i.e. spatial extent) V, is of the order of $3.3\xi(0) \approx 330/W^2$ for weak disorder and unity in the limit of strong disorder, $W \to \infty$ [207]. The average spacing of eigenvalues of NMEs within the range of a localization volume is given by $d \approx \Delta/V$, with Δ being the spectrum width. The two frequency scales $d \leq \Delta$ determine the packet evolution details in the presence of nonlinearity.

In order to write the equations of motion of Hamiltonian (7.20) in the normal mode space of the system, we insert $\psi_l = \sum_\nu A_{\nu,l} \phi_\nu$ in (7.21), with $|\phi_\nu|^2$ denoting the time-dependent amplitude of the νth mode. Then, using (7.18) and the orthogonality of NMEs the equations of motion (7.21) read

$$i\dot{\phi}_\nu = \lambda_\nu \phi_\nu + \beta \sum_{\nu_1, \nu_2, \nu_3} I_{\nu, \nu_1, \nu_2, \nu_3} \phi_{\nu_1}^* \phi_{\nu_2} \phi_{\nu_3} \tag{7.24}$$

with $I_{\nu, \nu_1, \nu_2, \nu_3} = \sum_l A_{\nu,l} A_{\nu_1,l} A_{\nu_2,l} A_{\nu_3,l}$ being the so-called overlap integral (see Exercise 7.4).

To study the spreading characteristics of wave packets let us order the NMEs by increasing value of the center-of-norm coordinate $X_\nu = \sum_l l A_{\nu,l}^2$ [131, 210, 311, 317]. Then, for DDNLS we follow the normalized norm density distributions $z_\nu \equiv |\phi_\nu|^2 / \sum_\mu |\phi_\mu|^2$, while for KG we monitor the normalized energy density distributions $z_\nu \equiv E_\nu / \sum_\mu E_\mu$ with $E_\nu = \dot{A}_\nu^2/2 + \omega_\nu^2 A_\nu^2/2$, where A_ν is the amplitude of the νth normal mode and ω_ν^2 its squared frequency. Usually these distributions are characterized by means of the second moment $m_2 = \sum_\nu (\nu - \bar{\nu})^2 z_\nu$ (which quantifies the wave packet's spreading width), with $\bar{\nu} = \sum_\nu \nu z_\nu$, and the participation number $P = 1/\sum_\nu z_\nu^2$, (i.e. the number of the strongest excited modes in z_ν). Another often used quantity is the compactness index $\zeta = P^2/m_2$, which quantifies the inhomogeneity of a wave packet. Thermalized distributions have $\zeta \approx 3$, while $\zeta \ll 3$ indicates very inhomogeneous packets, e.g. sparse (with many holes) or partially selftrapped ones (see [317] for more details).

The frequency shift δ of a single site oscillator induced by the nonlinearity for the DDNLS model is $\delta_D = \beta|\psi_l|^2$, while for the KG system the squared frequency shift of a single site oscillator is $\delta_K \sim E_l$, E_l being the energy of the oscillator. Since all NMEs are exponentially localized in space, each effectively couples to a finite number of neighbor modes. The nonlinear interactions are thus of finite range; however, the strength of this coupling is proportional to the norm (energy) density for the DDNLS (KG) model. If the packet spreads far enough, we can generally define two norm (energy) densities: one in real space, $n_l = |\psi_l|^2$ (E_l) and the other in normal mode space, $n_\nu = |\phi_\nu|^2$ (E_ν).

Typically, in order to obtain a statistical description of the wave packet dynamics averaging over hundreds of realizations is performed. As a result, no strong difference is seen between the norms (energies) in real and normal mode space, and therefore, we treat them generally as some characteristic norm (n) or energy (E) density. The frequency shift due to nonlinearity is then $\delta_D \sim \beta n$ for the DDNLS model, while the square frequency shift is $\delta_K \sim E$ for the KG lattice.

7.4.2.2 Regimes of Wave Packet Spreading

Let us now discuss in more detail the different wave packet evolutions. For strong nonlinearities a substantial part of the wave packet is self-trapped. This is due to nonlinear frequency shifts, which will tune the excited sites immediately out of resonance with the non-excited neighborhood [125, 127]. The existence of the selftrapping regime was theoretically predicted for the DDNLS model in [202] (see also [317] for more details). According to the theorem stated in [202], for large enough nonlinearities ($\delta_D > \Delta_D$), single site excitations cannot uniformly spread over the entire lattice. Consequently, a part of the wave packet will remain localized, although the theorem does not prove that the location of this inhomogeneity is constant in time. In fact, partial self-trapping will occur already for $\delta_D \geq 2$ ($\delta_K \gtrsim 1/W$) since at least some sites in the packet may be tuned out of resonance. The selftrapping regime has been numerically observed for single-site excitations [131, 311, 317] and for extended excitations [47, 210], both for the DDNLS and the KG models, despite the fact that the KG system conserves only the total energy H, and the selftrapping theorem can not be applied there.

When selftrapping is avoided for $\delta_D < 2$ ($\delta_K \lesssim 1/W$), two different spreading regimes were predicted in [124] having different dynamical characteristics: an asymptotic weak chaos regime, and a potential intermediate strong chaos one. The basic assumption for this prediction is that the spreading of the wave packet is due to the chaoticity inside the packet. This practically means that a normal mode in a layer of width V in the cold exterior of the wave packet—which borders the packet but will belong to the core of the spreading packet at later times—is incoherently heated by the packet. Numerical verifications of the existence of these regimes were presented in [47, 210, 311].

Let us take a closer look at these behaviors by considering initial "block" wave packets, where L central oscillators of the lattice are excited having the same norm (energy). In the weak chaos regime, for $L \geq V$ and $\delta < d$ most of the NMEs are weakly interacting with each other. Then the subdiffusive spreading of the wave packet is characterized by $m_2 \sim t^{1/3}$. If the nonlinearity is weak enough to avoid selftrapping, yet strong enough to ensure $\delta > d$, the strong chaos regime is realized. Wave packets in this regime initially spread faster than in the case of weak chaos, with $m_2 \sim t^{1/2}$. Note that a spreading wave packet launched in the regime of strong chaos will increase in size, drop its norm (energy) density, and therefore a crossover into the asymptotic regime of weak chaos will occur at later times.

We turn our attention now to the case $L < V$. In this case the wave packet initially spreads over the localization volume V during a time interval $\tau_{in} \sim 2\pi/d$, even in the absence of nonlinearities [124, 210]. The initial average norm (energy) density n_{in} (E_{in}) of the wave packet is then lowered to $n(\tau_{in}) \approx n_{in}L/V$ ($E(\tau_{in}) \approx E_{in}L/V$). Further spreading of the wave packet in the presence of nonlinearities is then determined by these reduced densities. Note that for single-site excitations ($L = 1$), the strong chaos regime completely disappears and the wave packet evolves either in the weak chaos or selftrapping regimes [131, 271, 317, 338].

7.4.2.3 Numerical Results

We present now some numerical results obtained in [210] showing the existence and the main characteristics of weak chaos, strong chaos, the crossover between them and the selftrapping regime (Fig. 7.10). These results were obtained by considering compact DDNLS wave packets at $t = 0$ spanning a width L centered in the lattice, such that within L there is a constant norm density and a random phase at each site (outside the volume L the norm density is zero). In the KG case, this corresponds to exciting each site in the width L with the same energy density, $E = H/L$, i.e. setting initial momenta to $p_l = \pm\sqrt{2E}$ with randomly assigned signs. Ensemble averages over disorder were calculated for $1,000$ realizations with $W = 4$, while $L = V = 21$, and system sizes of $1,000 - 2,000$ sites were considered. For the DDNLS, an initial norm density of $n_{in} = 1$ was taken, so that $\delta_D = \beta$. The values of β (E for KG) were chosen to give the three expected spreading regimes, respectively $\beta \in \{0.04, 0.72, 3.6\}$ and $E \in \{0.01, 0.2, 0.75\}$.

Let us describe the results of Fig. 7.10 referring mainly to the DDNLS model. In the regime of weak chaos we see a subdiffusive growth of m_2 according to $m_2 \sim t^\alpha$ with $\alpha \approx 1/3$ at large times. In the regime of strong chaos we observe exponents $\alpha \approx 1/2$ for $10^3 \lesssim t \lesssim 10^4$ (KG: $10^4 \lesssim t \lesssim 10^5$). Time averages in these regions over the green curves yield $\alpha \approx 0.49 \pm 0.01$ (KG: 0.51 ± 0.02). With spreading continuing in the strong chaos regime, the norm density in the packet decreases, and eventually satisfies $\delta_D \leq d_D$ ($\delta_K \leq d_K$). This results in a dynamical *crossover* to the slower weak chaos subdiffusive spreading. Fits of this decay suggest that $\alpha \approx 1/3$ at $10^{10} \lesssim t \lesssim 10^{11}$ for both models. In the regimes of weak and strong chaos, the

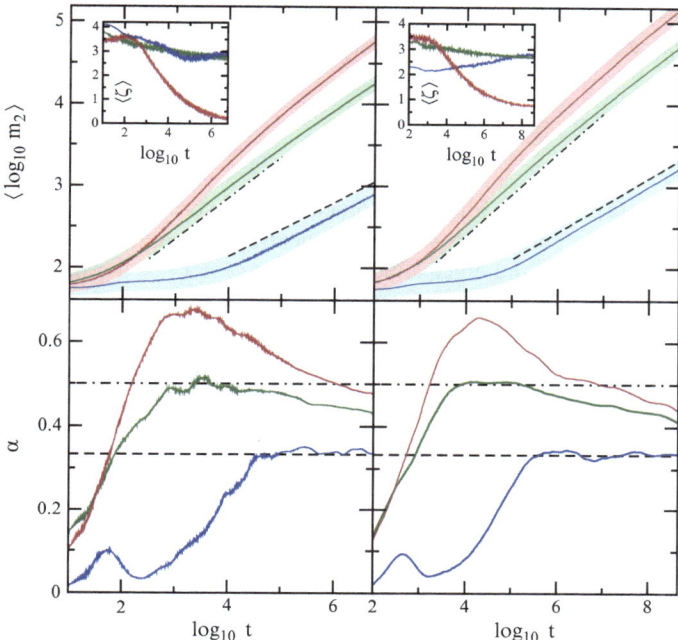

Fig. 7.10 *Upper row*: Average log of second moments (inset: average compactness index) vs. log time for the DDNLS (KG) on the left (*right*), for $W = 4$, $L = 21$. Colors correspond to the three different regimes: (1) weak chaos—*blue*, $\beta = 0.04$ ($E = 0.01$), (2) strong chaos—*green*, $\beta = 0.72$ ($E = 0.2$), (3) self-trapping—*red*, $\beta = 3.6$ ($E = 0.75$). The respective lighter surrounding areas show one-standard-deviation error. *Dashed lines* are to guide the eye to $\sim t^{1/3}$, while *dot-dashed lines* are guides for $\sim t^{1/2}$. *Lower row*: Finite difference derivatives $\alpha(\log t) = d \langle \log m_2 \rangle / d \log t$ for the smoothed m_2 data respectively from above curves (after [210])

compactness index at largest computational times is $\zeta \approx 2.85 \pm 0.79$ (KG: 2.74 ± 0.83), as seen in the blue and green curves of the insets of Fig. 7.10. This means that the wave packet spreads, but remains rather compact and thermalized ($\zeta \approx 3$).

Numerical studies of several additional models of disordered nonlinear one-dimensional lattices demonstrate that in the presence of nonlinearities subdiffusive spreading is always observed, so that the second moment grows initially as $m_2 \sim t^\alpha$ with $\alpha < 1$, showing signs of a crossover to the asymptotic $m_2 \sim t^{1/3}$ law at larger times [47, 210]. Remarkably, subdiffusive spreading was also observed for large disorder strengths, when the localization volume (which defines the number of interacting partner modes) tends to one [47]. Such results support the conjecture that the wave packets, once they spread, do so up to infinite times in a subdiffusive way, bypassing Anderson localization of the linear wave equations. Nevertheless, the validity of this conjecture is still an open issue.

It is worth-mentioning that when the nonlinearity strength tends to very small values, waiting times for wave packet spreading of compact initial excitations increase beyond the detection capabilities of current computational tools. The corresponding question of whether a KAM regime can indeed be approached at finite nonlinearity strength was addressed in [178], but remains still unanswered.

Exercises

Exercise 7.1. (a) Identify all fixed points of the two-dimensional map (7.5) as functions of the parameters ω and γ. Linearize the equations of the map about them and compute the eigenvalues and eigenvectors of the linearized map in every case, thus determining the stability of each point.
(b) Choose ω and γ values such that the origin is a saddle, place many initial conditions on the corresponding eigenvectors and trace out numerically the stable and unstable manifolds emanating out of the point $(0, 0)$ in that case (note that to follow the stable manifold you will need to use the equations of the inverse map).
(c) Computing as accurately as possible one of their primary intersections locate one homoclinic point, whose (forward and backward) iterations should enable you to compute one of the simple breathers of the DNLS equation.

Exercise 7.2. (a) Substituting (7.2) in the KG equation of motion (7.1) derive the infinite dimensional map satisfied by the Fourier coefficients $A_n(k)$ for a periodic solution of the system with frequency ω. Using these equations prove that the origin is of the saddle type provided the following condition is satisfied

$$C(k) = 2 + \frac{K - \omega^2 k^2}{\alpha} > 2, \tag{7.25}$$

thus showing that breathers can exist in the KG system with frequency $\omega_b = \omega$ if $\omega^2 k^2 > K + 4\alpha$ for all $k = 1, 2, \ldots$. It follows, therefore, that if this condition is satisfied for $k = 1$ it will also hold for all other k.

(b) Referring to (7.7), with $C = C(1)$, consult reference [61] and use the method proposed in Exercise 7.1 to compute homoclinic orbits for the cubic map (7.7) that approximate simple breathers of the KG system. Consult also Problem 7.1 below to see how to do this in a systematic way.

Exercise 7.3. Prove Proposition 7.1 of Sect. 7.3.1 reasoning as follows: Let $\mathbf{x} = \hat{\mathbf{x}}(t)$ denote a periodic solution of

$$\ddot{\mathbf{x}} = f(\mathbf{x}). \tag{7.26}$$

We wish to show that there exists an L_c such that for $L > L_c$, $\mathbf{x} = \hat{\mathbf{x}}(t)$ is an asymptotically stable solution of the system

$$\ddot{\mathbf{x}} = f(\mathbf{x}) + L \frac{\mathrm{d}}{\mathrm{d}t}(\hat{\mathbf{x}} - \mathbf{x}). \tag{7.27}$$

To this end, linearize (7.26), writing $\mathbf{x} = \hat{\mathbf{x}}(t) + \boldsymbol{\varepsilon}(t)$ and perform the change of variables $\boldsymbol{\varepsilon}(t) = \boldsymbol{\xi}(t)\exp(-\frac{Lt}{2})$, transforming the linear system into a Hill's type equation for $\boldsymbol{\xi}(t)$, involving the Jacobian matrix $Df(\hat{\mathbf{x}}(t))$, whose coefficients are periodic in time. Now use the fact that the solution of Hill's equation can be written as a linear combination of (n-dimensional) vector periodic functions $\mathbf{P}_k(t), k = 1, 2, \ldots, n$ multiplied by exponentials in the Floquet exponents, β_k. Examining the nature of these exponents, show that the solution $\mathbf{x} = \hat{\mathbf{x}}(t)$ is stable if the control parameter $L > 2|\beta_{\max}|$, where $\beta_{\max}$ is the largest in absolute value of all real β_k, since then the function $\boldsymbol{\varepsilon}(t)$ will tend to zero as time goes to infinity. Hint: Consult [240, 250, 344].

Exercise 7.4. Substituting $\psi_l = \sum_v A_{v,l}\phi_v$ in (7.21), and using (7.18) and the orthogonality of NMEs, prove that the equations of motion of the DDNLS system (7.20) in normal mode space is given by (7.24).

Problems

Problem 7.1. (a) Consider the "soft spring" KG potential

$$V(x) = \frac{1}{2}Kx^2 - \frac{1}{4}x^4. \tag{7.28}$$

Use this form of $V(x)$ in the equations of motion (7.1) to search for periodic solutions of the form (7.2) with $\omega_b = \omega$, equating coefficients of $exp(ik\omega t)$. Thus, derive the following algebraic system for the coefficients $A_n(k)$

$$A_{n+1}(k) + A_{n-1}(k) = C(k)A_n(k) - \frac{1}{\alpha}\sum_{k_1}\sum_{k_2}\sum_{k_3}A_n(k_1)A_n(k_2)A_n(k_3),$$

$$k_1 + k_2 + k_3 = k, \tag{7.29}$$

for every $k = 1, 2, \ldots, M$, with

$$C(k) = 2 + \frac{K - k^2\omega^2}{\alpha} \tag{7.30}$$

assuming that the Fourier coefficients $A_n(k)$ decrease sufficiently rapidly with n, so that M of them are enough to describe the dynamics. As explained in Sect. 7.2, a necessary condition for the existence of breathers is that the trivial solution of (7.29), i.e. $A_n(k) = 0$, for all n, k, be a saddle point of the $2M$-dimensional map (7.29).

(b) Linearize (7.29) near the trivial point and prove that it is a saddle with an M-dimensional stable and an M-dimensional unstable manifold, if and only if $|C(k)| > 2$ for all $k = 1, 2, \ldots, M$. Show that the smallest value of

ω that makes it possible for a breather with this frequency to exist is $\omega = \sqrt{K - \alpha(C(1) - 2)}$. Consider now the simple approximation that the breather is represented by a single mode only, i.e. $u_n(t) = 2A_n(1)cos\omega t$, for $-M < n < M$, and scale the Fourier coefficients by $A_n(1) = \sqrt{\alpha} A_n$ to obtain the map

$$A_{n+1} + A_{n-1} + C(1)A_n = -3A_n^3, \tag{7.31}$$

(see also (7.7)).

(c) Analyzing (7.31), show that the only possibility for breathers is $C(1) > 2$, with frequency $\omega < \sqrt{K}$ *below* the phonon band. Explain why the parameter range for breathers of (7.28) is considerably more limited than the "hard spring" case (see (7.1)), where $\omega > \sqrt{K + 4\alpha}$. Write the above map in the two-variable form $A_{n+1} = -B_n - C(1)A_n - 3A_n^3$, $B_{n+1} = A_n$ and determine the eigenvectors of its linearized equations about the origin $(0,0)$. Now use the method described in Exercise 7.1 to trace out numerically the invariant manifolds of the origin, locate their intersections and compute some of the breathers and multibreathers. Start with accurate representations of their $u_n(0) = 2A_n(1)$ and $\dot{u}_n(0) = 0$ initial conditions and integrate the equations of motion (see (7.1)) for long times to determine numerically the stability of these solutions, under small changes of their initial conditions.

Problem 7.2. (a) Use the approach outlined in Sect. 7.2.1 to compute homoclinic orbits of the map (7.5), corresponding to symmetric breathers and multi-breathers of the DNLS (7.4), as follows: Solve (7.10), noting that, once $\mathbf{x}_{-N}$ is known as an initial condition, the values of $\mathbf{x}_1$, $\mathbf{x}_0$ and $\mathbf{x}_{-1}$ are uniquely obtained by direct iteration of the map (7.8). Observe next that setting $\mathbf{x}_1 = F(\mathbf{x}_0)$ and $\mathbf{x}_{-1} = F^{-1}(\mathbf{x}_0)$, (7.10) can be solved for $\mathbf{x}_0$, by looking for solutions of the (7.11), whose number of unknowns is d, since $Z(\mathbf{x}_0) = 0$ implies $\mathbf{x}_0 = M\mathbf{x}_0$ (i.e. the second component of $\mathbf{x}_0$ is zero, $\mathbf{w}_0 = 0$).

(b) Express now the desired solution $\mathbf{x}_{-N}$, lying on the d-dimensional unstable manifold, by a coordinate σ satisfying $Z(\sigma) = 0$. Since $\mathbf{x}_0 = F^N(\mathbf{x}_{-N})$ this also completely determines all the unknowns in (7.11) and hence every solution of this equation defines an orbit of (7.8) having the desired asymptotic behavior. One convenient way to define the unknown $\sigma = (\sigma_1, \ldots, \sigma_d)$ is to approximate the state $\mathbf{x}_{-N}$ by a point lying on the Euclidean unstable manifold of the linearized equations and express it as a linear combination of the eigenvectors $\mathbf{E}_i^u$, writing $\mathbf{x}_{-N} = \varepsilon \sum_{i=1}^{d} \sigma_i \mathbf{E}_i^u + \mathbf{x}_{eq}$, where $|\lambda_i| > 1$ are the corresponding eigenvalues and $|\epsilon| \ll 1$ a scalar parameter denoting the accuracy of the computation. Thus, you may now substitute this expression in $Z(\sigma) = 0$ and solve for the resulting d equations with d unknowns, using an appropriate Newton iterative method.

Chapter 8
The Statistical Mechanics of Quasi-stationary States

Abstract This chapter adopts an altogether different approach to the study of chaos in Hamiltonian systems. We consider, in particular, probability distribution functions (pdfs) of sums of chaotic orbit variables in different regions of phase space, aiming to reveal the statistical properties of the motion in these regions. If the orbits are strongly chaotic, these pdfs tend to a Gaussian and the system quickly reaches an equilibrium state described by Boltzmann-Gibbs statistical mechanics. There exist, however, many interesting regimes of weak chaos characterized by long-lived quasi-stationary states (QSS), whose pdfs are well-approximated by q-Gaussian functions, associated with nonextensive statistical mechanics. In this chapter, we study such QSS for a number of N dof FPU models, as well as 2D area-preserving maps, to locate weakly chaotic QSS, investigate the complexity of their dynamics and discover their implications regarding the occurrence of dynamical phase transitions and the approach to thermodynamic equilibrium, where energy is equally shared by all degrees of freedom of the system.

8.1 From Deterministic Dynamics to Statistical Mechanics

As the reader has undoubtedly noticed, we have regarded so far in this book the solutions of Hamiltonian systems as individual trajectories (or orbits), evolving in a deterministic way, according to Newton's equations of motion. In other words, we have systematically chosen initial conditions locally in various areas of phase space and have sought to characterize the resulting orbits as "ordered" or "chaotic", aiming to understand the dynamics more globally in phase space. By order, we mean that the corresponding domain is predominantly occupied by invariant tori, while chaos implies that extremely sensitive dependence on initial conditions prevails in the region under study.

Clearly, however, if we seriously wish to shift our attention from local to global dynamics, such characterizations are too simplistic. As we have already seen in

T. Bountis and H. Skokos, *Complex Hamiltonian Dynamics*,
Springer Series in Synergetics, DOI 10.1007/978-3-642-27305-6_8,
© Springer-Verlag Berlin Heidelberg 2012

earlier chapters, order and chaos are globally much more complex than what their local manifestations might entail. In Chap. 6, for example, we realized that s-dimensional invariant tori of N dof Hamiltonians, with $s \ll N$, owing to their localization properties in Fourier space, turn out to be very important in our attempt to understand the (global) phenomenon of FPU recurrences. Indeed, quasiperiodic orbits of as few as $s = 2, 3, \ldots$ (rationally independent) frequencies were found to dominate the dynamics at low energies to the extent that, even when these orbits become unstable and the corresponding tori break down, they still represent quasi-stationary states, which preclude the onset of energy equipartition for very long times.

A different kind of localized solutions (this time in configuration space), known as discrete breathers, was discussed in Chap. 7. They were also found to hinder energy equipartition, through the persistence of stable periodic oscillations, in which very few particles participate with appreciable amplitudes! Indeed, discrete breathers are observed in many cases to be surrounded by low-dimensional tori and clearly constitute one more example where local order influences global dynamics in multi-dimensional Hamiltonian systems.

One may, therefore, well wonder: since there do exist hierarchical levels of order, could there also exist hierarchies of disorder, where chaotic orbits are confined, for very long times, in limited domains of phase space? Such regimes may be characterized, for example, by small LCEs and weakly chaotic dynamics, compared with regimes of strong chaos, which are larger and possess higher LCEs. As we will show in the present chapter, such distinction can indeed be made, but we must be prepared to combine our deterministic methods with the probabilistic approach of statistical mechanics.

As is well-known, the statistical analysis of dynamical systems has a long history. Probability density functions (pdfs) of chaotic orbits have been studied for many decades and by many authors, aiming to understand the transition from deterministic to stochastic (or ergodic) behavior [11, 20, 117, 184, 185, 266, 268, 289–291, 308]. In this context, the fundamental question that arises concerns the existence of an appropriate invariant probability density (or ergodic measure), characterizing phase space regions where solutions generically exhibit chaotic dynamics. If such an invariant measure can be established for almost all initial conditions (i.e. except for a set of measure zero), one has a firm basis for studying the system at hand from a statistical mechanics point of view.

If, additionally, this invariant measure turns out to be a continuous and sufficiently smooth function of the phase space coordinates, one can invoke the Boltzmann-Gibbs (BG) microcanonical ensemble and attempt to evaluate all relevant quantities at thermal equilibrium, like partition function, free energy, entropy, etc. On the other hand, if the measure is absolutely continuous (as e.g. in the case of the so-called Axiom A dynamical systems), one might still be able to use the formalism of ergodic theory and Sinai-Ruelle-Bowen measures to study the statistical properties of the model [117].

In all these cases, viewing the values of one, or a linear combination of components of a chaotic trajectory at discrete times t_n, $n = 1, \ldots, N$ as realizations of several independent and identically distributed random variables X_n and calculating the distribution of their sums in the context of the Central Limit Theorem (CLT) [282] one expects to find a Gaussian, whose mean and variance are those of the X_n's. This is indeed what happens in many chaotic dynamical systems studied to date which are ergodic, i.e. almost all their orbits (except for a set of measure zero) pass arbitrarily close to any point of the constant energy manifold, after sufficiently long times. What is also true is that, in these cases, at least one LCE is positive, stable periodic orbits are absent and the constant energy manifold is covered uniformly by chaotic orbits, for all but a (Lebesgue) measure zero set of initial conditions.

But then, what about chaotic regions of limited extent, at energies where the MLE is "small" and stable periodic orbits are present, whose islands of invariant tori and sets of cantori around them occupy a positive measure subset of the energy manifold? In such regimes of weak chaos, it is known that many orbits "stick" for long times to the boundaries of these islands and chaotic trajectories diffuse slowly through multiply connected regions in a highly non-uniform way [8, 89, 249]. Such examples occur in many physically realistic systems studied in the current literature (see e.g. [131, 177, 316, 317]).

These are cases where exponential separation of nearby solutions is not uniform, exponential decay of correlations is not generic and chaotic orbits can no longer be viewed as independent and/or identically distributed random variables. What kind of pdfs would we expect from a computation of their sums, if not Gaussians? This type of Hamiltonian dynamics is strongly reminiscent of many examples of physical systems governed by long range interactions, like self-gravitating systems of finitely many mass points, interacting black holes and ferromagnetic spin models, in which power laws are dominant over exponential decay [332].

8.1.1 Nonextensive Statistical Mechanics and q-Gaussian pdfs

As we also discussed in Sect. 1.4, multi-particle systems belong to different universality classes, according to their statistical properties at equilibrium. In the most widely studied class, if the system can be at any one of $i = 1, 2, \ldots, W$ states with probability p_i, its entropy is given by the BG formula

$$S_{BG} = -k \sum_{i=1}^{W} p_i \ln p_i, \tag{8.1}$$

where k is Boltzmann's constant, under the constraint

$$\sum_{i=1}^{W} p_i = 1. \tag{8.2}$$

As is well-known, the BG entropy is *additive* in the sense that, for any two independent systems A and B, the entropy of their sum is the sum of the individual entropies, i.e. $S_{BG}(A + B) = S_{BG}(A) + S_{BG}(B)$. It is also *extensive*, as it grows linearly with the number of dof N, as $N \to \infty$. These properties are associated with the fact that different parts of BG systems are highly uncorrelated and their dynamics is statistically uniform in phase space.

There is, however, an abundance of physical systems characterized by strong correlations, for which the assumptions of extensivity and additivity are not generally valid [332]. In fact, as we explain in this chapter, Hamiltonian systems provide a wealth of examples governed by such complex statistics, especially near the boundaries of islands of ordered motion, where orbits "stick" for very long times and the dynamics becomes very weakly chaotic. It is for this kind of situations that Tsallis proposed the entropy formula

$$S_q = k \frac{1 - \sum_{i=1}^{W} p_i^q}{q - 1} \text{ with } \sum_{i=1}^{W} p_i = 1, \qquad (8.3)$$

that depends on an index q, for a set of W states with probabilities p_i $i = 1, \ldots, W$, obeying the constraint (8.2). The S_q entropy is not additive, since $S_q(A + B) = S_q(A) + S_q(B) + k(1-q)S_q(A)S_q(B)$ and generally not extensive. The pdf replacing the Gaussian in this type of nonextensive statistical mechanics is the q-Gaussian distribution

$$P(s) = a \exp_q(-\beta s^2) \equiv a \left[1 - (1 - q)\beta s^2 \right]^{\frac{1}{1-q}} \qquad (8.4)$$

obtained as an extremum (maximum for $q > 0$ and minimum for $q < 0$) of the Tsallis entropy (8.3), under appropriate constraints [332]. The q index satisfies $1 < q < 3$ to make (8.4) normalizable, β is an arbitrary parameter and a a normalization constant. Note that in the limit $q \to 1$ (8.4) tends to the Gaussian distribution, i.e. $\exp_q(-\beta x^2) \to \exp(-\beta x^2)$.

The above approach does not constitute, of course, the only possible choice for analyzing the statistics of strongly correlated systems at thermal equilibrium. As Tsallis points out in his book [332], various entropic forms have been proposed by different authors as alternatives to the BG entropy. S_q, however, turns out to enjoy a number of important properties, also shared by S_{BG}, that appear to render it superior to other choices. For example, it satisfies uniqueness theorems analogous to those of Shannon and Khinchin obeyed by S_{BG} and is Lesche stable (i.e. robust under small variations of the state probabilities p_i) and concave for all $q > 0$. By contrast, the Rényi entropy, for example, defined by [281]

$$S_q^R = k \frac{\ln \sum_{i=1}^{W} p_i^q}{1 - q}, \qquad (8.5)$$

although additive, fails to satisfy many other important requirements of entropic forms like concavity and Lesche stability [332].

What we wish to do in this chapter is study the complex statistics of several multi-dimensional Hamiltonian systems involving either nearest neighbor interactions or long range forces. We focus on weakly chaotic regimes and demonstrate by means of numerical experiments, in the spirit of the CLT, that pdfs of sums of orbit components *do not* rapidly converge to a Gaussian, but are well approximated, for long integration times, by the q-Gaussian distribution (8.4). At longer times, of course, chaotic orbits generally leak out of smaller regions to larger chaotic seas, where obstruction by islands and cantori is less dominant and the dynamics is more uniformly ergodic. This transition is signaled by the q-index of the distribution (8.4) decreasing towards $q = 1$, which represents the limit at which the pdf becomes a Gaussian.

Thus, in our models, q-Gaussian distributions represent *quasi-stationary states* (QSS) that are often very long-lived, especially inside "thin" chaotic layers near periodic orbits that have just turned unstable. This suggests that it might be useful to study these pdfs (as well as their associated q values) for sufficiently long times and try to derive useful information regarding these QSS directly from the chaotic orbits, at least for time intervals accessible by numerical integration. QSS have also been studied in coupled standard maps and the so-called Hamiltonian Mean Field (HMF) model in [22, 23], but not from the viewpoint of sum distributions.

One must be very careful, however, with regard to the kind of functions one uses to approximate these pdfs. While it is true that q-Gaussians offer, in general, a quite accurate representation of QSS sum distributions, they are *not* the only possible choice. In fact, it has already been shown in certain systems that another so-called "crossover" function [77, 332, 333] can describe the same data with better accuracy. As pointed out in many references [108, 166, 167], there are cases where other functions can be used to approximate better sum distributions of weakly chaotic QSS.

8.2 Statistical Distributions of Chaotic QSS and Their Computation

Let us start again with an autonomous N dof Hamiltonian function of the form

$$H \equiv H(\mathbf{q}(t), \mathbf{p}(t)) = H(q_1(t), \dots, q_N(t), p_1(t), \dots, p_N(t)) = E \qquad (8.6)$$

where $(q_k(t), p_k(t))$ are the positions and momenta respectively representing the solutions in phase space at time t. What we wish to study here is the statistical properties of these solutions in regimes of weakly chaotic motion, where the Lyapunov exponents [31, 32, 117, 310] are positive but very small. Such situations arise often when one considers orbits which diffuse into thin chaotic layers and wander through a complicated network of "islands", sticking for very long times to the boundaries of these islands on a surface of constant energy.

There are several interesting questions one would like to ask here: How long do these weakly chaotic states last? Assuming they are quasi-stationary, can we describe them statistically by observing some of their chaotic orbits? What type of distributions characterize these QSS and how could one connect them to the actual dynamics of the corresponding Hamiltonian system? Can we use these statistical considerations to understand important *physical* properties of system (8.6) like energy equipartition? Is there any connection between the statistics of these QSS and the breakdown of certain localization phenomena discussed in earlier chapters?

The approach we shall follow is in the spirit of the well-known Central Limit Theorem [282] and is described in detail in [18]. In particular, Hamilton's equations of motion will be solved numerically for a large set of initial conditions to construct distributions of suitably rescaled sums of M values of a generic observable $\eta_i = \eta(t_i)$, $i = 1, \ldots, M$, which depends linearly on the components of the solution. If these are viewed as different random variables (in the limit $M \to \infty$), we may evaluate their sum

$$S_M^{(j)} = \sum_{i=1}^{M} \eta_i^{(j)} \tag{8.7}$$

for $j = 1, \ldots, N_{\text{ic}}$ initial conditions. Thus, we can analyze the statistics of these sums, centered about their mean value and rescaled by their standard deviation σ_M, writing them as

$$s_M^{(j)} \equiv \frac{1}{\sigma_M} \left(S_M^{(j)} - \langle S_M^{(j)} \rangle \right) = \frac{1}{\sigma_M} \left(\sum_{i=1}^{M} \eta_i^{(j)} - \frac{1}{N_{\text{ic}}} \sum_{j=1}^{N_{\text{ic}}} \sum_{i=1}^{M} \eta_i^{(j)} \right) \tag{8.8}$$

over the N_{ic} initial conditions.

Plotting now the normalized histogram of the probabilities $P(s_M^{(j)})$ as a function of $s_M^{(j)}$, we compare our pdfs with various functional forms found in the literature. If the variables are independent and identically distributed, as explained in the previous section, we expect a Gaussian. What we find, however, in many cases, is that the data is much better approximated by a q-Gaussian function of the form

$$P(s_M^{(j)}) = a \exp_q(-\beta s_M^{(j)2}) \equiv a \left[1 - (1-q)\beta s_M^{(j)2} \right]^{\frac{1}{1-q}} \tag{8.9}$$

(see (8.4)). The normalization of this pdf is achieved by setting

$$\beta = a\sqrt{\pi} \frac{\Gamma\left(\frac{3-q}{2(q-1)}\right)}{(q-1)^{\frac{1}{2}} \Gamma\left(\frac{1}{q-1}\right)}, \tag{8.10}$$

where Γ is the Euler Γ function, showing that the allowed values of q are $1 < q < 3$.

Let us now describe the numerical procedure one may use to calculate these pdfs. First, we specify an observable denoted by $\eta(t)$ as one (or a linear combination) of the components of the position vector $\mathbf{q}(t)$ of a chaotic solution of Hamilton's equations of motion, located initially at $(\mathbf{q}(0), \mathbf{p}(0))$. Assuming that the orbit visits all parts of a QSS during the integration interval $0 \le t \le t_{\mathrm{f}}$, we divide this interval into N_{ic} equally spaced, consecutive time windows, which are long enough to contain a significant part of the orbit. Next, we subdivide each such window into a number M of equally spaced subintervals and calculate the sum $S_M^{(j)}$ of the values of the observable $\eta(t)$ at the right edges of these subintervals (see (8.7)). In this way, we treat the point at the beginning of every time window as a new initial condition and repeat this process N_{ic} times to obtain as many sums as required to have reliable statistics. Consequently, at the end of the integration, we compute the average and standard deviation of the sums (8.7), evaluate the N_{ic} rescaled quantities $s_M^{(j)}$ and plot the histogram $P(s_M^{(j)})$ of their distribution.

As we shall see in the next sections, in regions of weak chaos these distributions are well-fitted by a q-Gaussian of the form (8.9) for fairly long time intervals. However, for longer times, the orbits often diffuse to wider domains of strong chaos and the well-known form of a Gaussian pdf is recovered.

8.3   FPU π-Mode Under Periodic Boundary Conditions

One very popular Hamiltonian to which we can apply our approach is the one-dimensional lattice of N particles with nearest neighbor interactions governed by the FPU-β Hamiltonian [121]

$$H(\mathbf{q}, \mathbf{p}) = \frac{1}{2} \sum_{j=1}^{N} p_j^2 + \sum_{j=1}^{N} \left[\frac{1}{2}(q_{j+1} - q_j)^2 + \frac{1}{4}\beta(q_{j+1} - q_j)^4 \right] = E \quad (8.11)$$

under periodic boundary conditions $q_{N+j}(t) = q_j(t)$, $j = 1, \ldots, N$, which we have studied extensively in this book. More specifically, we shall first concentrate on orbits starting near a well-known NNM of this system called the π-mode, which has been studied in detail in several publications [16, 66, 67, 105, 218, 272]. This simple periodic solution is defined by (3.22)

$$q_j(t) = -q_{j+1}(t) \equiv q(t), \quad j = 1, \ldots, N \quad (8.12)$$

with N even.

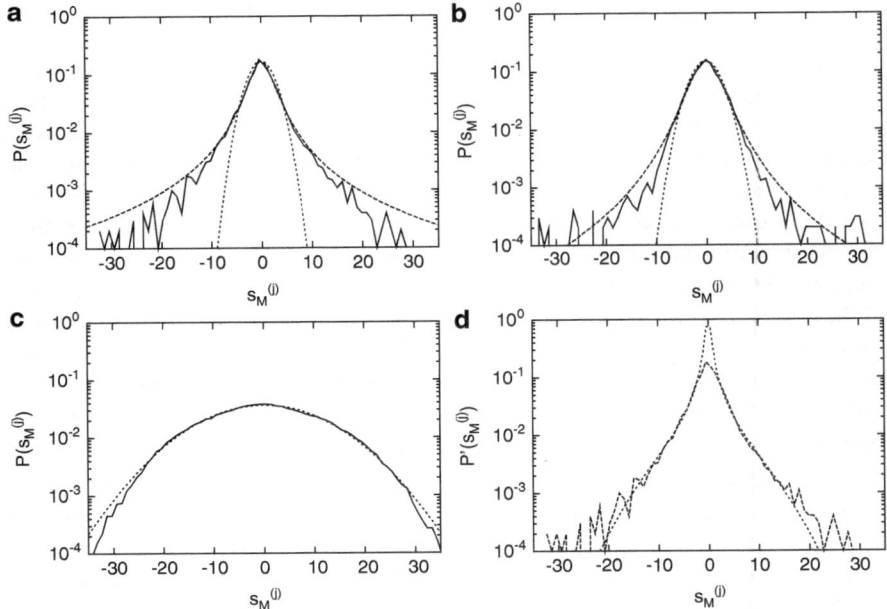

Fig. 8.1 Plot in linear-log scale of numerical (*solid curve*), q-Gaussian (*dashed curve*) and Gaussian (*dotted curve*) distributions for the FPU π-mode with periodic boundary conditions, with $N = 128$, $\beta = 1$ and $E = 0.768 > E_u \approx 0.0256$. Panel (**a**) corresponds to final integration time $t_f = 10^5$ using $N_{ic} = 10^4$ time windows and $M = 10$ terms in the computation of the sums. Here, the numerical fitting with a q-Gaussian gives $q \approx 1.818$ with $\chi^2 \approx 0.0007$. Panel (**b**) corresponds to $t_f = 10^6$, $N_{ic} = 10^4$ and $M = 100$ and the numerical fitting gives $q \approx 1.531$ with $\chi^2 \approx 0.0004$. Panel (**c**) corresponds to $t_f = 10^8$, $N_{ic} = 10^5$ and $M = 1,000$. It is evident that the numerical distribution (*solid curve*) has almost converged to a Gaussian (*dotted curve*). Panel (**d**) compares the same pdf as in panel (**a**) with the $\tilde{P}$ function of (8.14) for $a_1 \approx 0.009$, $a_q \approx 2.849$ and $q \approx 2.179$ with $\chi^2 \approx 0.00008$ (*dashed curve*) (after [18])

Our aim here is to investigate chaotic states near this orbit at energies where it has just become unstable. To this end, let us choose as our observable the quantity

$$\eta(t) = q_{\frac{N}{2}}(t) + q_{\frac{N}{2}-1}(t) \tag{8.13}$$

which satisfies $\eta(t) = 0$ at the π-mode. Thus, starting close to (8.12), $\eta(t)$ remains near zero at energies where the mode is stable and grows in magnitude at energies where the π-mode has destabilized, i.e. $E > E_u$. To compare with results published in the recent literature (see e.g. [218]), we first consider the case $N = 128$ and $\beta = 1$, for which $E_u \approx 0.0257$ [16] and take as our total energy $E = 0.768$ (i.e. $\varepsilon = E/N = 0.006$), at which the π-mode is certainly unstable. The accuracy of the integration performed in [18] is determined at each time step by requiring that $H(\mathbf{q}(t), \mathbf{p}(t))$ is within 10^{-5} from the energy value set initially at time $t = 0$.

As we see in Fig. 8.1, when the total integration time t_f is increased, the pdfs (solid curves) approach closer and closer to a Gaussian with q tending to 1.

Moreover, this seems to be independent of the values of N_{ic} and/or M, at least up to the final integration time $t_{\mathrm{f}} = 10^8$. For example, when the parameters N_{ic} and M in Fig. 8.1b are varied, one obtains q-Gaussians of very similar shape with q between 1.51 and 1.67. It is important to note, however, that the same data may be better fitted by other similar looking functions: For example, in Fig. 8.1d the numerical distribution (solid curve) is more accurately approximated by the so-called "crossover" function [77, 332, 333]

$$\tilde{P}(s_M^{(j)}) = \frac{1}{\left\{1 - \frac{a_q}{a_1} + \frac{a_q}{a_1} \exp[(q-1)a_1 s_M^{(j)2}]\right\}^{\frac{1}{q-1}}}, \quad a_1, a_q \geq 0 \text{ and } q > 1, \quad (8.14)$$

where $a_1 \approx 0.009$, $a_q \approx 2.849$ and $q \approx 2.179$ with $\chi^2 \approx 0.00008$, in contrast to the $\chi^2 \approx 0.0007$ obtained by fitting the same distribution by a q-Gaussian with $q \approx 1.818$ (see Fig. 8.1a). We note that χ^2 denotes the well known chi-square test of reliability of a given statistical hypothesis, see e.g. [157]. Equation 8.14 represents a "crossover" between q-Gaussians and Gaussians and takes into account finite size (and time) effects reflected in the lowering of the tails of the corresponding distributions.

Regarding this QSS, it is interesting to note that when the final integration time is increased beyond $t_f \approx 4 \times 10^7$, one observes that the LCEs monotonically increase and attain bigger values than those computed up to $t_f \simeq 4 \times 10^7$ (see Fig. 8.2b). This may signify that the trajectories drift away from the neighborhood of the π-mode and enter a larger chaotic subspace of the energy manifold.

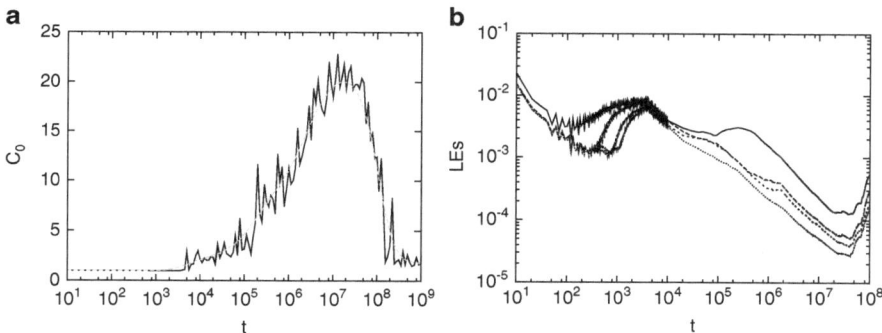

Fig. 8.2 (**a**) Plot of C_0, (8.15), as a function of time for the unstable ($E = 0.768 > E_u \approx 0.0257$) π-mode with $\beta = 1$ and $N = 128$. The *grey curve* corresponds to the time average of the *solid curve*. (**b**) Log-log plot of the four biggest LCEs as a function of time for the same parameters as in panel (**a**) (after [18])

Similar results for this system have also been obtained by other authors using an analogous methodology [218]. They too find that if one takes $t_f = 10^6$ as the longest integration time, one obtains pdfs that are well approximated by q-Gaussians. Their transitory character, however, is revealed when one integrates up to $t_f = 10^7$ or longer, when the pdfs are clearly seen to approach a Gaussian.

8.3.1 Chaotic Breathers and the FPU π-Mode

As we discovered in Chap. 6, physical properties like energy equipartition among all degrees of freedom have been at the center of many investigations of FPU systems. In one such study, focusing on the π-mode under periodic boundary conditions [105], the authors studied the time evolution of an FPU-β chain towards equipartition using initial conditions close to that mode. They observed that at energies well above its threshold of destabilization, a remarkable localization phenomenon occurs: A large amplitude excitation spontaneously occurs (strongly resembling a discrete breather), which has a finite lifetime and moves chaotically along the chain, keeping all the energy restricted among very few particles. Moreover, numerical results suggest that these "chaotic" breathers break down just before the system reaches energy equipartition.

As pointed out in [105], this phenomenon can be monitored by evaluating the function

$$C_0(t) = N \frac{\sum_{n=1}^{N} E_n^2}{(\sum_{n=1}^{N} E_n)^2}, \qquad (8.15)$$

where E_n denotes the energy per site

$$E_n = \frac{1}{2} p_n^2 + \frac{1}{2} V(q_{n+1} - q_n) + \frac{1}{2} V(q_n - q_{n-1}) \qquad (8.16)$$

with $1 \leq n \leq N$ and $V(x) = \frac{1}{2}x^2 + \frac{\beta}{4}x^4$. Since $C_0(t) = 1$, if $E_n = E/N$ at each site n and $C_0(t) = N$ if the energy is localized at only one site, it can serve as an efficient indicator of energy localization in the chain.

In their experiments, these authors used $N = 128$, $\beta = 0.1$, $E \approx 42.2707$ well above the lowest destabilization energy of the π-mode $E_u \approx 0.25725$ and plotted C_0 versus t. Distributing evenly the energy among all sites of the π-mode at $t = 0$, they observed that C_0 initially grows to relatively high values, indicating that the energy localizes at a few sites. After a certain time, however, C_0 reaches a maximum and decreases towards an analytically derived asymptotic value $\bar{C}_0 \approx 1.795$ [105], which is associated with the breakdown of the chaotic breather and the onset of energy equipartition in the chain.

In [18], the same study of the π-mode was performed at a lower energy, $E = 0.768 > E_u \approx 0.0257$ (keeping $\beta = 1$, $N = 128$) and QSS were approximated by q-Gaussian distributions, which are related to the lifetime of chaotic breathers as follows: In Fig. 8.2a, C_0 is plotted as a function of time, verifying indeed that it grows over a long interval ($t \approx 1.8 \times 10^8$) after which it starts to decrease and eventually tends to the asymptotic value $\bar{C}_0 \simeq 1.795$ associated with energy equipartition. In Fig. 8.2b it is shown that the four biggest LCEs decrease towards zero until about $t \approx 4 \times 10^7$, but then start to increase towards positive values indicating the unstable character of the π-mode.

In Fig. 8.3 we present the instantaneous energies E_n of all $N = 128$ sites at different times, together with their associated sum distributions. As shown in Fig. 8.3a, the energy becomes localized at only a few sites, demonstrating the occurrence of a chaotic breather at times of order $10^4 \lesssim t \lesssim 10^8$. Next, Fig. 8.3b shows that when the time is further increased (e.g. to $t = 6 \times 10^8$), the chaotic breather is destroyed and the system reaches equipartition.

Comparing Figs. 8.1, 8.2a and 8.3, we deduce that while the chaotic breather still exists, a QSS is observed fitted by q-Gaussians with q well above unity, as seen in Fig. 8.3c. However, as the chaotic breather breaks down for $t > 10^8$ and energy equipartition is reached (see Fig. 8.3b), $q \to 1$ and q-Gaussian distributions rapidly converge to Gaussians (see Fig. 8.1c) in full agreement with what is expected from BG statistical mechanics.

In Fig. 8.3d we plot estimates of the q values obtained, when one computes chaotic orbits near the π-mode at this energy density $\epsilon = E/N = 0.006$. Remarkably enough, even though these values have an error bar of about $\pm 10\%$ due to the different statistical parameters M, N_{ic}, used in the computation, they exhibit a clear tendency to fall closer to 1 for $t > 10^7$, where energy equipartition is expected to occur.

It is indeed a hard and open problem to determine exactly how equipartition times T_{eq} scale with the energy density $\epsilon = E/N$ and other parameters (like β), particularly in the thermodynamic limit, even in one-dimensional FPU lattices. Although it is a question that has long been studied in the literature [25, 35, 37, 105, 111], the precise scaling exponents by which T_{eq} depends on ϵ, β, etc. are not yet precisely known for general mode excitations.

It would be very interesting if our q values could help in this direction. In fact, carrying out more careful calculations, as in Fig. 8.3d, at other values of the specific energy, e.g. $\epsilon = 0.04$ (see Fig. 8.4) and $\epsilon = 0.2$, one finds q plots that exhibit a clear decrease to values close to 1, at $T_{eq} \approx 7.5 \times 10^5$ and $T_{eq} \approx 7.5 \times 10^4$ respectively, approximately where the corresponding chaotic breathers collapse. Still, even though these results are consistent with what is known in the literature, the limited accuracy of the above approach does not allow one to say something meaningful about scaling laws, as the system tends to equilibrium.

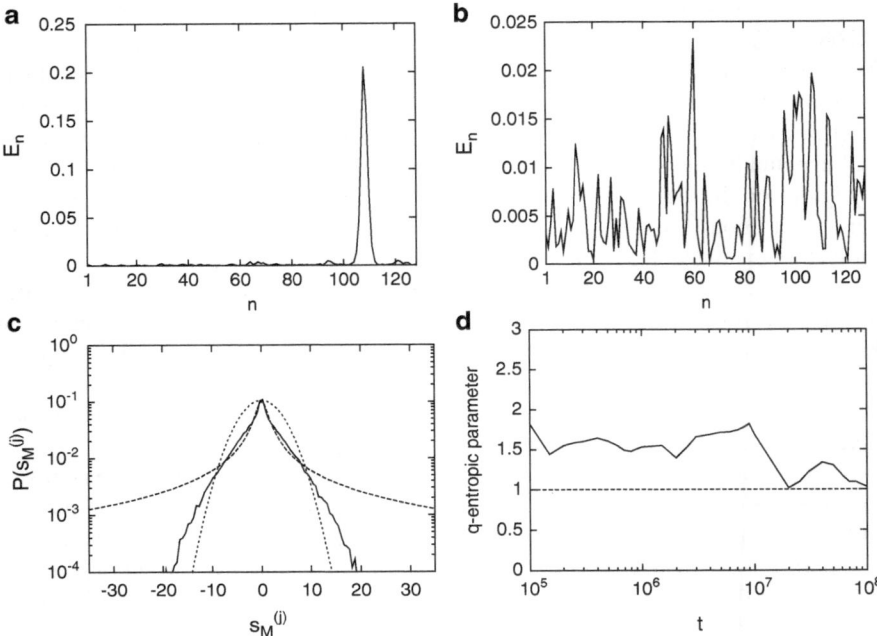

Fig. 8.3 In panel (**a**) at $t = 10^7$, near the maximum of $C_0(t)$ (see Fig. 8.2a), we see a chaotic breather. In panel (**b**) at $t = 6 \times 10^8$, this breather has collapsed and the system has reached a state whose distribution is very close to a Gaussian, as implied already by the pdf shown in Fig. 8.1c at $t = 10^8$. Note the scale difference in the vertical axes. Panel (**c**) at $t = 10^7$ shows that the sum distribution, near the maximum of the chaotic breather, is still quite close to a q-Gaussian with $q \approx 2.6$. Panel (**d**) presents an estimate of the q index at different times, which shows that its values on the average fall significantly closer to 1 for $t > 10^7$ (after [18])

Fig. 8.4 Plot of q values as a function of time for energy density $\epsilon = 0.04$, i.e. $E = 5.12$ (*dashed curve*) for the π-mode of an FPU periodic chain with $N = 128$ particles and $\beta = 1$. The *dotted* and *solid curves* show error bars in the form of plus and minus one standard deviation. In agreement with other studies the transition to values close to $q = 1$ occurs here at $T_{eq} \approx 7.5 \times 10^5$ (after [18])

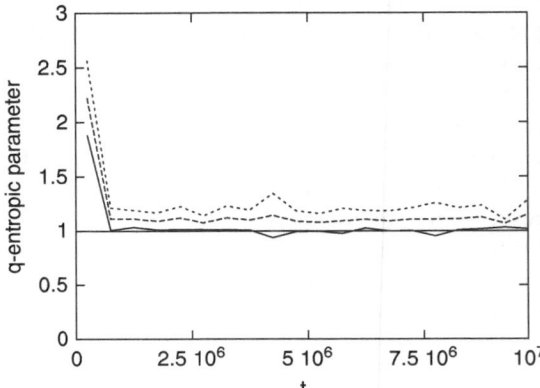

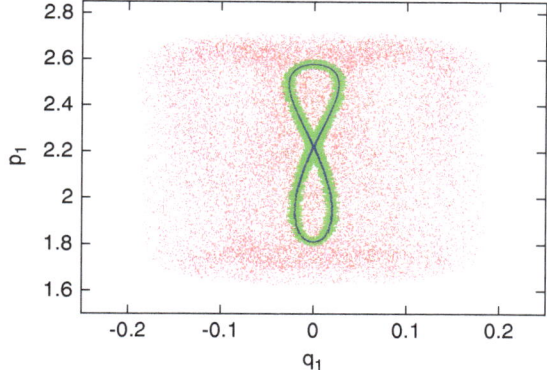

Fig. 8.5 The "figure eight" chaotic region (*blue color*) is observed for an initial condition at a distance close to the unstable SPO1 mode (depicted as the saddle point at $q_3 = 0$ and $p_3 > 0$). A slightly more extended "figure eight" region (*green points*) occurs for an initial condition a little further away and a large scale chaotic region (*red points*) arises for an initial condition even more distant on the surface of section (q_1, p_1) computed at times when $q_3 = 0$. Orbits are integrated up to $t_f = 10^5$ using $E = 7.4$, $N = 5$ and $\beta = 1.04$ (after [18])

8.4   FPU SPO1 and SPO2 Modes Under Fixed Boundary Conditions

Let us now examine the chaotic dynamics near NNMs of the FPU system under fixed boundary conditions, i.e. $q_0(t) = q_{N+1}(t) = 0$. In particular, we first evaluate pdfs of sums of chaotic orbit components near the SPO1 mode (see (3.26)), which keeps one particle fixed for every two adjacent particles oscillating with opposite phase. This mode is defined for N odd by

$$q_{2j}(t) = 0, \qquad q_{2j-1}(t) = -q_{2j+1}(t), \ j = 1, \dots, \frac{N-1}{2}. \qquad (8.17)$$

As shown in Fig. 8.5, the chaotic region close to this solution (when it has just become unstable) appears for a long time isolated in phase space from other chaotic domains. In fact, one finds several such domains, embedded one into each other. For example, in the case $N = 5$ and $\beta = 1.04$, a "figure eight" chaotic region appears on the surface of section (q_1, p_1) of Fig. 8.5 computed at times when $q_3 = 0$ (and $p_3 > 0$) at the energy $E = 7.4$. Even though the SPO1 mode is unstable (depicted as the saddle point at the middle of this surface of section) orbits starting sufficiently nearby remain in its vicinity for very long times, forming eventually the thin blue "figure eight" at the center of the figure. Starting, however, at points a little further away a more extended chaotic region is observed, plotted by green points, which still resembles a "figure eight". Choosing even more distant initial conditions, a large scale chaotic region plotted by red points becomes evident in Fig. 8.5.

It is, therefore, reasonable to regard these different dynamical behaviors near the SPO1 mode as QSS and characterize them by pdfs of sum distributions, as explained in Sect. 8.2. The idea behind this is that orbits starting initially at the immediate vicinity of the unstable SPO1 mode behave differently than those lying further away, since the latter orbits have the ability to explore more uniformly parts of the constant energy surface. Thus, q-Gaussian-like distributions with $1 < q < 3$ are expected near SPO1, while for orbits starting sufficiently far the distributions we expect to find are Gaussians with $q \to 1$.

To test the validity of these ideas, let us choose the quantity

$$\eta(t) = q_1(t) + q_3(t) \tag{8.18}$$

as our observable, which is exactly equal to zero at the SPO1 orbit. Following what was presented in Sect. 8.2, we now study the motion related to *three* different initial conditions as a result of integration over longer and longer times, during which an orbit passes through all the different stages depicted in Fig. 8.5. In particular, in Fig. 8.6a, we see the surface of section created by the trajectory starting close to the NNM and integrated up to $t_f = 10^5$ while in the following two panels we see the same surface of section computed for final integration times of $t_f = 10^7$ and $t_f = 10^8$ respectively. The parameters are the same as in Fig. 8.5.

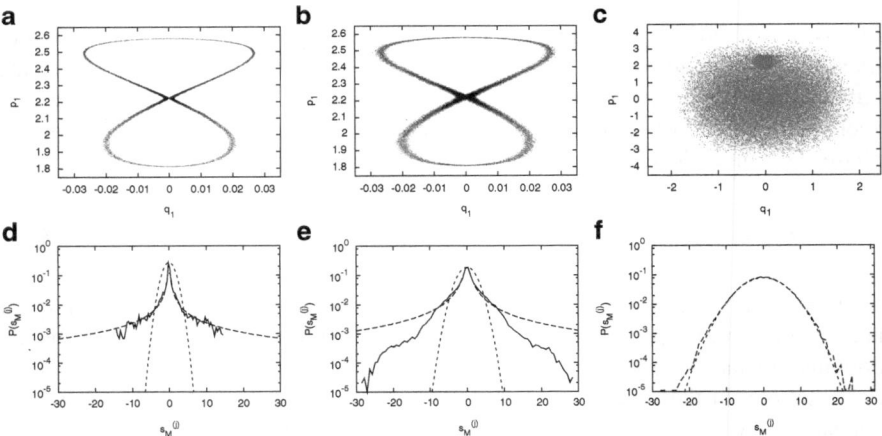

Fig. 8.6 (a) The (q_1, p_1) surface of section of anf orbit integrated up to $t_f = 10^5$ and starting close to the unstable SPO1 orbit for $N = 5$ and $\beta = 1.04$ at energy $E = 7.4$. (b) and (c) are same as (a) but for $t_f = 10^7$ and $t_f = 10^8$ respectively. (d)–(f) Plots in linear-log scale of numerical (*solid curve*), of q-Gaussian (*dashed curve*) and Gaussian (*dotted curve*) for th einitial conditions of (a)–(c) respectively. In particular: (d) is for $t_f = 10^5$, $N_{ic} = 10^4$ and $M = 10$, (e) for $t_f = 10^7$, $N_{ic} = 10^5$ and $M = 1,000$ terms, and (f) for $t_f = 10^8$, $N_{ic} = 10^5$ and $M = 1,000$ (after [18])

Clearly, as the integration time increases, orbits starting close to the unstable SPO1 mode, eventually wander over a more extended part of phase space, covering gradually all of the energy surface when the integration time is sufficiently large (e.g. $t_f = 10^8$).

An important question arises here: Are these behaviors reflected in the statistics associated with these trajectories? The answer to this question is presented in Fig. 8.6d–f. In particular, Fig. 8.6d shows in linear-log scale the numerical (solid curve), q-Gaussian (dashed curve) and Gaussian (dotted curve) distributions for the initial condition located closest to SPO1, using $t_f = 10^5$, $N_{ic} = 10^4$ time windows and $M = 10$ terms in the computations of the sums. In this case, a q-Gaussian fitting of the data gives $q \approx 2.785$ with $\chi^2 \approx 0.000\,31$. This distribution corresponds to the surface of section shown in Fig. 8.6a. If one now increases t_f by two orders of magnitude (see panel (f)) using $N_{ic} = 10^5$, $M = 1,000$ and performs the same kind of fit one gets $q \approx 2.483$ with $\chi^2 \approx 0.000\,47$.

It is important to emphasize, however, that the lower parts (tails) of the solid curve distribution of Fig. 8.6e are not fitted well by a q-Gaussian. This suggests that by increasing the integration time, the initial pdf takes a form that may well be approximated by other types of functions, like e.g. (8.14). This distribution corresponds to the surface of section of Fig. 8.6b. By increasing the time further to $t_f = 10^8$ and using $N_{ic} = 10^5$ and $M = 1,000$ terms, we observe that the solid curve of panel (g) is indeed very close to a Gaussian ($q \approx 1.05$), characterizing the chaotic regime plotted in the surface of section of Fig. 8.6c.

Next, the authors of [18] turned to another nonlinear mode of the FPU Hamiltonian with fixed boundary conditions called the SPO2 mode (see (3.27)). This is a NNM which keeps every third particle fixed, while the two in between move in exact out-of-phase motion. What is important about this NNM is that it becomes unstable at much lower energies (i.e. $E_u/N \propto N^{-2}$) compared to SPO1 ($E_u/N \propto N^{-1}$) [13], much like the low $k = 1, 2, 3, \ldots$ mode periodic orbits connected with the breakdown of FPU recurrences [64, 94]. Thus, it is expected that near SPO2, orbits will be more weakly chaotic than SPO1 and hence QSS should persist for longer times. This is exactly what happens. As Fig. 8.7 clearly shows, the dynamics in a close vicinity of SPO2 has the features of what we might call "edge of chaos": Orbits wander in a regime of very small (positive) LCEs, tracing a kind of "banana"-shaped region much different than the "figure eight" we had observed near SPO1. Remarkably, the pdfs in this case (at least up to $t_f = 10^{10}$), actually converge to a smooth function, never deviating towards a Gaussian, as in the QSS of other FPU systems.

More specifically, let us set $N = 5$, $\beta = 1$, $E = 0.5$ and choose an orbit located initially close to the SPO2 solution, which has just turned unstable (at $E_u \approx 0.4776$). As we can see in Fig. 8.7a, the dynamics here yields banana-like orbits at least up to $t_f = 10^{10}$. The weakly chaotic nature of the motion is plainly depicted in Fig. 8.7b, where the four positive LCEs are plotted. Note that, although they do decrease towards zero for a very long time, at about $t_f \approx 10^9$, the largest one of them tends to converge to a very small value (about 10^{-8}), indicating that the orbit is chaotic, sticking perhaps to an "edge of chaos" region around SPO2.

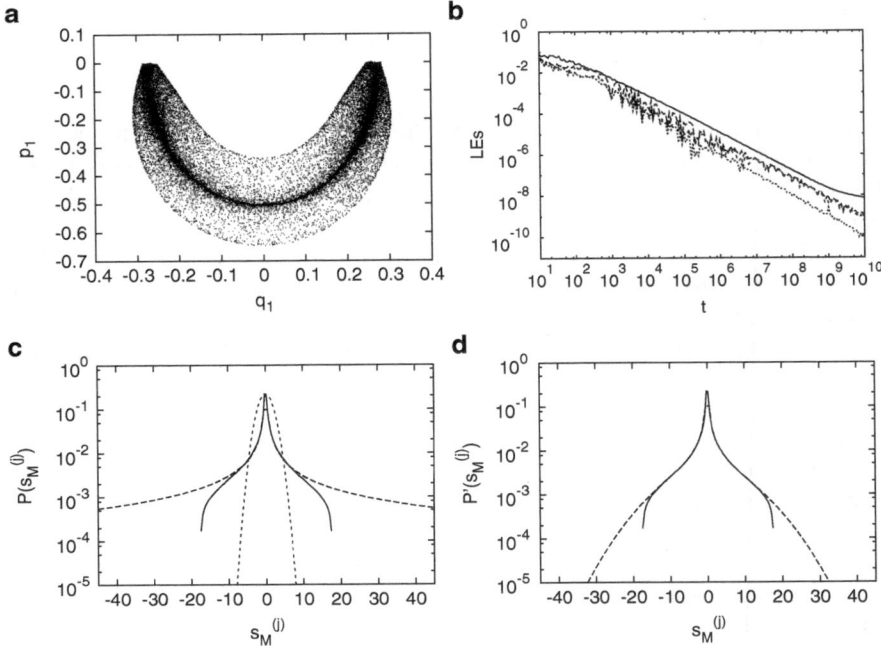

Fig. 8.7 (**a**) The (q_1, p_1) surface of section of an orbit integrated up to $t_f = 10^{10}$ starting in the vicinity of the unstable SPO2 mode. (**b**) The corresponding four biggest LCEs. (**c**) Linear-log scale plot of the numerical (*solid curve*), q-Gaussian (*dashed curve*) and Gaussian (*dotted curve*) distributions. (**d**) The *solid curve* distribution of panel (**c**) is better fitted by the *dashed curve* of the $\tilde{P}$ function of (8.14) (after [18])

In Fig. 8.7c, the corresponding pdf is plotted at time $t_f = 10^{10}$ (whose shape does not change after $t_f = 10^7$). An extremely long-lasting QSS is formed, whose distribution is well-fitted by a q-Gaussian with $q \approx 2.769$ and $\chi^2 \approx 4.44 \times 10^{-5}$. The "legs" of this distribution away from the center deviate from the q-Gaussian shape, but remain far from the Gaussian plotted as a dotted curve in the figure. Performing a similar fitting of our data with the function (8.14), in Fig. 8.7d, as in the case of Fig. 8.1d, one finds that the numerical distribution (solid curve) of Fig. 8.7c is better approximated by (8.14) where $a_1 \approx 0.006$, $a_q \approx 170$ and $q \approx 2.82$ with $\chi^2 \approx 2.06 \times 10^{-6}$, compared with the $\chi^2 \approx 4.44 \times 10^{-5}$ obtained by fitting the same distribution by a q-Gaussian with $q \approx 2.769$ in Fig. 8.7c.

Thus, in "thin" chaotic layers of multi-dimensional Hamiltonian systems with small positive LCEs it is possible to find non-Gaussian QSS that persist for very long times as in the SPO2 case. Numerical evidence suggests that in these regimes chaotic orbits stick for long time intervals to a complex network of islands, where

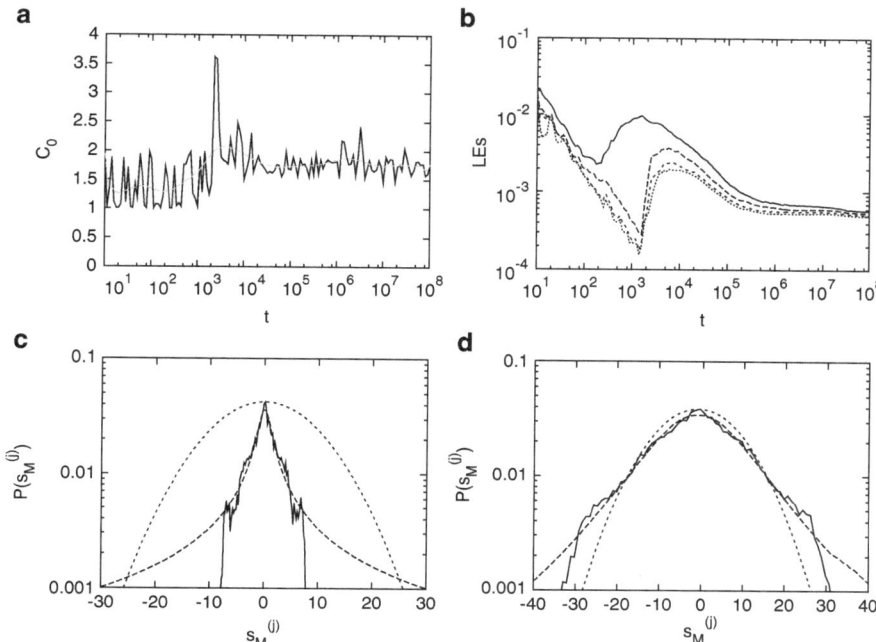

Fig. 8.8 (**a**) Plot of C_0 (8.15) (*solid curve*) as a function of time for a perturbation of the unstable ($E = 1.5 > E_u \approx 1.05226$) SPO1 mode with $\beta = 1.04$ and $N = 129$. The *grey curve* corresponds to the time average of the *solid curve*. (**b**) Log-log plot of the four biggest LCEs as a function of time. Plots in linear-log scale of numerical (*solid curve*), q-Gaussian (*dashed curve*) and Gaussian (*dotted curve*) distributions: (**c**) for final integration time $t_f = 4 \times 10^5$, $N_{ic} = 10^4$ time windows and $M = 20$ terms in the sums, (**d**) for $t_f = 2 \times 10^6$, $N_{ic} = 5 \times 10^4$ and $M = 20$ (after [18])

their statistics is well approximated by q-Gaussian distributions connected with nonextensive statistical mechanics.

It is interesting to study the dynamics near the unstable SPO1 and SPO2 modes using an analysis similar to the one carried out for the π-mode, in Sect. 8.3.1, following [105]. In particular, focusing on small perturbations of both modes, one may wish to investigate how C_0 (8.15) depicts the evolution towards energy equipartition and whether this transition will again be preceded by the appearance of chaotic breathers.

To find out, consider the SPO1 mode and follow neighboring orbits starting at a small distance from the mode when it has just turned unstable. As explained in detail in [18], C_0 is found to grow on the average without a clear maximum, settling down to a value near 1.75 (see Fig. 8.8), indicating that the system reaches equipartition at about $t \approx 10^6$, where the four biggest LCEs begin to converge to their final values, as shown in Fig. 8.8b. Interestingly, during this period, the corresponding distribution in Fig. 8.8c (using the observable $\eta = q(64) + q(65) + q(66)$) is well fitted by a q-Gaussian with $q \approx 2.843$ at $t = 4 \times 10^5$. For longer times, distributions quickly tend to Gaussians as shown in Fig. 8.8d for $t = 2 \times 10^6$. Thus, even though (unlike the π-mode) no chaotic breather is observed in the time C_0 takes to relax to its limiting value, sum distributions are well approximated by q-Gaussians and only converge to Gaussians when energy equipartition has occurred.

Repeating this process now for an orbit starting close to the SPO2 mode, when it has just destabilized, one finds that C_0 also does not exhibit a distinct high maximum, but grows on the average more slowly than in the SPO1 case (see [18]). Thus, it takes longer to approach its limit, indicating that the system reaches equipartition at $t \approx 4 \times 10^8$, where the four biggest Lyapunov exponents cease to decrease towards zero and tend to small positive values. The corresponding sum distribution is well fitted by a q-Gaussian with a high $q = 1.943$ value, while we need to increase the time to 5×10^8 to see pdfs that are much closer to a Gaussian ($q \approx 1.0$), indicating that this is a lower bound for energy equipartition.

8.5 q-Gaussian Distributions for a Small Microplasma System

We now turn our attention to a Hamiltonian system very different than the FPU systems of the previous sections and discuss a number of results presented in [18], concerning a system of few degrees of freedom characterized by long range interactions of the Coulomb type. As we will see, the approach described in this chapter will allow us to study statistically the dynamics of its transition from a crystal-like to a liquid-like phase (the so-called "melting transition") [17, 168] at small energies, as well as identify a transition from the liquid-like to a gas-like phase, as the total energy increases further [145].

We shall focus, in particular, on a microplasma system consisting of N ions of equal mass $m = 1$ and electric charge Q moving in a Penning trap in the presence of an electrostatic potential [17, 145]

$$\Phi(x, y, z) = V_0 \frac{2z^2 - x^2 - y^2}{r_0^2 + 2z_0^2} \tag{8.19}$$

(r_0, z_0 are physical parameters of the trap), and a constant magnetic field in the z direction, whose vector potential is

$$\mathbf{A}(x, y, z) = \frac{1}{2}(-By, Bx, 0). \tag{8.20}$$

The system is described by the Hamiltonian

$$\mathcal{H} = \sum_{i=1}^{N} \left\{ \frac{1}{2m} [\mathbf{p}_i - q\mathbf{A}(\mathbf{r}_i)]^2 + Q\Phi(\mathbf{r}_i) \right\} + \sum_{1 \le i < j \le N} \frac{Q^2}{4\pi\varepsilon_0 r_{ij}}, \qquad (8.21)$$

where $\mathbf{r}_i$ is the position of the ith ion, r_{ij} is the Euclidean distance between ith and jth ions and ε_0 is the vacuum permittivity. In the Penning trap, the ions are subjected to a harmonic confinement in the z direction with frequency

$$\omega_z = \left[\frac{4QV_0}{m(r_0^2 + 2z_0^2)} \right]^{\frac{1}{2}} \qquad (8.22)$$

while, in the perpendicular direction they rotate (due to the cyclotron motion) with frequency $\omega_C = QB/m$. Thus, in a frame rotating about the z axis with Larmor frequency $\omega_L = \omega_C/2$, the ions are subjected to an overall harmonic potential with frequency $\omega_x = \omega_y = (\omega_C^2/4 - \omega_z^2/2)^{1/2}$ in the direction perpendicular to the magnetic field. In the rescaled time $\tau = \omega_C t$, position $\mathbf{R} = \mathbf{r}/a$ and energy $H = \mathcal{H}/(m\omega_C^2 a^2)$ with $a = [Q^2/(4\pi\varepsilon_0 m\omega_C^2)]^{1/3}$, our Hamiltonian (8.21) reads

$$H = \frac{1}{2} \sum_{i=1}^{N} \mathbf{P}_i^2 + \sum_{i=1}^{N} \left[\left(\frac{1}{8} - \frac{\gamma^2}{4} \right)(X_i^2 + Y_i^2) + \frac{\gamma^2}{2} Z_i^2 \right] + \sum_{i<j} \frac{1}{R_{ij}} = E \qquad (8.23)$$

where E is the total constant energy, $\mathbf{R}_i = (X_i, Y_i, Z_i)$ and $\mathbf{P}_i = (P_{X_i}, P_{Y_i}, P_{Z_i})$ are the positions and momenta in $\mathbb{R}^3$ respectively of the N ions, R_{ij} is the Euclidean distance between different ions i, j given by

$$R_{ij} = [(X_i - X_j)^2 + (Y_i - Y_j)^2 + (Z_i - Z_j)^2]^{\frac{1}{2}} \qquad (8.24)$$

and $\gamma = \omega_z/\omega_C$.

Due to the form of the potential in (8.23), the ions perform bounded motion provided $|\gamma| < 1/\sqrt{2}$. The Penning trap is called prolate if $|\gamma| < 1/\sqrt{6}$, isotropic if $|\gamma| = 1/\sqrt{6}$ and oblate if $1/\sqrt{6} < |\gamma| < 1/\sqrt{2}$. Thus, the motion is quasi one-dimensional in the limit $\gamma \to 0$ and quasi two-dimensional in the limit $\gamma \to 1/\sqrt{2}$. The Z direction is a symmetry axis and hence, the Z component of the angular momentum, $L_Z = \sum_{i=1}^{N} X_i P_{Y_i} - Y_i P_{X_i}$, is conserved, being a second integral of the motion. We may, therefore, set from here on the angular momentum equal to zero (i.e. $L_Z = 0$) and study the motion in the Larmor rotating frame.

In [17], the authors demonstrate the occurrence of a dynamical regime change in a microplasma system composed of $N = 5$ ions and confined in a prolate quasi one-dimensional configuration of $\gamma = 0.07$. More specifically, in the lower energy regime, a transition from crystal-like to liquid-like behavior was observed, called the

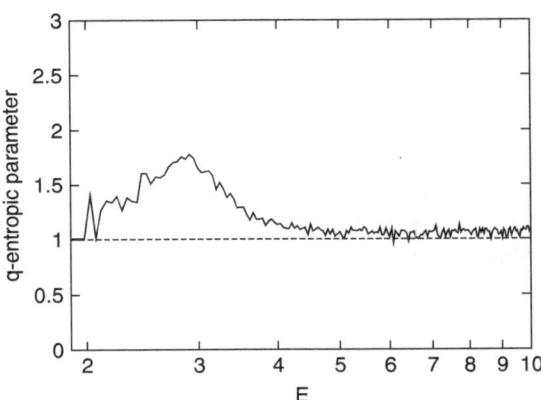

Fig. 8.9 Plot in log-linear scale of the q parameter (*solid curve*) as a function of the energy E of the microplasma (8.23) for $\gamma = 0.07$ (prolate trap) and $N = 5$. Plotted here is also the line $q = 1$ for comparison with the Gaussian case (after [18])

"melting" phase transition. First, it was shown that this transition is not associated with a sharp increase of the temperature at some critical energy, as might have been expected at first sight. Furthermore, the positive LCEs achieve their maximum at energies much higher than the "melting" regime. Thus, it appears that no clear macroscopic reasoning is available for identifying and analyzing this process.

For this reason, the SALI method [309, 313, 314] described in Chap. 5 was invoked to study the local microscopic dynamics in detail [17]. It was discovered that there exists indeed an energy range of weakly chaotic behavior, i.e. $E \in \Delta E_{mt} = (2, 2.5)$, where the positive LCEs are very small and the SALI exhibits a stair-like decay to zero with varying decay rates. This suggests the presence of long-lived "sticky" components executing a multi-stage diffusion process near the boundaries of resonance islands [316]. Thus, it was concluded that it is in this energy interval that the above "melting" transition occurs.

Motivated by these results, the investigation was pursued further in [18], for the same γ and N, in the regime where the minimum energy at which ions start moving appreciably about their equilibria positions is $E_0 \approx 1.8922$. The aim was to study the melting transition as the energy E of the system increases above E_0, using pdfs associated with chaotic trajectories to relate microscopic to macroscopic observables. Based on the results described earlier in this chapter, one might expect to find q-Gaussian approximations with $1 < q < 3$ in the vicinity of ΔE_{mt}, where the positive LCEs L_i, $i = 1, \ldots, 3N$ are quite small compared to their maximum value of $L_1 \approx 0.0558$ attained at $E \approx 5.95$.

Figure 8.9 shows the results of this study for the interval $(E_0, 10]$. Choosing as an observable the quantity $\eta(t) = X_1(t)$ and setting $N_{ic} = 2 \times 10^4$, $M = 1,000$ the equations of motion were integrated for a total time $t_f = 2 \times 10^7$. Remarkably, in the energy range ΔE_{mt} of the "melting" transition, the values of the entropic index of the q-Gaussian pdfs approximating the data, were found to be well above $q = 1$, indicating that the statistics is certainly not Gaussian. In fact, the detected

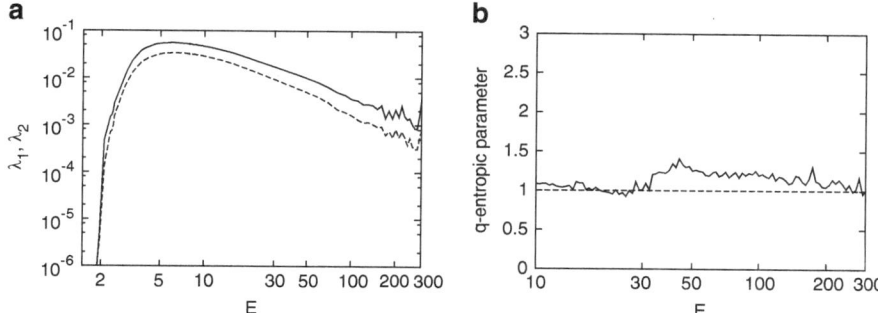

Fig. 8.10 (**a**) The two biggest positive LCEs (L_1 (*solid curve*) and L_2 (*dashed curve*), denoted by λ_1 and λ_2 respectively. as a function of the energy E in log-log scale. (**b**) Plot in log-linear scale of the q parameter (*solid curve*) as a function of the energy E of the microplasma Hamiltonian (8.23) for $\gamma = 0.07$ and $N = 5$. Plotted here is also the line $q = 1$ for comparison with the Gaussian case (after [18])

q-Gaussian shape persists up to $E \approx 4$, implying that the energy interval of the transition may be slightly bigger than was originally estimated.

Next, in [18] the second transition of system (8.23) was studied, from a liquid-like to a gas-like phase. Here, the system is studied over an energy range where the largest positive LCEs decrease towards zero following (8.25), as shown in Fig. 8.10a. Thus, as one moves to higher energies, the system passes from a strongly chaotic regime at energies where the LCEs are maximal (i.e. for $E \gtrsim 6$) to energies where the motion is much less chaotic.

As demonstrated in [145], if all N^2 inter-particle interaction terms of the Coulomb part of Hamiltonian (8.23) at distances R_{ij} contribute to the MLE L_1, it is possible to derive, for sufficiently large values of $T \to \infty$ (and therefore of $E \to \infty$), that

$$L_1 \sim \left\langle \frac{N^2}{R_{ij}^3} \right\rangle \sim N \frac{(\ln T)^{\frac{1}{2}}}{T^{\frac{3}{4}}} \tag{8.25}$$

where T is related to the temperature of the system. This formula explains the asymptotic power law decay of the biggest LCE observed in Fig. 8.10a. It can also be shown that the second largest LCE L_2 (dashed curve) obeys a similar formula as a function of E.

The important result obtained in [18] is that, in this case also, the q indices of the corresponding pdfs remain well above unity for $E \in (30, 200)$, as shown in Fig. 8.10b. This suggests that in this range also a dynamical change to a weaker form of chaos occurs, associated with small positive LCEs. Most probably, the N^2 inter-particle terms do not contribute significantly to the transition between the fully developed chaotic regime of a liquid-like state and the more ordered dynamics of a gas-like regime.

It is indeed interesting to compare the plots of the index q as a function of energy shown in Figs. 8.9 and 8.10b, with the behavior of q as a function of time shown in Figs. 8.3d and 8.4 for the chaotic regime near the π-mode of the FPU system. These results demonstrate that, in very different situations, studying the statistics of chaotic QSS and monitoring the values of the entropic index of q-Gaussian approximations can prove very useful in revealing important physical phenomena like the onset of energy equipartition, or the occurrence of dynamical phase transitions.

8.6   Chaotic Quasi-stationary States in Area-Preserving Maps

Two-dimensional area-preserving maps are excellent models for studying the qualitative dynamics of Hamiltonian systems of two degrees of freedom. The Poincaré maps that we extensively analyzed in Chaps. 2 and 3 are all area-preserving maps. As such, they generically possess families of invariant closed curves (representing quasiperiodic orbits), which form complete barriers to all motion evolving inside "islands" centered around stable periodic orbits in the two-dimensional phase space. At the boundaries of these islands, a complex network of smaller islands and invariant Cantor sets (often called cantori) exists, to which chaotic orbits are observed to "stick" for very long times. It is here that trajectories often get trapped into QSS that can be very long-lived. Such phenomena have so far been thoroughly studied in terms of a number of dynamical mechanisms responsible for chaotic transport (see e.g. [236, 238, 347]).

It would, therefore, be very natural to wonder: Could we also study the dynamics in these "sticky" regions using the probabilistic techniques of nonextensive statistical mechanics? What would pdfs of sums of iterates reveal that would help us understand the complexity of motion in these domains? Will we uncover interesting connections between the "geometry" of phase space dynamics and the time-evolving statistics of chaotic orbits, as we did in earlier sections through a similar analysis of multi-dimensional Hamiltonian systems? As we describe in what follows, it is indeed possible to find in such "weakly chaotic" domains of area-preserving maps long-lived QSS, whose pdfs do not rapidly converge to a Gaussian, but pass through several stages, some of which are very well approximated by q-Gaussian distributions.

Let us start by recalling a number of studies of such pdfs in 1D maps [7, 292, 327, 328], as well as higher-dimensional conservative maps [115], in similar "edge of chaos" domains, where the MLE either vanishes or is very close to zero. These studies have already provided ample evidence of the usefulness of q-Gaussian distributions, in the context of the CLT. Despite the controversy that some of these findings have generated [68, 158], it is worth pointing out that for one-dimensional maps, the situation is a lot clearer. Indeed, as shown convincingly in [328, 333], when one approaches the critical point of the period-doubling route to chaos taking into account a proper scaling relation that involves the Feigenbaum constant δ and the location of the critical point, the pdfs of sums of iterates of the logistic map are approximated by a q-Gaussian far better than the Lévy distribution proposed in [158]. This suggests the need for a more thorough investigation of low-dimensional maps, within a nonextensive statistical mechanics approach, in which q-Gaussian distributions represent *metastable states*, or QSS of the dynamics [22, 23, 251, 285].

Since we are going to work in the context of the CLT, it is important to note that such a theorem has been verified for deterministic systems [29, 44, 237]. Attempts to generalize the CLT have also been published demonstrating that, for certain classes of strongly correlated random variables, their rescaled sums approach a q-Gaussian limit distribution [161, 335, 336]. As stated earlier, however, objections have also been raised by some authors [108, 166, 167], questioning whether q-Gaussians can indeed constitute attractors, as in some mathematical models where this was thought to be the case, other functions were shown to describe the sum distributions in the CLT limit.

To understand the situation better, therefore, it is instructive to probe deeper into the complex dynamics of conservative maps in weakly chaotic domains and investigate pdfs of rescaled sum of M iterates, in the large M limit, and for many different initial conditions. We will thus be able to shed more light on possible connections between "geometric" properties of chaotic regions and their statistics expressed by pdfs of long-lived QSS in these domains. As shown recently in [293], we will discover that, in general, as M grows, these pdfs pass from a q-Gaussian to an exponential form (having a triangular shape in semi-log plots), ultimately tending to become true Gaussians, as "stickiness" to an 'edge of chaos' subsides in favor of more uniformly chaotic motion. Still, we will also demonstrate that there are cases where the orbits evolve in such convoluted pathways that q-Gaussian approximations persist for as long as it was possible to iterate the equations of motion.

8.6.1 Time-Evolving Statistics of pdfs in Area-Preserving Maps

Let us, therefore, consider in what follows 2D maps of the form

$$x_{n+1} = f(x_n, y_n), \quad y_{n+1} = g(x_n, y_n) \tag{8.26}$$

and treat their chaotic orbits as generators of random variables, as we did for Hamiltonian systems earlier in this chapter. We will thus compute long sequences produced by many initial conditions and examine whether these can be considered as independent and identically distributed quantities as required by the classical CLT. In this regard, it is important to recall that the well known CLT assumption about the independence of identically distributed random variables can be replaced by the weaker property of asymptotic statistical independence, as shown in [237].

Thus, we may proceed to compute the generalized rescaled sums of one of the iterates, say x_n of the map (8.26) in the sense of the CLT [29, 44, 237]:

$$Z_M = M^{-\gamma} \sum_{n=1}^{M} (x_n - \langle x \rangle) \tag{8.27}$$

where $\langle \cdot \rangle$ implies averaging over a large number of iterations M *and* a large number of randomly chosen initial conditions N_{ic}. For fully chaotic systems $\gamma = 1/2$ and the distribution of (8.27) in the limit ($M \to \infty$) is expected to be a Gaussian [237]. Alternatively, however, we may define the non-rescaled variable

$$z_M = \sum_{n=1}^{M} [x_n - \langle x \rangle] \tag{8.28}$$

(absorbing the factor $M^{-\gamma}$) and analyze its pdf normalized by its variance as follows (see [293] and Sect. 8.2 of this chapter): First, we construct the sums $S_M^{(j)}$ obtained by adding iterates x_n ($n = 0, \ldots, M$) of the map (8.26), where (j) represents the dependence on the randomly chosen initial conditions $x_0^{(j)}$, with $j = 1, 2, \ldots, N_{ic}$. Next, we focus on the s_m^j variables, which are the S_m^js centered about their mean and rescaled by their standard deviation σ, and follow the procedure outlined in Sect. 8.2 (see (8.8)).

More specifically, we compute the pdfs of $s_M^{(j)}$ and plot, in the sections that follow, the corresponding histograms of $P(s_M^{(j)})$ for sufficiently small increments (e.g. $\Delta s_M^{(j)} = 0.05$) to smoothen out fine details and analyze their functional form. Following [293], we express in our plots the independent variable as $z/\sigma \equiv s_M^{(j)}$.

Before proceeding, however, to investigate in detail these phenomena in a family of area-preserving maps, we refer the reader to Problem 8.1 below, where he (or she) is asked to apply the above methodology to a 2D map, which possesses a strange attractor when it is dissipative and an annular region with complex chaotic dynamics when it is area-preserving. It is indeed remarkable how the Gaussian that dominates the statistics in the presence of a strange attractor, yields its place to a different function, which is well-approximated by a q-Gaussian in the conservative case.

8.6.2   The Perturbed MacMillan Map

Let us consider the so-called perturbed MacMillan map, introduced in [151] to study the onset of chaos via Mel'nikov theory, in the form:

$$\begin{cases} x_{n+1} = y_n \\ y_{n+1} = -x_n + 2\mu \dfrac{y_n}{1 + y_n^2} + \varepsilon \left(y_n + \beta x_n \right) \end{cases} \tag{8.29}$$

where ε, β, μ are physically important parameters. For $\varepsilon = 0$, (8.29) is known to be integrable possessing a simple polynomial invariant that allows it to have a saddle point at the origin, for $\mu > 1$ (see Exercise 8.1). The Jacobian determinant is $J = 1 - \varepsilon\beta$, so that (8.29) is area-preserving for $\beta = 0$, and dissipative for $\varepsilon\beta > 0$. Here, we only consider the area-preserving case $\beta = 0$, so that the only relevant parameters are (ε, μ).

As discussed in Exercise 8.1, for $\mu > 1$ the unperturbed map possesses a "figure eight" invariant curve with a self-intersection at the origin that is a fixed point of saddle type. For $\varepsilon > 0$, the invariant manifolds of this saddle split and a thin chaotic layer appears surrounding two large islands in the (x_n, y_n) plane. The MLE L_1 for $\mu = 1.6$ and $0.2 \leq \epsilon \leq 1.8$ is found to vary between 0.0875 and 0.03446. By analyzing the histogram of the normalized sums of (8.28) within this chaotic layer for a wide range of parameters ε and μ, we find that, in many cases, the pdfs begin by being well approximated by q-Gaussians and then turn to exponentials $\sim e^{-k|z|}$, whose triangular shape on semi-log plots eventually starts to approach an inverted parabola representing the Gaussian function. Thus, monitoring chaotic orbits in this region for increasingly large numbers of iterations M, we observe the occurrence of different QSS described by these distributions. One thus obtains some very interesting results, which are described in detail in [293] and are briefly discussed in the sections that follow.

Let us focus, in particular, on the time-evolving statistics of two examples of the MacMillan map, which represent respectively: (1) One set of cases with a "figure eight" chaotic domain whose distributions pass through a succession of pfds before converging to an ordinary Gaussian, and (2) a set with more complicated domains extending around many islands, where q-Gaussian pdfs dominate the statistics for very long times and no tendency to a Gaussian is observed.

8.6.2.1 The $\varepsilon = 0.9$, $\mu = 1.6$ Class of Examples

The case $\varepsilon = 0.9$, $\mu = 1.6$ is a typical example characterized by time-evolving pdfs. As shown in Fig. 8.11b, the corresponding phase space plot yield a seemingly simple chaotic region in the form of a "figure eight", while the corresponding pdfs do *not* converge to a single distribution, passing from a q-Gaussian-looking function to an exponential distribution.

Analyzing carefully the time evolution of chaotic orbits in this example, we observe that there exist at least three long-lived QSS. In particular, for $i = 1, \ldots, M = 2^{16}$, a QSS is produced whose pdf is close to a pure $q = 1.6$-Gaussian. Figure 8.12 shows some pdfs for different numbers of iterates M. Note that these pdfs correspond to a "figure eight" chaotic region that evolves essentially around two large islands (Fig. 8.11(a)). However, for $M > 2^{16}$, a more complex structure emerges: Iterates stick around new islands, and a transition is evident from q-Gaussian to exponentially decaying pdfs [293].

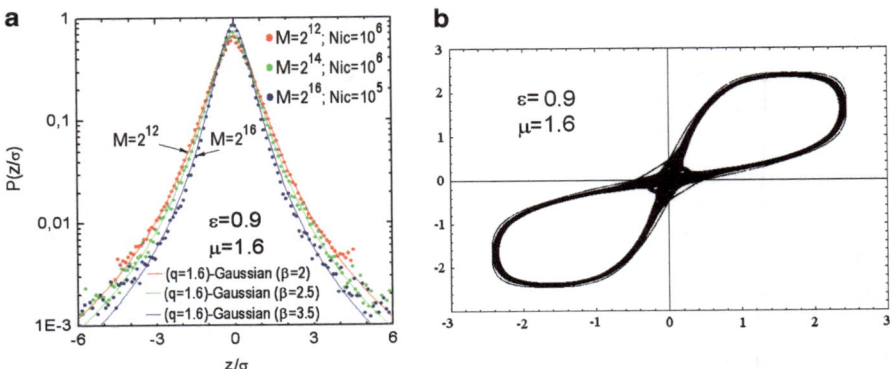

Fig. 8.11 (**a**) pdfs of the renormalized sums of M iterates of the MacMillan map (8.29), for $\varepsilon = 0.9$, $\mu = 1.6$, $M \leq 2^{16}$ and N_{ic} initial conditions randomly chosen within a square $(0, 10^{-6}) \times (0, 10^{-6})$ about the origin. (**b**) Phase space plot for $M = 2^{16}$ (after [293])

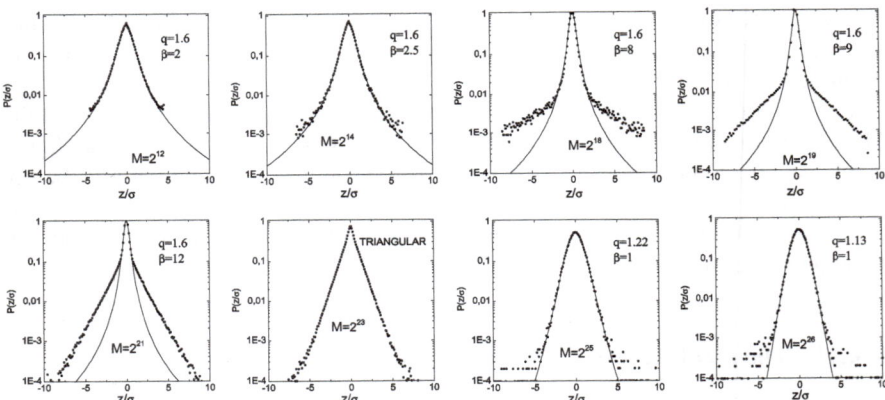

Fig. 8.12 Detailed evolution of the pdfs of the MacMillan map for $\varepsilon = 0.9$, $\mu = 1.6$, as M increases from 2^{12} to 2^{26}, respectively (after [293])

Clearly, therefore, for $\varepsilon = 0.9$ (and other cases with $\varepsilon = 0.2, 1.8$) more than one QSS coexist whose pdfs are well-approximated by a sequence of q-Gaussians. In fact, for $10^{18} \leq M \leq 2^{21}$, the central part of the distribution is still well-fitted by a q-Gaussian with $q = 1.6$. However, as we continue to iterate the map to $M = 2^{23}$, intermediate states are observed, which are better described by an exponential distribution. From here on, as $M > 2^{23}$, the central part of the pdfs is close to a Gaussian (see Fig. 8.12) and a true Gaussian is expected in the limit ($M \to \infty$). The evolution of this sequence of successive QSS as M increases is shown in detail in Fig. 8.12.

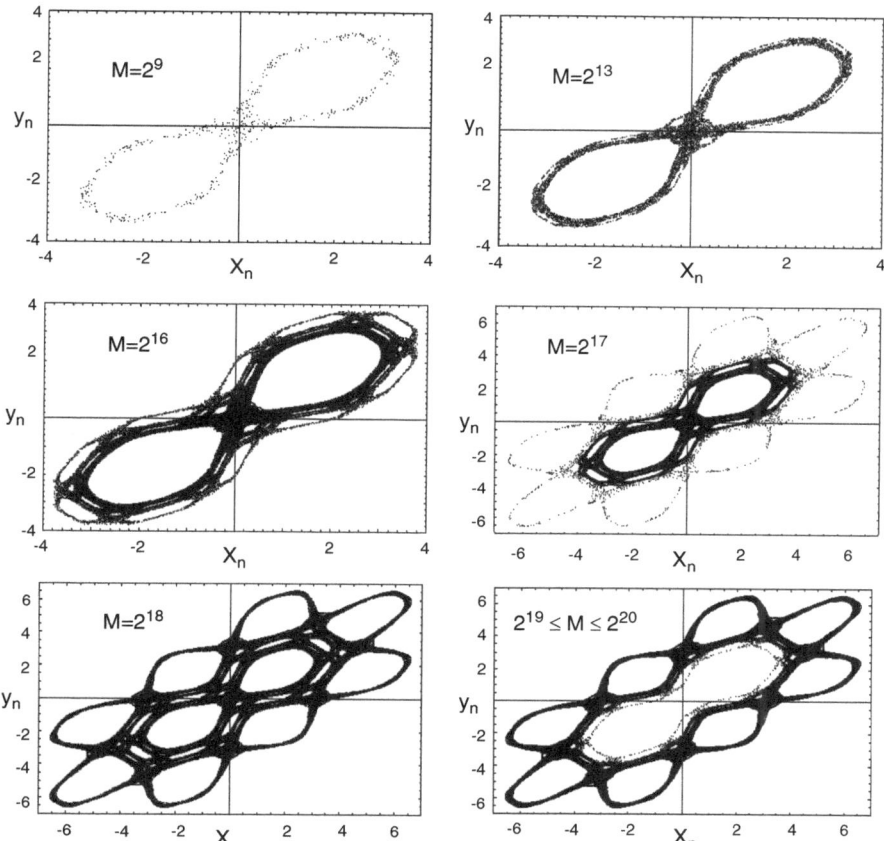

Fig. 8.13 Structure of phase space plots of the MacMillan map (8.29) for parameter values $\varepsilon = 1.2$ and $\mu = 1.6$, starting from a random set of initial conditions chosen randomly within a square $(0, 10^{-6}) \times (0, 10^{-6})$ about the origin and iterating the map M times (after [293])

8.6.2.2   The $\epsilon = 1.2$, $\mu = 1.6$ Class of Examples

Let us now turn to the behavior of the $\varepsilon = 1.2$, $\mu = 1.6$ class, whose MLE is $L_1 \approx 0.05$, hence smaller than that of the $\varepsilon = 0.9$ case (where $L_1 \approx 0.08$). As clearly seen in Fig. 8.13, a diffusive behavior sets in here that extends outward in phase space, enveloping a chain of eight large islands to which the orbits "stick" as the number of iterations grows to $M = 2^{19}$.

Some representative pdfs of this evolution are seen in Fig. 8.14. In this class of MacMillan maps, the chaotic domain under study extends around several large islands and is apparently richer in "stickiness" phenomena. This higher complexity of the dynamics may very well be the reason why the corresponding chaotic states

Fig. 8.14 Pdfs of the
renormalized sums of M
iterates of the MacMillan
map (8.29), for $\varepsilon = 1.2$,
$\mu = 1.6$, $M = 2^{18}$, 2^{19} and
2^{20}. Note the apparent
convergence to a q-Gaussian,
due to the complicated
chaotic dynamics around the
large islands of Fig. 8.13
(after [293])

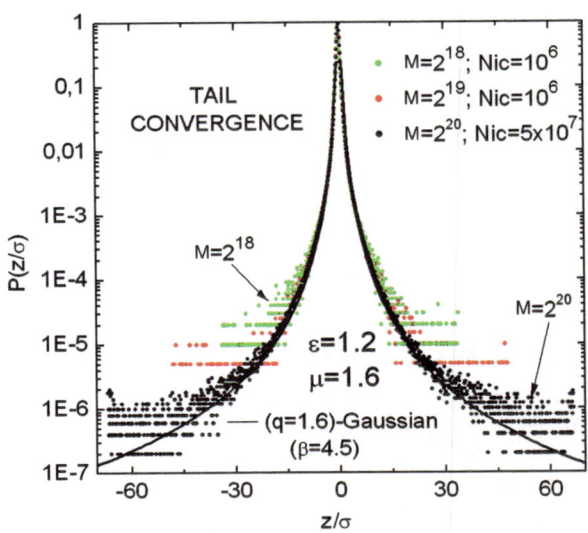

possess pdfs that are well approximated by q-Gaussians with $q > 1$ and persist for
extremely large numbers of iterations of the MacMillan map [293].

Exercises

Exercise 8.1. Consider the perturbed MacMillan map (8.29) of Sect. 8.6.1.

(a) Prove that it is integrable for $\varepsilon = 0$, by demonstrating that it possesses the
polynomial invariant $I(x_n, y_n) = x_n^2 y_n^2 + x_n^2 + y_n^2 - 2\mu x_n y_n$. Investigate the
form of the invariant curves foliating the plane for different values of μ and
show that when $|\mu| > 1$ the origin is a saddle point. In what way is the dynamics
for $\mu > 1$ different from $\mu < -1$?

(b) Locate the stable and unstable manifolds of the linearized equations about the
origin, for $\mu = 1.6$ and $\varepsilon = 0.001$. Now follow them numerically in the full 2-
dimensional plane starting with many initial conditions placed densely on these
manifolds, very close to $(0,0)$. Begin with $\beta = 0$ and plot the intersections of the
manifolds, delineating the "figure eight" region that you observe numerically at
the central part of the plane. Increasing by small steps the value of β, can you
locate numerically the critical value β_c beyond which the manifolds cease to
intersect? Hint: This phenomenon of *homoclinic tangency* in two-dimensional
mappings is discussed in [151].

Problems

Problem 8.1. Consider the example of the Ikeda map [141]

$$\begin{cases} x_{n+1} = R + u(x_n \cos \tau - y_n \sin \tau) \\ y_{n+1} = u(x_n \sin \tau + y_n \cos \tau) \end{cases} \tag{8.30}$$

where $\tau = C_1 - C_2/(1 + x_n^2 + y_n^2)$, R, u, C_1, C_2 are free parameters. Its Jacobian determinant is $J(R, u, \tau) = u^2$, hence (8.30) is dissipative for $|u| < 1$ and area-preserving for $|u| = 1$.

(a) Fix the values of $C_1 = 0.4$, $C_2 = 6$ and $R = 1$ and show by plotting the iterates of (8.30) in the (x_n, y_n) plane that, for $u = 0.9$, all orbits converge on a strange attractor. Now set $u = 1$ and demonstrate numerically the existence of a chaotic annular region surrounding a central domain about the origin, inside which the motion is predominantly quasiperiodic.

(b) Write a code implementing the procedure outlined in Sect. 8.6.1 and evaluate pdfs of the normalized variables $s_M^{(j)}$ for the parameter values $u = 0.9, 1$, for a high number of iterations M and a large number of initial conditions N_{ic}. Thus, show that the pdf of the strange attractor of the Ikeda map can be well fitted by a Gaussian.

(c) Consider now the area-preserving case $u = 1$, and select initial conditions in the outer chaotic annulus surrounding the origin. Show that in this case the pdfs do *not* converge to a Gaussian, but to a very different function, whose central part is well-fitted by a q-Gaussian with $q = 5.3$, even for very large M. Can you detect any dynamical features on the boundary of this annular region that might justify its characterization as an "edge of chaos" regime?

Problem 8.2. It is possible to extend the study of Sect. 8.6 to higher dimensional conservative maps and obtain results on their chaotic QSS and complex statistics. One such example is a 4D symplectic mapping model of accelerator dynamics [54, 59] (see (5.21)). After some appropriate scaling, the equations of this map can be written as follows

$$\begin{cases} x_{n+1} = 2c_x x_n - x_{n-1} - \rho x_n^2 + y_n^2 \\ y_{n+1} = 2c_y y_n - y_{n-1} + 2x_n y_n \end{cases} \tag{8.31}$$

where $\rho = \beta_x s_x / \beta_y s_y$, $c_{x,y} \equiv \cos(2\pi q_{x,y})$ and $s_{x,y} \equiv \sin(2\pi q_{x,y})$, $q_{x,y}$ are the so-called betatron frequencies and $\beta_{x,y}$ are the betatron functions of the accelerator. Following [54], set $q_x = 0.21$, $q_y = 0.24$ and assume that $\beta_{x,y}$ are proportional to $q_{x,y}^{-1}$.

(a) Investigate weak diffusive phenomena in the y-direction of the four-dimensional phase space, as follows: Start by verifying first that for $y_1 = y_0 = 0$ the point $(x_0, x_1) = (-0.0049, -0.5329)$ is located within a thin chaotic layer surrounding five large islands in the (x_n, x_{n-1}) plane. Then, keeping the same (x_0, x_1), choose (y_1, y_0) very close to zero and observe the evolution of the y_ns, indicating the growth of the beam in the vertical direction as the number of iterations M increases. Plot the projections of the iterates separately on the (x_n, x_{n+1}) and (y_n, y_{n+1}) planes.

(b) Write a code to evaluate the pdfs associated with these orbits, following the procedure outlined in Sect. 8.6.1. Starting always with the same (x_0, x_1), use different initial values for y_0 ($y_1 = 0$) to compute pdfs of the normalized sums of iterates of the y_n-variable. Show that, just as in the case of two-dimensional maps, these distributions are initially of the q-Gaussian type, evolving into exponential distributions and finally turning into Gaussians. Note, for example, that one such QSS with a maximum amplitude of about 0.00001 is apparent up to $N = 2^{19}$, when its amplitude is suddenly *tripled* in the y-direction.

(c) Demonstrate that the closer one starts to $y_0 = y_1 = 0$ the more the resulting pdf resembles a q-Gaussian, while the larger the y_0, y_1 values the faster the pdfs tend towards a Gaussian-like shape.

Chapter 9
Conclusions, Open Problems and Future Outlook

Abstract The final chapter first summarizes and discusses the main conclusions described in the book. We then list a number of open problems, which we feel should be further pursued in continuation of what we have presented in earlier chapters. We start with some recent results that extend the mathematical theory of integrability from the viewpoint of singularity analysis and continue with some directions that further develop the topics of nonlinear normal modes, localization, diffusion and the complex statistical properties of nonlinear lattices. Finally, regarding the future outlook of research in Hamiltonian dynamics, we briefly review three topics of great current interest that were not treated in the book, but are extremely important in view of their far-reaching experimental applications: (1) anomalous heat conduction and the discovery of mechanisms that control heat flow based on the dynamics of Hamiltonian lattices, (2) soliton dynamics in nonlinear photonic structures and (3) kinetic theory of Hamiltonian systems with applications to plasma physics.

9.1 Conclusions

We hope that the preceding eight chapters have given the reader a fairly complete picture of what we believe are some of the most exciting and physically relevant results of recent research in the field of Hamiltonian dynamics. Our purpose was clearly not to dwell on the rigorous mathematical aspects of this research. What we wanted to do is present a number of fascinating developments, which demonstrate that multi-dimensional Hamiltonian systems still hold many secrets in their behavior that cannot be explained by some celebrated theorem and are not tractable by one of the classical approaches of local analysis or straightforward application of perturbation theory.

Hamiltonian systems are *not* complex in the sense that one speaks of complexity in present day scientific literature. They are not described by complex networks, adaptive systems or agent based models and do not display self-organization, pattern formation, or emergence of collective behavior. They are characterized, however, by

T. Bountis and H. Skokos, *Complex Hamiltonian Dynamics*,
Springer Series in Synergetics, DOI 10.1007/978-3-642-27305-6_9,
© Springer-Verlag Berlin Heidelberg 2012

varying degrees of order and chaos at different scales and exhibit fascinating out-of-equilibrium phenomena, long lived metastable states, emergence and breakdown of localized motion and surprising diffusion and transport properties. It is in this sense that we have called Hamiltonian dynamics complex.

One important feature that characterizes the Hamiltonian systems treated in this book is their relevance to specific physical applications. The state of complete order, expressed mathematically by the property of integrability (see Chap. 2) has puzzled researchers since the times of Poincaré and Kowalevskaya at the turn of the twentieth century. Despite the great activity inspired by the work of Painlevé, from the early 1900s until today [98], very few examples have been found that are completely ordered, possessing a full set of integrals. On the other hand, the celebrated KAM theorem of the 1960s and Nekhoroshev's work on the stability of near-integrable Hamiltonian systems [256] demonstrated that close to complete order there is a vast number of physically important examples (the solar system being the most famous), whose behavior is *globally* ordered for "most" initial conditions in phase space and exponentially long intervals in time.

What happens then as one departs from integrability by varying the system's parameters, or waits patiently to find out whether certain weakly chaotic orbits will reach an equilibrium state, where all fundamental modes share the same energy? The fundamental modes we speak of here are none other than the NNMs of Chaps. 3 and 4, formed by the continuation of the linear normal modes of harmonic oscillator systems with nearest-neighbor interactions. NNMs play a very important role in the study of local and global stability of Hamiltonian dynamics. They have been especially helpful in the investigation of the approach to "thermal" equilibrium in Hamiltonian lattices ever since the discovery of FPU recurrences by Fermi, Pasta and Ulam in the mid 1950s [71], which remains to this day a topic of great research interest, as discussed in Chap. 6 of the book.

Chapter 5 discusses local dynamic indicators widely used in distinguishing ordered from chaotic dynamics. More specifically, it reviews the topic of variational equations and tangent dynamics of a reference orbit and focuses on a set of particularly efficient indices called SALI and GALI, which utilize more than one variational vector to identify the nature of the orbit. Analytical results regarding the long time asymptotic behavior of these indices were presented and a great number of numerical experiments were described, verifying the validity of the asymptotic formulas and demonstrating their usefulness in multi-dimensional Hamiltonian systems. Our main conclusion here is that the GALI method is indeed superior to most other similar tools used in the literature for several reasons: First, it detects chaotic behavior faster than other methods, due to the exponential decay of the $GALI_k$s observed more rapidly in the plots of the higher k indices. It is also perhaps the only method that can determine the dimensionality of a torus of quasiperiodic motion and accurately predict its destabilization threshold.

One of the most important issues treated in this book to which the GALI method applies is the breakdown of FPU recurrences that constitutes the main theme of the results described in Chap. 6. In that chapter, we approach the problem of FPU recurrences for the viewpoint of localization in Fourier (or modal q) space.

Using techniques of Poincaré-Linstedt perturbation theory combined with ample numerical evidence, we demonstrate that FPU recurrences can be understood in terms of q-tori which reconcile the approach of q-breathers [128, 129] with of the "natural packet" scenario proposed in [37, 38]. Since these recurrences are related to the presence of stable q tori, the GALI method can be invoked to determine conditions under which q tori become unstable. This leads to a weakly diffusive motion of nearby orbits and eventually to the collapse of recurrences and the onset of energy equipartition in the FPU lattice.

These topics are closely related to the phenomena of localization and diffusion discussed in Chap. 7, this time in configuration (rather than Fourier) space. Indeed, the surprising occurrence of localized oscillations in nonlinear Hamiltonian lattices, pointed out as early as 1988 [306] and studied profusely in the literature ever since, has severe consequences for energy transport and diffusive processes arising in a multitude of applications, like Josephson junction networks, nonlinear optical waveguide arrays, Bose-Einstein condensates and antiferromagnetic layered structures [125]. Furthermore, the breakdown of localization due to disorder and the associated diffusive spread of wavepackets is at present a subject of intense investigation [131, 317], in view of their many applications to light propagation in spatially random nonlinear optical media and related Bose-Einstein condensate phenomena.

Finally, we come to the remarkable complex statistical distributions of weakly chaotic motion in Hamiltonian systems discussed at length in Chap. 8. As with complete order, it turns out that complete disorder is also an exception in Hamiltonian dynamics. Widespread global chaos, giving rise to Gaussian pdfs according to BG statistics, is known to prevail in the so-called Anosov or Sinai systems [11, 12, 307, 308] characterized by completely chaotic motion, such as elastic particles in a box or a periodic array of scatterers, motion on surfaces of everywhere negative curvature, etc. To be sure, Gaussian pdfs and BG statistics were also discovered by many researchers in large scale chaotic regions featured in mixed systems, where domains of order and chaos coexist.

Complexity, however, in the sense discussed in the present book, is manifested in domains of weak chaos, near the boundaries of ordered motion or in cases where stable periodic motion first destabilizes as the total energy is increased. It is there that we find "thin layers" of chaotic orbits, whose pdfs in the sense of the Central Limit Theorem do *not* converge to Gaussians, even after very long times. It is in these cases where complex statistics arises associated with what we have called metastable or quasistationary states in Chap. 8. Remarkably enough, our investigations of these QSS share many common features with analogous studies of dynamical systems characterized by long range interactions and strongly correlated motion [332].

The pdfs of sums of variables associated with this type of QSS are found to be very well described by a class of functions called q-Gaussians related to the notions of Tsallis' entropy and nonextensive statistical mechanics, where the index q satisfies $1 < q < 3$ and $q = 1$ corresponds to the Gaussian distribution. A detailed numerical analysis of these pdfs demonstrates that their occurrence is far from rare. Indeed, in many of the physically important Hamiltonian systems explored in this way, we have discovered that these QSS are closely connected to energy equipartition in one-dimensional lattices, dynamical transitions in a microplasma Hamiltonian and other related phenomena [18]. Particularly with regard to the characterization of energy transport and diffusion in Hamiltonian lattices, it would be of crucial importance to use the index q as a measure of how far our pdfs lie from the Gaussian equilibrium state of BG statistical mechanics.

9.2 Open Problems

As we all know, a book devoted to issues of active scientific interest cannot provide the final word on any of the topics treated in its chapters. Instead, what we have tried to do is communicate to the reader our strong conviction that a subject so classical, so thoroughly studied analytically, numerically and experimentally as Hamiltonian dynamics, still has many surprises to offer. The main reason for this, in our view, lies in the remarkable capacity of Hamiltonian dynamics to arise in many novel and exciting physical applications, where it continues to generate new directions of research.

We would like to close this volume, therefore, by mentioning a number of open problems, which we feel should be pursued, firstly because this will advance our knowledge and expertise in the field of Hamiltonian systems. More importantly, however, progress in these issues will most likely provide answers to present day questions of great physical concern. Naturally, our point of departure will be the topics we have presented in this book and hence the list of problems outlined below is far from being comprehensive. Nevertheless, even in this limited context, we believe that the reader will find our problems interesting enough to try them based on the methods and information contained in the present volume.

9.2.1 Singularity Analysis: Where Mathematics Meets Physics

As we have seen in Chap. 2, the topic of integrability has been of great interest in the study of Hamiltonian systems for centuries. One of the reasons, of course, is that in this way one could study models, where the dynamics is perfectly ordered and understood. As the integrable examples, however, turned out to be a precious few, many physicists were discouraged and integrability gradually became a favourite playground for mathematicians. This situation was drastically changed in the late 1960s, when soliton equations were discovered, which were not only physically interesting [352] but could also be solved analytically by the method of Inverse Scattering Transforms (see e.g. [2, 3]). In this way, the abundance of applications of soliton equations in water waves, nonlinear optics, Josephson junctions and even field theory has kept the subject of integrable PDEs alive for the past 40 years.

Such was not the case, however, with integrable systems of ODEs describing Hamiltonian dynamics. After the celebrated discovery of the Toda lattice [329] and the Ablowitz-Ladik model [4], the search for integrable Hamiltonian systems did not produce many physically interesting examples [60]. Despite a thorough exploration using the tools of Painlevé analysis [98, 279] the harvest of new integrable Hamiltonians has not been particularly satisfying.

Nevertheless, an interesting conclusion did arise concerning the case of *near-integrable* Hamiltonians, where the Painlevé requirement allowing only poles as movable singularities was lifted and *multi-valued* solutions emerged in the complex domain. Ziglin's theory of non-integrability [152] showed that small perturbations of a Painlevé integrable system produced *infinitely branched* solutions, typically containing logarithmic singularities. More importantly, however, the coefficients of the logarithmic terms in the singular expansions were shown to be related to the Mel'nikov integral [160], which is known to provide an estimate of the width of the chaotic layer in nearly-integrable systems [152, 287, 288].

The next question, of course, is what happens when the solutions of a Hamiltonian system are locally *finitely branched*. As was demonstrated in a number of papers [51, 113, 132], local finite branching is *not* sufficient to ensure integrability. It is important that the global Riemann surface be *finitely sheeted*. What often happens in non-integrable systems is that, as one repeatedly integrates along a closed loop in the complex time plane, new singularities appear, which preclude the return to the initial values and produce an infinitely sheeted solution surface, even though each singularity is *by itself* finitely branched. Every time a system of ODEs with finitely branched singularities turned out to also possess a finitely sheeted Riemann surface, a known integrable system was recovered which can be transformed to one that has the full Painlevé property after appropriate variable transformations [152].

Thus, an open problem is to discover a new integrable ODE (or systems of ODEs), whose solutions are locally finitely branched with a finitely sheeted Riemann surface, but which has not yet been identified by the Painlevé analysis. An attempt in that direction was made in [132] in the framework of perturbation theory. Unfortunately, in all the cases tried, the finitely sheeted property established at first order was not preserved at higher orders.

A related problem is to connect the structure of the finitely sheeted Riemann surface of the solutions of a problem with finetely branched singularities to the *dynamical properties* of the orbits. An interesting start in this direction was made in [69,70], where an integrable model of this type was analysed and certain remarkable properties of its solutions (such as the isochrony of its periodic orbits) were found to be related quantitatively to the structure of its Riemann surface. A next step might be to vary slightly the parameters of the model and investigate the onset of chaotic motions in connection with changes in the geometry of the Riemann surface.

Another important problem is to use the analysis described in Chap. 7 of [152] and compute the Mel'nikov vector of a nearly integrable Hamiltonian system using information about the singularities of its solutions on the invariant manifolds associated with its saddle points. The usefulness of this approach lies in the fact that one does not need to know explicitly the corresponding homoclinic (or heteroclinic) solutions of the integrable system to perform this computation. Estimating the magnitude of the components of the Mel'nikov vector, the challenging question is to extract some information about the *width* of the chaotic layer of the system under different *projections* of its dynamics in the multi-dimensional phase space.

9.2.2 Nonlinear Normal Modes, Quasiperiodicity and Localization

Let us recall that in Chap. 3 the problem of stability of motion was approached starting from certain very simple periodic solutions of N dof Hamiltonians like the FPU$-\beta$ one-dimensional lattice, under periodic and fixed boundary conditions. One reason for proceeding in this way is the fact that these periodic solutions are expressed in terms of known functions and the study of their (linear) stability leads to the analysis of a single second order ODE of the Lamé type.

Concerning in particular the solutions called SPO1 (3.26) and SPO2 (3.27) of the FPU-β model under fixed boundary conditions, we observed in Sect. 3.2 that SPO1 destabilizes at much higher energies than SPO2. In fact, the destabilization threshold of SPO2 as a function of the number of particles N was seen to follow the law (3.39), $E_c \propto N^{-2}$, derived in [128] for the low $q = 1, 2, 3, ..$ modes, where q numbers the NNM continuations of the linear modes of the system. And yet our SPO2 solution corresponds to a much higher mode with $q = 2(N + 1)/3$! Why is that so? How can the instability threshold of such a high mode—even in its asymptotic form—coincide with the one satisfied by the low q modes? We do not know.

The SPO1 solution on the other hand, is identified with the $q = (N + 1)/2$ mode and satisfies a very different asymptotic destabilization law of the form $E_c \propto N^{-1}$. The open question, therefore, is: What is the theory that governs the stability of the higher q NNMs as $N \to \infty$? We do not speak here of course of the *highest* q modes at the right end of the spectrum ($q = ... N - 2, N - 1, N$), for which the analysis is very similar to that of the lowest ones, at least for the FPU-β chain under fixed boundary conditions.

It is well-known, after all, that most studies carried out to date on the stability of NNMs in these FPU lattices concern the low q modes, since they are the ones that play the main role in the phenomenon of FPU recurrences (see Chap. 6). What then is the significance of the higher modes? Are they responsible for some other physically important phenomenon? Does their instability threshold obey an asymptotic law different than those we have mentioned so far and what is the deeper meaning of these laws? Again we don't know.

One way to approach this problem, at least for the case of periodic boundary conditions is provided by the theory and results of Chap. 4 (see e.g. Sect. 4.4.2). In that theory, symmetry properties of the equations of motion are exploited to form *bushes* of orbits consisting of linear combinations of fundamental NNMs. Studying then the interactions among the modes belonging to each bush, it is possible to study analytically the linear stability of the bush itself, which generally represents a quasiperiodic solution of the system whose frequencies are those of the modes participating in the bush. Thus, it becomes possible to explore stability properties of these Hamiltonians more globally, as has been done e.g. in [82, 87] (see a review of this approach in [64]).

An open problem, therefore, is to combine the results of the bush theory with the study of stability of low-dimensional tori (called q-tori) discussed in Chap. 6 to investigate more deeply the phenomenon of FPU recurrences. This effort may meet with limited success for lattices with fixed boundaries, due to the small number of symmetries present in that case. Our suggestion, therefore, is to study the FPU recurrence problem for *periodic* boundary conditions from the point of view of bush theory and compare the results with the q-tori approach of Chap. 6. Note that care must be taken to deal with the twofold degeneracy of the linear spectrum in periodic models, which modifies somewhat the analysis of the recurrence problem [272].

Let us now come to the question of stability of tori of quasiperiodic motion and its connection with the phenomenon of localization in nonlinear lattices. As the reader recalls, in Chap. 6 we interpreted the phenomenon of FPU recurrences as a result of localization in the Fourier *modal* space and studied in Sect. 6.3.1 the stability of the associated q-tori using the GALI methods of Chap. 5. What about localization in *configuration* space as exemplified e.g. by the occurrence of discrete breathers, like those investigated in Chap. 7? Do we also find low-dimensional tori in their neighborhood?

In [63] this question was answered affirmatively in the case of a lattice model described by the Hamiltonian

$$H(t) = \sum_{n=1}^{N} \left\{ \frac{1}{2}\dot{x}_n^2 + \frac{1}{2}[1 - \epsilon \cos(\omega_d t)]x_n^2 - \frac{1}{4}x_n^4 + \frac{K}{4}(x_{n+1} - x_n)^4 \right\}, \quad (9.1)$$

in which harmonic nearest-neighbor interactions are absent and a periodic forcing term has been added with amplitude ϵ and frequency ω_d. This Hamiltonian was studied in detail in [241], where among other phenomena chaotic breathers were also observed that remain localized on a small number of sites for very long times. As was explicitly shown in [63] via the GALI criterion, in the neighborhood of certain fundamental discrete breathers of (9.1), low-dimensional tori are seen to exist, which become unstable as the parameters of the problem are varied.

The following question arises therefore: Do these localization phenomena depend on the fact that (9.1) is *free* from linear modes that can destroy localized oscillations through resonances with the phonon spectrum? It would be very interesting to investigate this question by adding to the above Hamiltonian a harmonic part of the form $\sum_{n=1}^{N} \{ \frac{L}{2}(x_{n+1} - x_n)^2 \}$, and slowly vary $L \geq 0$ to determine how these terms affect the behavior of the dynamics.

As we also found out in Chaps. 5 and 8, another class of models that one can study in connection with Hamiltonian dynamics are *symplectic maps*, which are numerically far easier to explore and arise in many physical applications [42, 50, 58,

180, 244, 269]. Thus, it might be quite illuminating to represent each oscillator by a two-dimensional symplectic map (e.g. of the standard map type [136,214,246]) and consider a one-dimensional "lattice" formed by N such maps coupled by nearest neighbor interactions as follows

$$x_{n+1}^j = x_n^j + y_{n+1}^j,$$

$$y_{n+1}^j = y_n^j + \frac{K_j}{2\pi} \sin(2\pi x_n^j) - \frac{\beta}{2\pi} \{\sin[2\pi(x_n^{j+1} - x_n^j)] + \sin[2\pi(x_n^{j-1} - x_n^j)]\}$$

$$(9.2)$$

where $j = 1, \ldots, N$, β is the coupling parameter between the maps and fixed boundary conditions $x_0 = x_{N+1} = 0$ are assumed.

This model was investigated in [63], where it was demonstrated numerically that it does possess discrete breather solutions, around which low-dimensional tori can be identified, as long as the discrete breather is stable. Furthermore, changing to linear normal mode coordinates, in which the linear parts of the maps are uncoupled, it is easy to show that when the energy is placed initially in $s = 1, 2, 3, \ldots$, of these modes, and the parameter β is small enough, FPU-like recurrences arise, accompanied by the presence of s-dimensional tori, just as we found in the FPU Hamiltonian models. Furthermore, the stability of these tori and the breakdown of recurrences as β increases can also be accurately monitored using the GALI method [63].

It would, therefore, be very interesting to extend these studies and investigate in greater detail dynamical phenomena in coupled standard maps, rather than pursuing them on Hamiltonian lattices, which is computationally far more time-consuming. For example, it would be very interesting to impose periodic boundary conditions to the coupled map system (9.2) and try to understand its bushes of orbits, FPU type recurrences and their effect on the onset of energy equipartition in such models.

Finally, it is important to mention certain open questions related with the stability of discrete breathers in Hamiltonian lattices. For instance, why is it that *simple* breathers (exhibiting only one exponentially localized central maximum) are so frequently found to be stable under small perturbations? More specifically, let us consider multibreathers, which represent localized oscillations with more than one major maximum (or minimum) and compute all of them for a one-dimensional lattice of N particles with harmonic nearest neighbor interactions and quartic on-site potential, using homoclinic orbit methods [39, 42]. Let us also associate them symbolically with sequences of $0, +, -$, corresponding to sites of small (or no) oscillation), large positive and large negative amplitude respectively [41]. Why is it that most of them are found to be *unstable* under small changes of their initial conditions [40]? More remarkably, how can one explain the observation that this instability characterizes all multibreathers which contain one (or more) of the combinations $+-$, $+0+$, or $+00+$ in their symbolic representation?

9.2.3 Diffusion, Quasi-stationary States and Complex Statistics

And now we come to one of the most engaging aspects of complex Hamiltonian dynamics, which concerns the phenomenon of diffusion, by which we mean slow and gradual transport of energy in weakly chaotic domains. This transport concerns either modes in Fourier space, as we found out while studying the breakdown of FPU recurrences in Chap. 6 (see Sect. 6.4), or particle motion in configuration space, resulting from the breakdown of localized disturbances in disordered lattices, as we discovered in Sect. 7.4.2.

Let us start with diffusion in modal space and recall the results presented in Sect. 6.4.1 on what we called stage II of the dynamics of the FPU-α lattice. We had shown first that the model exhibits a slow transport of its total energy E from an initial "packet" of low modes to the tail modes, on its way to energy equipartition as E is increased. Studying the rate of this process, we observed that it is *sub-diffusive* ($\propto t^{\gamma}$, with $\gamma < 1$) and had attempted to estimate the time it takes for the system to reach equipartition, using the theoretical prediction of the diffusion rate of the motion described in Sect. 6.4.2.

More specifically, we defined a quantity $\eta(t)$ (6.44) as the sum of normalized harmonic energies of the upper third of the mode spectrum and plotted in Fig. 6.10 $\eta(t)$ as a function of t, for $E = 0.01$, $E = 0.1$ and $E = 1$. We thus observed a diffusion law of the form $\eta \propto Dt^{\gamma}$, demonstrating first that the process indeed is one of subdiffusion. Attempting then to estimate the time interval T^{eq} required for energy equipartition in Sect. 6.4.3, we found $T^{eq} \sim E^{-3}$, in agreement with an analogous formula given in [112], for energies $E \in [0.4, 2]$. This prediction, however, disagrees with the results of numerical integration for $0.1 < E < 0.4$, while for $E \leq 0.1$, it is impossible to check its validity by direct computation. Indeed, for $E = 0.01$ the value of the exponent γ is practically zero.

The open problem, therefore, is to extend the studies we have described for the FPU-α model to energy regimes where all attempts of the kind mentioned above have failed. To date we do not know of a theoretical estimate that allows us to reliably predict, even for one-dimensional lattices of the FPU type, the time T^{eq} required for equipartition when E becomes too small. Thus, we can only guess: Either equipartition times become exponentially long, or the system remains trapped on the neighborhood of a q-torus for all time and equipartition *never* occurs below a certain energy threshold.

This brings us to a similar question regarding diffusion in one-dimensional Hamiltonian lattices in the presence of disorder, discussed in Sect. 7.4 of Chap. 7: What happens if nonlinearity is introduced to the system? How can we understand its effect on the localization properties of wave packets in disordered systems? This challenging problem has been investigated in recent years by many researchers [47, 48, 124, 131, 144, 202, 210, 252–255, 271, 311, 317, 338, 339]. Most of these papers consider the evolution of an initially localized wave packet and show that wave packets spread subdiffusively for sufficiently strong nonlinearities.

On the other hand, if the nonlinearities are weak enough, wave packets appear to be frozen, in the sense of Anderson localization, at least for finite integration times. In fact, an alternative interpretation regarding wave packet evolution has been conjectured in [178] claiming that these states may be localized on some KAM torus for *infinite* times! Which of these interpretations is correct? Do wave packets continue to spread subdiffusively in some weakly chaotic domain for all time, or do they eventually settle down on a finite (or infinite) dimensional torus?

We suggest here that this question may also be studied by a *statistical* approach, similar to the one we have described in Chap. 8 of this book. As we have explained in that chapter, chaotic orbits in weakly chaotic domains of multi-dimensional Hamiltonian systems are often seen to form QSS that are extremely long lived. This is exemplified by pdfs of sums of their variables, which, in the sense of the CLT, do not converge to Gaussians, but are well-described by a class of so-called q-Gaussian functions, whose index lies in the interval $1 < q < 3$ and tends to $q = 1$ when the QSS reaches the Gaussian state of strong chaos.

An interesting open problem, therefore, may be formulated as follows: Consider the QSS formed by the spreading of a wave packet resulting from the delocalization of a central excitation in a disordered KG or DNLS one-dimensional lattice. Evaluate the sum of the position coordinates of all particles and study the time evolution of its normalized pdf as it evolves in a presumably weakly chaotic region of phase space. In the event that this pdf is well approximated by a q-Gaussian, estimate the index q and plot its behavior as a function of time. If its values show a tendency to approach $q = 1$ as time progresses then the process can be characterized as chaotically diffusing, while if the q values remain far from unity for all time, this would constitute evidence that the dynamics eventually "sticks" to the boundary of some high-dimensional torus of the system.

Finally, we mention some additional open problems that concern the complex statistics of weakly chaotic motion in Hamiltonian systems. In Chap. 8, we saw that it was possible to find QSS that can be fitted well by q-Gaussian pdfs for a number of Hamiltonian systems mainly of the FPU type [18]. This has been primarily achieved in the neighborhood of certain NNMs of the FPU-β model, at energies where these modes first turn unstable as the total energy E is increased. Examples of such NNMs were the π-mode in the case of periodic boundary conditions and the SPO1 and SPO2 modes for fixed boundary conditions [18].

The obvious question then is: Where else in the multi-dimensional phase space can one find weakly chaotic regimes, where similar QSS phenomena are observed? Are these important enough physically to justify the use of nonextensive statistical mechanics to characterize the dynamics in the corresponding regimes? One such possibility already mentioned above concerns the spreading of wave packets in disordered nonlinear lattices. Another one refers to the role of q-Gaussian approximations of pdfs for chaotic orbits associated with the breakdown of recurrences in the FPU-α model, which was recently studied in [15]. Are there any other examples of Hamiltonian dynamics, where this type of complex statistics might be relevant?

9.3   Future Outlook

9.3.1   Anomalous Heat Conduction and Control of Heat Flow

The problem of heat conduction in dielectric crystals has been an important issue in mathematical physics for many years. The first celebrated approach to this problem is due to none other than Rudolf Peierls (1907–1905), who proposed in the 1930s an explanation based on the fundamental assumption that the nonlinearity of phonon interactions can lead to a state of thermal equilibrium, at which normal diffusion can occur with constant heat conductivity [264].

Ever since the famous FPU experiments of the 1950s, however, there has been an impressive amount of analytical and numerical work by many authors, who showed on a wide variety of lattice models that nonlinearity is *not* sufficient to ensure energy equipartition between the phonon modes and normal thermal conduction. Indeed, even in cases where normal heat conductivity was observed, the heat conduction coefficient often depends strongly on the model and differs significantly from what is measured in actual physical experiments [219].

As noted by Casati and Li in [74], one-dimensional FPU lattices exhibit anomalous heat conduction, whose thermal conductivity coefficient κ is found to diverge with the system size L as $\kappa \sim L^{2/5}$, while when transverse motions are included in the model one finds $\kappa \sim L^{1/3}$ [343]. In fact, a simple formula has been proposed relating heat conductivity with anomalous diffusion in a one dimensional system [221], as follows: Let us recall that Fourier's law of heat conduction states that the heat flux j is proportional to the temperature gradient ∇T as

$$j = -\kappa \nabla T, \tag{9.3}$$

this implies that j is *independent* of the size of the system. Denoting energy diffusion by $< \sigma^2 > = 2Dt^{\alpha}$, $(0 < \alpha \leq 2)$, it has been demonstrated in [221] that $\kappa \propto L^{\beta}$, with $\beta = 2 - 2/\alpha$!

This latter relation, corresponding to subdiffusion for $\alpha < 1$ and superdiffusion for $\alpha > 1$, was found to be in good agreement with many results available from one-dimensional lattices, as well as existing data from many physical systems [74]. On the other hand, the situation regarding the divergence of the conductivity coefficient with system size in two and three spatial dimensional FPU models is still a topic of ongoing investigation [220, 302].

Another major development in this field has been the realization that exponential dynamical instability is a sufficient but *not necessary* condition for the validity of Fourier's law (9.3) [9, 76, 222–224]. Indeed, in a model of a Lorentz gas "channel", represented by two parallel lines and a series of semicircles of radius R placed along the channel in a "triangular" way so that no particle can move in the horizontal direction without colliding with the disks, accurate numerical evidence has shown [9] that heat conduction *does* obey Fourier's law. As the dynamics of the Lorentz gas is rigorously known to be mixing with positive Kolmogorov-Sinai entropy, this result clearly suggests that mixing with positive Lyapunov exponents is a "sufficient" condition for the validity of Fourier's law of heat conduction.

On the other hand, when the semi-circles in the above model are replaced by triangles, with all angles irrational multiples of π, it has been shown [75] that the Lyapunov exponent is zero and yet the model exhibits unbounded Gaussian diffusive behavior. This strongly suggests that normal heat conduction can take place even without the strong requirement of exponential instability.

In another series of studies, researchers considered the possibility of actually being able to *control* the heat flow, in a variety of one-dimensional nonlinear lattices. Based on a model proposed in [326], in which heat was numerically shown to flow in one preferential direction, Li, Wang and Casati [225] considered an improved version described by the Hamiltonian

$$ H = \sum_{i=1}^{N} \left(\frac{p_i^2}{2m} + \frac{1}{2}k(x_i - x_{i+1} - a)^2 - \frac{V}{(2\pi)^2}cos(2\pi x_i) \right), \qquad (9.4) $$

and divided the corresponding lattice in three parts: On the left part (L) the particles are assigned a spring constant $k = k_L$ and on site parameter $V = V_L$, while on the right part (R) $k = k_R = \lambda k_L$ and $V = V_R = \lambda V_L$. The two ends of the lattice are connected to heat reservoirs at temperatures $T = T_L = T_0(1 + \Delta)$, $T = T_R = T_0(1 - \Delta)$ and in the middle part of the lattice the spring constant was set to $k = k_{int}$. Fixing now the values of $a = m = 1$, $V_L = 5$, $k_L = 1$, one may adjust the parameters $\Delta, \lambda, k_{int}, T_0$ and plot the heat current as a function of Δ for different values of T_0.

As is well-known, (9.4) is the Hamiltonian of the Frenkel-Kontorova model, which has been shown to exhibit normal heat conduction [172] in the homogeneous case $k_L = k_{int} = k_R = k$, $V_L = V_R = V$. However, setting $k_{int} = 0.05, \lambda = 0.2$ and $N = 100$ a remarkable phenomenon was observed: For $\Delta > 0$ the heat current j_+ was found to increase with Δ, while in the region $\Delta < 0$ the heat current j_- is almost zero, implying that the system behaves as a thermal insulator! In fact, by increasing the value of $T_0 = 0.05, 0.07, 0.09$ the rectifying efficiency, defined as $|j_+/j_-|$ can increase by as much as a few hundred times [225]!

Thus the study of these models has opened up the way for a multitude of applications of great importance such the designing of novel thermal materials and devices like thermal rectifiers [225, 326] and thermal transistors [230]. Indeed, the possibility of building such devices has been demonstrated in many laboratory experiments [79, 192, 270, 348], whose discoveries may eventually lead to practically useful products.

9.3.2 Complex Soliton Dynamics in Nonlinear Photonic Structures

The behavior of light in media with nonlinear optical response is one more topic of intense research interest, where Hamiltonian dynamics is involved. The main reason for this is the fact that nonlinear effects in light propagation such as self-localization and nonlinear interactions have far-reaching technological applications in optical telecommunications, medicine and biotechnology. The invention of the laser along with recent developments in material science and the advent of novel nonlinear optical materials have led to an explosion of experimental and theoretical studies in nonlinear optics. More specifically, nonlinear effects in optical devices and the complexity of the associated wave dynamics have been recognized as effective potential mechanisms for light control.

One of the most fundamental phenomena in nonlinear optical wave propagation is the self-trapping of spatially localized beams, resulting from the balance between *diffraction* and nonlinearity. The concept of the soliton, as a solitary wave which remains intact under mutual interactions (see Sect. 6.1.1), has played a crucial role in the development of nonlinear optics [190,330]. Solitons are obtained as solutions of PDEs corresponding to infinite-dimensional integrable Hamiltonian systems and are remarkably robust under small perturbations [2,3]. In the context of nonlinear optics, wave propagation is accurately modeled by the Nonlinear Schrödinger Equation (NLS) under the so-called paraxial approximation [215].

Integrability, however, as we have so often said, is the exception. Once spatial *inhomogeneities* are allowed in the system the equations describing wave propagation become non-integrable due to symmetry breaking. This excludes the existence of pure solitons in the strict mathematical sense, but gives rise to the occurrence of a plethora of solitary waves, which do not have a counterpart in the homogeneous case [91, 182, 183, 216]. More importantly, it results in complex and interesting wave dynamics, allowing for the control and routing of light beams. For example, an unstable stationary solution could evolve into a stable one under certain circumstances, providing a transition phenomenon with many potential applications. Furthermore, many studies have shown that different types of inhomogeneity can be very useful for designing photonic structures having desirable features and properties. The latter are expected to provide optical devices with an all-optical circuitry capable of switching, routing, information processing and computing beyond the limitations of electronics-based systems. It is also important to emphasize that all the above concepts and effects are relevant to many other domains of nonlinear physics such as the formation of spin waves in magnetic films and the rapidly developing field of coherent matter waves and nonlinear atom optics [65, 260].

Now, wave propagation of the electric field variable $u = u(z, x)$ in an inhomogeneous medium with Kerr-type nonlinearity is described by the perturbed NLS

$$i \frac{\partial u}{\partial z} + \frac{\partial^2 u}{\partial x^2} + 2|u|^2 u + \varepsilon \left[\Delta n_0(x, z) + \Delta n_2(x, z)|u|^2 \right] u = 0 \qquad (9.5)$$

where z and x are the normalized propagation distance and transverse coordinate respectively and the potential functions $\Delta n_i(x, z)$, $i = 0, 2$ model the spatial variation of the linear and nonlinear refractive indices. The dimensionless parameter ε indicates the strength of the potential and is "small" for relatively weakly inhomogeneous media, which is the case of interest in technological applications. The functions $\Delta n_i(x,z)$, $i = 0, 2$ can be either periodic or nonperiodic in x and z.

As is well-known, the unperturbed NLS ((9.5) with $\varepsilon = 0$), has a fundamental soliton solution of the form

$$u(x,z) = A\,\text{sech}[A(x - x_0)]e^{i(\frac{v}{2}x + 2\sigma)} \tag{9.6}$$

where $x_0 = x_0(z) = vz$, $\sigma = \sigma(z) = -zv^2/8 + zA^2/2$, A is the amplitude or the inverse width of the soliton solution, x_0 is its center (called "center of mass" due to an analogy between solitons and particles), v the constant velocity, and σ the nonlinear phase shift.

The longitudinal evolution of the center x_0 under the lattice perturbation is obtained by applying the effective-particle method [191]. According to this method one assumes that the functional form as well as other properties of the soliton (width, power) are conserved in the case of the weakly perturbed NLS. This assumption has to be verified, however, through numerical integration of (9.5) at least to a good approximation. The motion of the center x_0 is equivalent to the motion of a particle with mass $m \equiv \int |u|^2 dx = 2A$ under the influence of an effective potential $V_{eff}(x_0, z) = 2\epsilon \int [\Delta n_0(x,z) + \Delta n_2(x,z)|u(x;x_0)|^2]\,|u(x;x_0)|^2 dx$ and is described by the nonautonomous Hamiltonian

$$H = \frac{mv^2}{2} + V_{eff}(x_0, z) \tag{9.7}$$

where z is considered as "time" and $v = \dot{x}_0$ (the dot denotes differentiation with respect to z) is the velocity of the soliton's "center of mass".

In the case of zero longitudinal modulation ($\partial V_{eff}/\partial z = 0$) the system is integrable and soliton dynamics is completely determined by the form of the z-independent effective potential. Stable and unstable solitons can be formed at the minima and maxima of the effective potential, while solitons can also be either trapped between two maxima of the effective potential or reflected by potential barriers. Soliton dynamics in several photonic structures can be described under this approach. For the simpler case of harmonic (sinusoidal) transverse modulation of the linear refractive index, the positions of the stable and unstable solitons are independent of the soliton "mass" (power) [119]. For more complex modulations of the linear and nonlinear refractive indices, as well as modulations with incommensurate wavelengths, soliton positions and stability depend strongly on the specific soliton power [196]. The latter results in additional functionality providing power-controlled discrimination of such structures. In addition, semi-periodic photonic structures can also be described such as interfaces between two periodic lattices with different characteristics as well as interfaces between periodic lattices and homogeneous media. Thus, it is shown that power-dependent soliton reflection, transmission, and trapping can be achieved by the formation of effective potential barriers and wells [197].

The presence of an explicit z dependence in the effective potential ($\partial V_{eff}/\partial z \neq 0$) results in the nonintegrability of the Hamiltonian system which describes the soliton motion and allows for a plethora of qualitatively different soliton evolution scenarios. The corresponding richness and complexity of soliton dynamics opens a large range of possibilities for interesting applications where the underlying inhomogeneity results in advanced functionality of the medium. Nonintegrability results in the destruction of the heteroclinic orbit (separating trapped from travelling solitons), allowing for dynamical trapping and detrapping of solitons. Therefore, we can have conditions for enhanced soliton mobility: Solitons with small initial energy can travel through the lattice as well as hop between adjacent potential wells and dynamically be trapped in a much wider area. Moreover, resonances between the frequencies of the internal unperturbed soliton motion and the z-modulation frequencies result in quasiperiodic trapping and symmetry breaking with respect to the velocity sign. It is worth mentioning that all these properties depend strongly on soliton characteristics, such as its amplitude A (see (9.6)), so that different solitons undergo qualitatively and quantitatively distinct dynamical evolution in the same inhomogeneous medium [263].

For strong modulations of the optical medium properties the effective particle approach is no longer valid. However, there are several cases where analytical solutions can still be found such as the case of a nonlinear Kronig-Penney model consisting of interlaced linear and nonlinear regions. The underlying systems are within a large class of *piecewise linear systems* for which analytical solutions can be constructed by utilizing the phase space of the system [194, 200]. Following this approach, solitary wave solutions have been obtained for periodic [193, 195] and nonperiodic [198] structures, which can be further utilized in perturbative techniques for the exploration of new classes of solutions.

Thus, the study of soliton dynamics in complex photonic structures is a challenging research direction, as it promises to provide light control functionality in appropriately designed media with numerous technological applications. It is a typical case of a field where complexity opens new possibilities with significant technological advantage, and where the concepts and methods of Hamiltonian dynamics provide excellent tools for understanding and designing practical systems with desirable properties.

9.3.3 Kinetic Theory of Hamiltonian Systems and Applications to Plasma Physics

The phase space evolution of particle distribution functions of non-integrable Hamiltonian systems is usually described by a *quasilinear theory* leading to an action diffusion equation, in which the perturbation terms are taken into account through a diffusion operator $\mathbf{D}$ that generally depends on both space and time [114, 337]. This diffusion equation may be derived starting with the Fokker Planck equation

$$\frac{\partial F}{\partial t} = -\frac{\partial}{\partial \mathbf{J}}(\mathbf{V}F) + \frac{1}{2}\frac{\partial^2}{\partial^2 \mathbf{J}}(\mathbf{D}F), \qquad (9.8)$$

where $\mathbf{J}$ is the action variable in x, y, z space. Noting now that [201, 226]

$$\mathbf{V} = \frac{1}{2} \frac{\partial \mathbf{D}}{\partial \mathbf{J}}, \tag{9.9}$$

the desired diffusion equation is directly obtained as

$$\frac{\partial F}{\partial t} = \frac{\partial}{\partial \mathbf{J}} \left(\mathbf{D} \frac{\partial F}{\partial \mathbf{J}} \right). \tag{9.10}$$

In the standard derivations of the above equations it is assumed that motion is randomized, with respect to the phase of the perturbation, after one period. This is related to the Markovian assumption, used for example in studying Brownian motion, according to which the dynamics is characterized by completely uncorrelated particle orbits, phase-mixing, loss of memory and ergodicity. These statistical properties naturally lead to the important simplification that the long time behavior of particle dynamics remains the same after one interaction time with the wave. The significant drawback is that the diffusion coefficient is singular, with a Dirac delta function singularity so that further considerations are necessary in order to utilize such operator in theoretical and numerical studies [114, 337].

Unfortunately, the Markovian assumption runs contrary to the dynamical behavior of particles interacting with coherent perturbations, since phase space is often a mixture of chaos and order with islands of coherent motion embedded within chaotic regimes [226]. Furthermore, near the boundaries of such islands particles are observed to "stick" and undergo coherent motion for times much longer than the interaction time. Even when the amplitude of the perturbations is assumed to be impractically large so that the entire phase is chaotic, the quasilinear theory fails to give an appropriate description of the evolution of the distribution function [280]. The persistence of long time correlations invalidates the Markovian assumption [36, 73, 301, 353].

Therefore, we need to develop a kinetic theory, based on first principles, in order to study a large variety of realistic systems for which the Markovian assumptions are simply not valid. Recently, such a theory was proposed using Hamiltonian perturbation theory and Lie transform techniques to derive a hierarchy of kinetic evolution equations that does not rely on any simplifying statistical assumptions [201]. The final kinetic equation in the hierarchy is obtained by averaging over all angles and represents an evolution equation for the distribution function in action space with an operator that is nonsingular and time-dependent. Moreover, this equation is time-reversible and is capable of describing the transient time evolution of the action distribution function as well as a range of dynamical phenomena much richer and wider than simple diffusion.

Let us consider the Hamiltonian corresponding to a generic system

$$H(\mathbf{J}, \boldsymbol{\theta}, t) = H_0(\mathbf{J}) + \epsilon H_1(\mathbf{J}, \boldsymbol{\theta}, t). \tag{9.11}$$

The Hamiltonian consists of two parts: the integrable part $H_0(\mathbf{J})$ that is a function of the constants of the motion $\mathbf{J}$ of a particle moving in a prescribed equilibrium field, and

$$H_1(\mathbf{J}, \boldsymbol{\theta}, t) = \sum_{\mathbf{m} \neq 0} A_{\mathbf{m}}(\mathbf{J}) e^{i(\mathbf{m} \cdot \boldsymbol{\theta} - \omega'_{\mathbf{m}} t)} \tag{9.12}$$

which includes perturbations to the equilibrium field. $\boldsymbol{\theta}$ are the angles canonically conjugate to the actions $\boldsymbol{J}$ and t is time. The complex frequency $\omega'_m = \omega_m + i\gamma_m$ allows for steady state ($\gamma_m = 0$), growing ($\gamma_m > 0$), or damped ($\gamma_m < 0$) waves. Note that for the case of a charged particle in a toroidal plasma, in the guiding center approximation for an axisymmetric equilibrium [186], the three actions are the magnetic moment, the canonical angular momentum, and the toroidal flux enclosed by a drift surface. The respective conjugate angles are the gyrophase, azimuthal angle, and poloidal angle. The perturbation terms correspond to electromagnetic waves and ϵ is an ordering parameter indicating that the effect of H_1 is perturbative.

According to the proposed kinetic theory [201], the operator $\mathbf{D}$ in (9.10) is given by

$$\mathbf{D}(\mathbf{J}, t) = \sum_{\mathbf{m} \neq 0} \frac{\mathbf{m}\mathbf{m} |A_{\mathbf{m}}(\mathbf{J})|^2 \, e^{2\gamma_{\mathbf{m}} t}}{\Omega_{\mathbf{m}}(\mathbf{J})^2 + \gamma_{\mathbf{m}}^2} \left\{ \gamma_{\mathbf{m}} \left[1 - e^{-\gamma_{\mathbf{m}} t} \cos\left(\Omega_{\mathbf{m}}(\mathbf{J}) t\right) \right] + \right.$$

$$\left. + \Omega_{\mathbf{m}}(\mathbf{J}) e^{-\gamma_{\mathbf{m}} t} \sin\left(\Omega_{\mathbf{m}}(\mathbf{J}) t\right) \right\} \tag{9.13}$$

where $\Omega_{\mathbf{m}}(\mathbf{J}) = \mathbf{m} \cdot \omega_0(\mathbf{J}) - \omega_m$ and $\mathbf{m}\mathbf{m}$ is a dyadic.

In the limit $t \to \infty$ and $\gamma_m t \to 0$, (9.13) leads to the standard time-independent quasilinear diffusion tensor

$$\mathbf{D}_{\mathbf{ql}}(\mathbf{J}) = \sum_{\mathbf{m} \neq 0} \mathbf{m}\mathbf{m} |A_{\mathbf{m}}(\mathbf{J})|^2 \, \delta\left(\Omega_{\mathbf{m}}(\mathbf{J})\right), \tag{9.14}$$

where δ is Dirac's delta function. The long time limit is justified only for statistically random, or Markovian processes [226], while the delta function excludes short time transient effects and is difficult to implement numerically. More importantly, the asymptotic time limit results in a time-irreversible Fokker Planck equation, while the time-dependent $\mathbf{D}$ in (9.13), being an odd function of time, leads to a time-reversible evolution equation. In contrast to the traditional quasilinear theory [114, 337], the kinetic evolution equation (9.10) with the time-dependent operator (9.13) does not distinguish between resonant and nonresonant particles and includes both growing and damped waves. In the vicinity of resonances given by $\Omega_{\mathbf{m}} = 0$, $\mathbf{D}$ is continuous and non-singular even when $\gamma_{\mathbf{m}} = 0$ and the width of the resonance decreases with time.

This kinetic theory is expected to provide an accurate description of many complex realistic systems of technological interest. Among the most important ones are fusion plasmas, where charged particles are confined by a static magnetic field and interact with either static magnetic perturbations or perturbations by radio frequency waves injected to heat the plasma, drive the current, and suppress local instabilities [186, 199]. In addition, the theory is directly applicable to space plasmas, particle acceleration studies and, in general, any Hamiltonian system where chaotic motion coexists with regular motion in a strongly inhomogeneous phase space.

References

1. F. Abdullaev, O. Bang, M.P. Sørensen (eds.), *Nonlinearity and Disorder: Theory and Applications*. NATO Science Series II: Mathematics, Physics and Chemistry, vol. 45 (Springer, Heidelberg, 2002)
2. M.J. Ablowitz, P.A. Clarkson, *Solitons, Nonlinear Evolution Equations and Inverse Scattering*. London Mathematical Society Lecture Note Series, vol. 149 (Cambridge University Press, Cambridge, 1991)
3. M.J. Ablowitz, H. Segur, *Solitons and the Inverse Scattering Transform* (SIAM, Philadelphia, 1981)
4. M.J. Ablowitz, B. Prinari, A.D. Trubatch, *Discrete and Continuous Nonlinear Schrödinger Systems* (Cambridge University Press, Cambridge, 2004)
5. E. Abrahams, P.W. Anderson, D.C. Licciardello, T.V. Ramakrishnan, Scaling theory of localization: absence of quantum diffusion in two dimensions. Phys. Rev. Lett. **42**, 673–676 (1979)
6. M. Abramowitz, I. Stegun, *Handbook of Mathematical Functions* (Dover, New York, 1965)
7. O. Afsar, U. Tirnakli, Probability densities for the sums of iterates of the sine-circle map in the vicinity of the quasiperiodic edge of chaos. Phys. Rev. E **82**, 046210 (2010)
8. Y. Aizawa, Symbolic dynamics approach to the two-dimensional chaos in area-preserving maps. Prog. Theor. Phys. **71**, 1419–1421 (1984)
9. D. Alonso, R. Artuso, G. Casati, I. Guarneri, Heat conductivity and dynamical instability. Phys. Rev. Lett. **82**, 1859–1862 (1999)
10. P.W. Anderson, Absence of diffusion in certain random lattices. Phys. Rev. **109**, 1492–1505 (1958)
11. D.V. Anosov, Geodesic flows on a compact Riemann manifold of negative curvature. Trudy Mat. Inst. Steklov **90**, 3–210 (1967). English translation, Proc. Steklov Math. Inst. **90**, 3–210 (1967)
12. D.V. Anosov, Y.G. Sinai, Some smooth Ergodic systems. Russ. Math. Surv. **22**(5), 103–167 (1967)
13. Ch. Antonopoulos, T. Bountis, Stability of simple periodic orbits and chaos in a Fermi-Pasta-Ulam lattice. Phys. Rev. E **73**, 056206 (2006)
14. Ch. Antonopoulos, T. Bountis, Detecting order and chaos by the linear dependence index (LDI) method. ROMAI J. **2**, 1–13 (2006)
15. Ch. Antonopoulos, H. Christodoulidi, Weak chaos detection in the Fermi-Pasta-Ulam-α system using q-Gaussian statistics. Int. J. Bifurc. Chaos **21**, 2285–2296 (2011)
16. Ch. Antonopoulos, T.C. Bountis, Ch. Skokos, Chaotic dynamics of N-degree of freedom Hamiltonian systems. Int. J. Bifurc. Chaos **16**, 1777–1793 (2006)

17. Ch. Antonopoulos, V. Basios, T. Bountis, Weak chaos and the "melting transition" in a confined microplasma system. Phys. Rev. E. **81**, 016211 (2010)
18. Ch. Antonopoulos, T. Bountis, V. Basios, Quasi-stationary chaotic states of multidimensional Hamiltonian systems. Phys. A **390**, 3290–3307 (2011)
19. V.I. Arnold, *Mathematical Methods of Classical Mechanics* (Springer, New York, 1989)
20. V.I. Arnold, A. Avez, Problèmes Ergodiques de la Mécanique Classique (Gauthier-Villars, Paris, 1967 / Benjamin, New York, 1968)
21. S. Aubry, Breathers in nonlinear lattices: existence, linear stability and quantization. Phys. D **103**, 201–250 (1997)
22. F. Baldovin, E. Brigatti, C. Tsallis, Quasi-stationary states in low-dimensional Hamiltonian systems. Phys. Lett. A **320**, 254–260 (2004)
23. F. Baldovin, L.G. Moyano, A.P. Majtey, A. Robledo, C. Tsallis, Ubiquity of metastable-to-stable crossover in weakly chaotic dynamical systems. Phys. A **340**, 205–218 (2004)
24. D. Bambusi, A. Ponno, On metastability in FPU. Commun. Math. Phys. **264**, 539–561 (2006)
25. D. Bambusi, A. Ponno, *Resonance, Metastability and Blow Up in FPU*. Lecture Notes in Physics, vol. 728 (Springer, New York/Berlin, 2008), pp. 191–205
26. R. Barrio, Sensitivity tools vs. Poincaré sections. Chaos Soliton Fract. **25**, 711–726 (2005)
27. R. Barrio, Painting chaos: a gallery of sensitivity plots of classical problems. Int. J. Bifurc. Chaos **16**, 2777–2798 (2006)
28. R. Barrio, W. Borczyk, S. Breiter, Spurious structures in chaos indicators maps. Chaos Soliton Fract. **40**, 1697–1714 (2009)
29. C. Beck, Brownian motion from deterministic dynamics. Phys. A **169**, 324–336 (1990)
30. G. Benettin, A. Ponno, Time-scales to equipartition in the Fermi-Pasta-Ulam problem: finite size effects and thermodynamic limit. J. Stat. Phys. **144**, 793–812 (2011)
31. G. Benettin, L. Galgani, A. Giorgilli, J.-M. Strelcyn, Lyapunov characteristic exponents for smooth dynamical systems and for Hamiltonian systems; a method for computing all of them. Part 1: Theory. Meccanica **15**, 9–20 (1980)
32. G. Benettin, L. Galgani, A. Giorgilli, J.-M. Strelcyn, Lyapunov characteristic exponents for smooth dynamical systems and for Hamiltonian systems; a method for computing all of them. Part 2: Numerical application. Meccanica **15**, 21–30 (1980)
33. G. Benettin, L. Galgani, A. Giorgilli, A proof of Nekhoroshev's theorem for the stability times in nearly integrable Hamiltonian systems. Celest. Mech. **37**, 1–25 (1985)
34. G. Benettin, A. Carati, L. Galgani, A. Giorgilli, The Fermi-Pasta-Ulam problem and the metastability perspective. Lecture Notes in Physics, vol. 728 (Springer, New York/Berlin, 2008), pp. 151–189
35. G. Benettin, R. Livi, A. Ponno, The Fermi-Pasta-Ulam problem: scaling laws vs. initial conditions. J. Stat. Phys. **135**, 873–893 (2009)
36. D. Benisti, D.F. Escande, Nonstandard diffusion properties of the standard map. Phys. Rev. Lett. **80**, 4871–4874 (1998)
37. L. Berchialla, A. Giorgilli, S. Paleari, Exponentially long times to equipartition in the thermodynamic limit. Phys. Lett. A, **321**, 167–172 (2004)
38. L. Berchialla, L. Galgani, A. Giorgilli, Localization of energy in FPU chains. Discret. Contin. Dyn. Syst. **11**, 855–866 (2004)
39. J.M. Bergamin, Numerical approximation of breathers in lattices with nearest-neighbor interactions, Phys. Rev. E **67**, 026703 (2003)
40. J.M. Bergamin, Localization in nonlinear lattices and homoclinic dynamics. Ph.D. Thesis, University of Patras, 2003
41. J.M. Bergamin, T. Bountis, C. Jung, A method for locating symmetric homoclinic orbits using symbolic dynamics. J. Phys. A-Math. Gen. **33**, 8059–8070 (2000)
42. J.M. Bergamin, T. Bountis, M.N. Vrahatis, Homoclinic orbits of invertible maps. Nonlinearity **15**, 1603–1619 (2002)
43. G.P. Berman, F.M. Izrailev, The Fermi-Pasta-Ulam problem: fifty years of progress. Chaos **15**, 015104 (2005)
44. P. Billingsley, *Convergence of Probability Measures* (Wiley, New York, 1968)

45. J. Billy, V. Josse, Z. Zuo, A. Bernard, B. Hambrecht, P. Lugan, D. Clément, L. Sanchez-Palencia, P. Bouyer, A. Aspect, Direct observation of Anderson localization of matter-waves in a controlled disorder. Nature **453**, 891–894 (2008)
46. G. Birkhoff, G.-C. Rota, *Ordinary Differential Equations* (Wiley, New York, 1978)
47. J.D. Bodyfelt, T.V. Laptyeva, Ch. Skokos, D.O. Krimer, S. Flach, Nonlinear waves in disordered chains: probing the limits of chaos and spreading. Phys. Rev. E **84**, 016205 (2011)
48. J.D. Bodyfelt, T.V. Laptyeva, G. Gligoric, D.O. Krimer, Ch. Skokos, S. Flach, Wave interactions in localizing media – a coin with many faces. Int. J. Bifurc. Chaos **21**, 2107–2124 (2011)
49. J. Boreux, T. Carletti, Ch. Skokos, M. Vittot, Hamiltonian control used to improve the beam stability in particle accelerator models. Commun. Nonlinear Sci. Numer. Simul. (2011) 17, 1725–1738 (2012)
50. J. Boreux, T. Carletti, Ch. Skokos, Y. Papaphilippou, M. Vittot, Efficient control of accelerator maps. Int. J. Bifurc. Chaos (2012, In Press) E-print arXiv:1103.5631
51. T. Bountis, Investigating non-integrability and Chaos in complex time. Phys. D **86**, 256–267 (1995)
52. T. Bountis, Stability of motion: From Lyapunov to the dynamics N-degree of freedom Hamiltonian systems. Nonlinear Phenomena and Complex Systems **9**, 209–239 (2006)
53. T. Bountis, J.M. Bergamin, *Discrete Breathers in Nonlinear Lattices: A Review and Recent Results*. Lecture Notes in Physics, vol. 626 (Springer, New York/Berlin, 2003)
54. T. Bountis, M. Kollmann, Diffusion rates in a 4-dimensional mapping model of accelerator dynamics. Phys. D **71**, 122–131 (1994)
55. T. Bountis, K.E. Papadakis, The stability of vertical motion in the N-body circular Sitnikov problem. Celest. Mech. Dyn. Astron. **104**, 205–225 (2009)
56. T. Bountis, H. Segur, in *Logarithmic Singularities and Chaotic Behavior in Hamiltonian Systems*, ed. by M. Tabor, Y. Treves. A.I.P. Conference Proceedings, vol. 88, 279–292 (A.I.P., New York, 1982)
57. T. Bountis, Ch. Skokos, Application of the SALI chaos detection method to accelerator mappings. Nucl. Instrum. Methods A **561**, 173–179 (2006)
58. T. Bountis, Ch. Skokos, Space charges can significantly affect the dynamics of accelerator maps. Phys. Lett. A **358**, 126–133 (2006)
59. T. Bountis, S. Tompaidis, Strong and weak instabilities in a 4-D mapping model of accelerator dynamics, in *Nonlinear Problems in Future Particle Accelerators*, ed. by W. Scandale, G. Turchetti (World Scientific, Singapore, 1991), pp. 112–127
60. T. Bountis, H. Segur, F. Vivaldi, Integrable Hamiltonian systems and the Painlevé property. Phys. Rev. A **25**, 1257–1264 (1982)
61. T. Bountis, H.W. Capel, M. Kollmann, J.C. Ross, J.M. Bergamin, J.P. van der Weele, Multibreathers and homoclinic orbits in one-dimensional nonlinear lattices. Phys. Lett. A **268**, 50–60 (2000)
62. T. Bountis, J.M. Bergamin, V. Basios, Stabilization of discrete breathers using continuous feedback control. Phys. Lett. A **295**, 115–120 (2002)
63. T. Bountis, T. Manos, H. Christodoulidi, Application of the GALI method to localization dynamics in nonlinear systems. J. Comput. Appl. Math. **227**, 17–26 (2009)
64. T. Bountis, G. Chechin, V. Sakhnenko, Discrete symmetries and stability in Hamiltonian dynamics. Int. J. Bifurc. Chaos **21**, 1539–1582 (2011)
65. V.A. Brazhnyi, V.V. Konotop, Theory of nonlinear matter waves in optical lattices. Mod. Phys. Lett. B **18**, 627–651 (2004)
66. N. Budinsky, T. Bountis, Stability of nonlinear modes and chaotic properties of 1D Fermi-Pasta-Ulam lattices. Phys. D **8**, 445–452 (1983)
67. A. Cafarella, M. Leo, R.A. Leo, Numerical analysis of the one-mode solutions in the Fermi-Pasta-Ulam system. Phys. Rev. E **69**, 046604 (2004)
68. P. Calabrese, A. Gambassi, Slow dynamics in critical ferromagnetic vector models relaxing from a magnetized initial state. J. Stat. Mech.-Theory Exp. **2007**, P01001 (2007)

69. F. Calogero, D. Gomez-Ullate, P.M. Santini, M. Sommacal, On the transition from regular to irregular motions, explained as travel on Riemann surfaces. J. Phys. A **38**, 8873–8896 (2005)
70. F. Calogero, D. Gomez-Ullate, P.M. Santini, M. Sommacal, Towards a theory of chaos explained as travel on Riemann surfaces. J. Phys. A **42**, 015205 (2009)
71. D.K. Campbell, P. Rosenau, G.M. Zaslavsky (eds.), The Fermi-Pasta-Ulam problem: the first 50 Years. Chaos, Focus Issue **15**, 015101 (2005)
72. R. Capuzzo-Dolcetta, L. Leccese, D. Merritt, A. Vicari, Self-consistent models of cuspy triaxial galaxies with dark matter haloes. Astrophys. J. **666**, 165–180 (2007)
73. J.R. Cary, D.F. Escande, A.D. Verga, Nonquasilinear diffusion far from the chaotic threshold. Phys. Rev. Lett. **65**, 3132–3135 (1990)
74. G. Casati, B. Li, Heat conduction in one dimensional systems: Fourier law, chaos, and heat control, in *Nonlinear Dynamics and Fundamental Interactions*. NATO Science Series, Springer, New York/Berlin, vol. 213, Part 1, 1–16 (2006)
75. G. Casati, T. Prosen, Mixing property of triangular billiards. Phys. Rev. Lett. **83**, 4729–4732 (1999)
76. G. Casati, J. Ford, F. Vivaldi, W.M. Visscher, One-dimensional classical many-body system having a normal thermal conductivity. Phys. Rev. Lett. **52**, 1861–1864 (1984)
77. A. Celikoglu, U. Tirnakli, S.M. Duarte Queirós, Analysis of return distributions in the coherent noise model. Phys. Rev. E **82**, 021124 (2010)
78. J. Chabé, G. Lemarié, B. Grémaud, D. Delande, P. Szriftgiser, J.-C. Garreau, Experimental observation of the Anderson metal-insulator transition with atomic matter waves. Phys. Rev. Lett. **101**, 255702 (2008)
79. C.W. Chang, D. Okawa, A. Majumdar, A. Zettl, Solid-state thermal rectifier. Science **314**, 1121 (2006)
80. G.M. Chechin, Computers and group-theoretical methods for studying structural phase transition. Comput. Math. Appl. **17**, 255–278 (1989)
81. G.M. Chechin, V.P. Sakhnenko, Interactions between normal modes in nonlinear dynamical systems with discrete symmetry. Exact results. Phys. D **117**, 43–76 (1998)
82. G.M. Chechin, K.G. Zhukov, Stability analysis of dynamical regimes in nonlinear systems with discrete symmetries. Phys. Rev. E **73**, 036216 (2006)
83. G.M. Chechin, T.I. Ivanova, V.P. Sakhnenko, Complete order parameter condensate of low-symmetry phases upon structural phase transitions. Phys. Status Solidi B **152**, 431–446 (1989)
84. G.M. Chechin, E.A. Ipatova, V.P. Sakhnenko, Peculiarities of the low-symmetry phase structure near the phase-transition point. Acta Crystallogr. A **49**, 824–831 (1993)
85. G.M. Chechin, N.V. Novikova, A.A. Abramenko, Bushes of vibrational modes for Fermi-Pasta-Ulam chains. Phys. D **166**, 208–238 (2002)
86. G.M. Chechin, A.V. Gnezdilov, M.Yu. Zekhtser, Existence and stability of bushes of vibrational modes for octahedral mechanical systems with Lennard-Jones potential. Int. J. Nonlinear Mech. **38**, 1451–1472 (2003)
87. G.M. Chechin, D.S. Ryabov, K.G. Zhukov, Stability of low-dimensional bushes of vibrational modes in the Fermi-Pasta-Ulam chains. Phys. D **203**, 121–166 (2005)
88. B.V. Chirikov, A universal instability of many-dimensional oscillator systems. Phys. Rep. **52**, 263–379 (1979)
89. B.V. Chirikov, D.L. Shepelyansky, Correlation properties of dynamical chaos in Hamiltonian systems. Phys. D **13**, 395–400 (1984)
90. S.-N. Chow, M. Yamashita, Geometry of the Melnikov vector, in *Nonlinear Equations in Applied Sciences*, ed. by W.F. Ames, C. Rogers (Academic Press, San Diego, 1991), pp. 79–148
91. D.N. Christodoulides, F. Lederer, Y. Silberberg, Discretizing light behavious in linear and nonlinear waveguide lattices. Nature **424**, 817 (2003)
92. H. Christodoulidi, Dynamics on low-dimensional tori and chaos in Hamiltonian systems. Ph.D. Thesis, University of Patras, 2010
93. H. Christodoulidi, T. Bountis, Low-dimensional quasiperiodic motion in Hamiltonian systems. ROMAI J. **2**, 37–44 (2006)

94. H. Christodoulidi, C. Efthymiopoulos, T. Bountis, Energy localization on q-tori, long-term stability, and the interpretation of Fermi-Pasta-Ulam recurrences. Phys. Rev. E **81**, 016210 (2010)
95. P.M. Cincotta, C. Simó, Simple tools to study global dynamics in non-axisymmetric galactic potentials-I. Astron. Astrophys. Suppl. **147**, 205–228 (2000)
96. P.M. Cincotta, C.M. Giordano, C. Simó, Phase space structure of multi-dimensional systems by means of the mean exponential growth factor of nearby orbits. Phys. D **182**, 151–178 (2003)
97. E.A. Coddington, N. Levinson, *Theory of Ordinary Differential Equations* (McGraw-Hill, New York, 1955)
98. R.M. Conte, M. Musette, *The Painlevé Handbook* (Springer, Heidelberg, 2008)
99. G. Contopoulos, *Order and Chaos in Dynamical Astronomy* (Springer, Heidelberg, 2002)
100. G. Contopoulos, B. Barbanis, Lyapunov characteristic numbers and the structure of phase-space. Astron. Astrophys. **222**, 329–343 (1989)
101. G. Contopoulos, P. Magnenat, Simple three-dimensional periodic orbits in a galactic-type potential. Celest. Mech. **37**, 387–414 (1985)
102. G. Contopoulos, N. Voglis, Spectra of stretching numbers and helicity angles in dynamical systems. Celest. Mech. Dyn. Astr. **64**, 1–20 (1996)
103. G. Contopoulos, N. Voglis, A fast method for distinguishing between ordered and chaotic orbits. Astron. Astrophys. **317**, 73–81 (1997)
104. G. Contopoulos, L. Galgani, A. Giorgilli, On the number of isolating integrals in Hamiltonian systems. Phys. Rev. A **18**, 1183–1189 (1978)
105. T. Cretegny, T. Dauxois, S. Ruffo, A. Torcini, Localization and equipartition of energy in the beta-FPU chain: chaotic breathers. Phys. D **121**, 109–126 (1998)
106. F. Dalfovo, S. Giorgini, L.P. Pitaevskii, S. Stringari, Theory of Bose-Einstein condensation in trapped gases. Rev. Mod. Phys. **71**, 463–512 (1999)
107. H.T. Davis, *Introduction to Nonlinear Differential and Integral Equations* (Dover, New York, 1962)
108. T. Dauxois, Non-Gaussian distributions under scrutiny. J. Stat. Mech.-Theory Exp. **2007**, N08001 (2007)
109. J. De Luca, A.J. Lichtenberg, Transitions and time scales to equipartition in oscillator chains: low-frequency initial conditions. Phys. Rev. E **66**, 026206 (2002)
110. J. De Luca, A.J. Lichtenberg, M.A. Lieberman, Time scale to ergodicity in the Fermi-Pasta-Ulam system. Chaos **5**, 283–297 (1995)
111. J. De Luca, A.J. Lichtenberg, S. Ruffo, Energy transitions and time scales to equipartition in the Fermi-Pasta-Ulam oscillator chain. Phys. Rev. E **51**, 2877–2885 (1995)
112. J. De Luca, A.J. Lichtenberg, S. Ruffo, Finite times to equipartition in the thermodynamic limit. Phys. Rev. E **60**, 3781–3786 (1999)
113. L. Drossos, T. Bountis, Evidence of natural boundary and nonintegrability of the mixmaster universe model. J. Nonlinear Sci. **7**, 1–11 (1997)
114. W.E. Drummond, D. Pines, Nonlinear stability of plasma oscillations. Nucl. Fusion Suppl. **3**, 1049–1057 (1962)
115. S.M. Duarte Queirós, The role of ergodicity and mixing in the central limit theorem for Casati-Prosen triangle map variables. Phys. Lett. A **373**, 1514–1518 (2009)
116. G. Duffing, *Erzwungene Schwingungen bei veränderlicher Eigenfrequenz und ihre technische Bedeutung* (Vieweg & Sohn, Braunschweig, 1918)
117. J.-P. Eckmann, D. Ruelle, Ergodic theory of chaos and strange attractors. Rev. Mod. Phys. **57**, 617–656 (1985)
118. J.T. Edwards, D.J. Thouless, Numerical studies of localization in disordered systems. J. Phys. C Solid **5**, 807–820 (1972)
119. N.K. Efremidis, D.N. Christodoulides, Lattice solitons in Bose-Einstein condensates. Phys. Rev. A **67**, 063608 (2003)
120. L.H. Eliasson, Absolutely convergent series expansions for quasi periodic motions. Math. Phys. Electron. J. **2**, 4 (1996)

121. E. Fermi, J. Pasta, S. Ulam, Studies of nonlinear problems. Los Alamos Sci. Lab. Rep. No. LA-1940 (1955), in *Nonlinear Wave Motion*, ed. by A.C. Newell. Lectures in Applied Mathematics, vol. 15 (Amer. Math. Soc., Providence, 1974), pp. 143–155

122. S. Flach, Conditions on the existence of localized excitations in nonlinear discrete systems. Phys. Rev. E **50**, 3134–3142 (1994)

123. S. Flach, Obtaining breathers in nonlinear Hamiltonian lattices. Phys. Rev. E **51**, 3579–3587 (1995)

124. S. Flach, Spreading of waves in nonlinear disordered media. Chem. Phys. **375**, 548–556 (2010)

125. S. Flach, A.V. Gorbach, Discrete breathers – Advances in theory and applications. Phys. Rep. **467**, 1–116 (2008)

126. S. Flach, A. Ponno, The Fermi-Pasta-Ulam problem: periodic orbits, normal forms and resonance overlap criteria. Phys. D **237**, 908–917 (2008)

127. S. Flach, C. Willis, Discrete breathers. Phys. Rep. **295**, 181–264 (1998)

128. S. Flach, M.V. Ivanchenko, O.I. Kanakov, q-Breathers and the Fermi-Pasta-Ulam problem. Phys. Rev. Lett. **95**, 064102 (2005)

129. S. Flach, M.V. Ivanchenko, O.I. Kanakov, q-breathers in Fermi-Pasta-Ulam chains: existence, localization, and stability. Phys. Rev. E **73**, 036618 (2006)

130. S. Flach, O.I. Kanakov, M.V. Ivanchenko, K. Mishagin, q-breathers in FPU-lattices – scaling and properties for large systems. Int. J. Mod. Phys. B **21**, 3925–3932 (2007)

131. S. Flach, D.O. Krimer, Ch. Skokos, Universal spreading of wavepackets in disordered nonlinear systems. Phys. Rev. Lett. **102**, 024101 (2009)

132. A.S. Fokas, T. Bountis, Order and the ubiquitous occurrence of Chaos. Phys. A **228**, 236–244 (1996)

133. J. Ford, The Fermi-Pasta-Ulam problem: paradox turns discovery. Phys. Rep. **213**, 271–310 (1992)

134. F. Freistetter, Fractal dimensions as chaos indicators. Celest. Mech. Dyn. Astron. **78**, 211–225 (2000)

135. C. Froeschlé, E. Lega, On the structure of symplectic mappings. The fast Lyapunov indicator: A very sensitive tool. Celest. Mech. Dyn. Astron. **78**, 167–195 (2000)

136. C. Froeschlé, Ch. Froeschlé, E. Lohinger, Generalized Lyapunov characteristic indicators and corresponding Kolmogorov like entropy of the standard mapping. Celest. Mech. Dyn. Astron. **56**, 307–314 (1993)

137. C. Froeschlé, E. Lega, R. Gonczi, Fast Lyapunov indicators. Application to asteroidal motion. Celest. Mech. Dyn. Astron. **67**, 41–62 (1997)

138. C. Froeschlé, R. Gonczi, E. Lega, The fast Lyapunov indicator: a simple tool to detect weak chaos. Application to the structure of the main asteroidal belt. Planet. Space Sci. **45**, 881–886 (1997)

139. F. Fucito, F. Marchesoni, E. Marinari, G. Parisi, L. Peliti, S. Ruffo, A. Vulpiani, Approach to equilibrium in a chain of nonlinear oscillators. J. Phys.-Paris **43**, 707–713 (1982)

140. L. Galgani, A. Scotti, Planck-like distributions in classical nonlinear mechanics. Phys. Rev. Lett. **28**, 1173–1176 (1972)

141. Z. Galias, Rigorous investigation of the Ikeda map by means of interval arithmetic. Nonlinearity **15**, 1759–1779 (2002)

142. G. Gallavotti, Twistless KAM tori. Commun. Math. Phys. **164**, 145–156 (1994)

143. G. Gallavotti, Twistless KAM tori, quasi flat homoclinic intersections, and other cancellations in the perturbation series of certain completely integrable Hamiltonian systems: a review. Rev. Math. Phys. **6**, 343–411 (1994)

144. I. García-Mata, D.L. Shepelyansky, Delocalization induced by nonlinearity in systems with disorder. Phys. Rev. E **79**, 026205 (2009)

145. P. Gaspard, Lyapunov exponent of ion motion in microplasmas. Phys. Rev. E **68**, 056209 (2003)

146. E. Gerlach, Ch. Skokos, Comparing the efficiency of numerical techniques for the integration of variational equations. Discr. Cont. Dyn. Sys.-Supp. September, 475–484 (2011)
147. E. Gerlach, S. Eggl, Ch. Skokos, Efficient integration of the variational equations of multi-dimensional Hamiltonian systems: application to the Fermi-Pasta-Ulam lattice. Int. J. Bifurc. Chaos (2012, In Press) E-print arXiv:1104.3127
148. A. Giorgilli, U. Locatelli, Kolmogorov theorem and classical perturbation theory. Z. Angew. Math. Phys. **48**, 220–261 (1997)
149. A. Giorgilli, U. Locatelli, A classical self-contained proof of Kolmogorov's theorem on invariant tori, in *Hamiltonian Systems of Three or More Degrees of Freedom*, ed. by C. Simó. NATO Advanced Study Institute, vol. 533 (Kluwer, Dordrecht, 1999), pp. 72–89
150. A. Giorgilli, D. Muraro, Exponentially stable manifolds in the neighbourhood of elliptic equilibria. Boll. Unione Mate. Ital. B **9**, 1–20 (2006)
151. M.L. Glasser, V.G. Papageorgiou, T.C. Bountis, Mel'nikov's function for two-dimensional mappings. SIAM J. Appl. Math. **49**, 692–703 (1989)
152. A. Goriely, *Integrability and Nonintegrability of Dynamical Systems* (World Scientific, Singapore, 2001)
153. G.A. Gottwald, I. Melbourne, A new test for chaos in deterministic systems. Proc. R. Soc. Lond. A Math. **460**, 603–611 (2004)
154. G.A. Gottwald, I. Melbourne, Testing for chaos in deterministic systems with noise. Phys. D **212**, 100–110 (2005)
155. E. Goursat, *Cours d' Analyse Mathématique vol. 2* (Gauthier-Villars, Paris, 1905)
156. B. Grammaticos, B. Dorizzi, R. Padjen, Painlevé property and integrals of motion for the Hénon-Heiles system. Phys. Lett. A **89**, 111–113 (1982)
157. P.E. Greenwood, M.S. Nikulin, A Guide to Chi-Squared Testing, (Wiley, New York, 1996)
158. P. Grassberger, Proposed central limit behavior in deterministic dynamical systems. Phys. Rev. E 79, 057201 (2009)
159. W. Greub, *Multilinear Algebra*, 2nd edn. (Springer, Heidelberg, 1978)
160. J. Guckenheimer, P. Holmes, *Nonlinear Oscillations, Dynamical Systems, and Bifurcations of Vector Fields* (Springer, New York, 1983)
161. M.G. Hahn, X. Jiang, S. Umarov, On q-Gaussians and exchangeability. J. Phys. A-Math. Theor. **43**, 165208 (2010)
162. E. Hairer, C. Lubich, G. Wanner, *Geometric Numerical Integration. Structure-Preserving Algorithms for Ordinary Differential Equations*. Springer Series in Comput. Math., vol. 31 (Springer, Berlin, 2002)
163. P. Hemmer, Dynamic and stochastic type of motion by the linear chain. Det Physiske Seminar i Trondheim **2**, 66 (1959)
164. M. Hénon, C. Heiles, The applicability of the third integral of motion: some numerical experiments. Astron. J. **69**, 73–79 (1964)
165. R.C. Hilborn, *Chaos and Nonlinear Dynamics* (Oxford University Press, New York, 1994)
166. H.J. Hilhorst, Note on a q-modified central limit theorem. J. Stat. Mech.-Theory Exp. **2010**, P10023 (2010)
167. H.J. Hilhorst, G. Schehr, A note on q-Gaussians and non-Gaussians in statistical mechanics. J. Stat. Mech.-Theory Exp. **2007**, P06003 (2007)
168. T.L. Hill *Thermodynamics of Small Systems* (Dover, New York, 1994)
169. E. Hille, *Lectures on Ordinary Differential Equations* (Addison-Wesley, Reading, 1969)
170. M.W. Hirsch, S. Smale, R.L. Devaney, *Differential Equations, Dynamical Systems and an Introduction to Chaos* (Elsevier, New York, 2004)
171. J.E. Howard, Discrete virial theorem. Celest. Mech. Dyn. Astron. **92**, 219–241 (2005)
172. B. Hu, B. Li, H. Zhao, Heat conduction in one-dimensional chains. Phys. Rev. E **57**, 2992 (1998)
173. H. Hu, A. Strybulevych, J. Page, S. Skipetrov, B. van Tiggelen, Localization of ultrasound in a three-dimensional elastic network. Nat. Phys. **4**, 945–948 (2008)
174. J.H. Hubbard, B.B. Hubbard, *Vector Calculus, Linear Algebra and Differential Forms: A Unified Approach* (Prentice Hall, Upper Saddle River, 1999)

175. M.C. Irwin, *Smooth Dynamical Systems* (Academic, New York, 1980)
176. N. Jacobson, *Lectures in Abstract Algebra*, vol. II (van Nostrand, Princeton, 1951)
177. M. Johansson, G. Kopidakis, S. Lepri, S. Aubry, Transmission thresholds in time-periodically driven nonlinear disordered systems. Europhys. Lett. **86**, 10009 (2009)
178. M. Johansson, G. Kopidakis, S. Aubry, KAM tori in 1D random discrete nonlinear Schrödinger model? Europhys. Lett. **91**, 50001 (2010)
179. O.I. Kanakov, S. Flach, M.V. Ivanchenko, K.G. Mishagin, Scaling properties of q-breathers in nonlinear acoustic lattices. Phys. Lett. A **365**, 416–420 (2007)
180. H. Kantz, P. Grassberger, Internal Arnold diffusion and chaos thresholds in coupled symplectic maps. J. Phys. A-Math. Gen. **21**, L127–L133 (1988)
181. G.I. Karanis, Ch.L. Vozikis, Fast detection of chaotic behavior in galactic potentials. Astron. Nachr. **329**, 403–412 (2008)
182. Y.V. Kartashov, V.A. Vysloukh, L. Torner, Soliton shape and mobility control in optical lattices. Prog. Opt. **52**, 63–148 (2009)
183. Y.V. Kartashov, B.A. Malomed, L. Torner, Solitons in nonlinear lattices. Rev. Mod. Phys. **83**, 247 (2011)
184. A. Katok, Lyapunov exponents, entropy and periodic orbits for diffeomorphisms. Publ. Math. IHÉS **51**, 137–173 (1980)
185. A. Katok, J.-M. Strelcyn, *Invariant Manifolds, Entropy and Billiards; Smooth Maps with Singularities*. Lecture Notes in Mathematics, vol. 1222 (Springer, Berlin, 1986)
186. A.N. Kaufman, Quasilinear diffusion of an axisymmetric toroidal plasma. Phys. Fluids **15**, 1063 (1972)
187. W. Ketterle, D.S. Durfee, D.M. Stamper-Kurn, Making, probing and understanding Bose-Einstein condensates, in *Bose-Einstein Condensation in Atomic Gases. Proceedings of the International School of Physics "Enrico Fermi"*, ed. by M. Inguscio, S. Stringari, C.E. Wieman (IOS Press, Amsterdam, 1999), pp. 67–176
188. P.G. Kevrekidis, *The Discrete Nonlinear Schrödinger Equation*. Tracts in Modern Physics, vol. 232 (Springer, Heidelberg, 2009)
189. Y.S. Kivshar, Intrinsic localized modes as solitons with a compact support. Phys. Rev. E **48**, R43–R45 (1993)
190. Y.S. Kivshar, G.P. Agrawal, *Optical Solitons. From Fibers to Photonic Crystals* (Academic, Amsterdam, 2003)
191. Y.S. Kivshar, B.A. Malomed, Dynamics of solitons in near-integrable systems. Rev. Mod. Phys. **61**, 763–915 (1989)
192. W. Kobayashi, Y. Teraoka, I. Terasaki, An oxide thermal rectifier. Appl. Phys. Lett. **95**, 171905 (2009)
193. Y. Kominis, Analytical solitary wave solutions of the nonlinear Kronig-Penney model in photonic structures. Phys. Rev. E **73**, 066619 (2006)
194. Y. Kominis, T. Bountis, Analytical solutions of systems with piecewise linear dynamics. Int. J. Bifurc. Chaos **20**, 509–518 (2010)
195. Y. Kominis, K. Hizanidis, Lattice solitons in self-defocusing optical media: analytical solutions of the nonlinear Kronig-Penney model. Opt. Lett. **31**, 2888–2890 (2006)
196. Y. Kominis, K. Hizanidis, Power dependent soliton location and stability in complex photonic structures. Opt. Expr. **16**, 12124–12138 (2008)
197. Y. Kominis, K. Hizanidis, Power-dependent reflection, transmission and trapping dynamics of lattice solitons at interfaces. Phys. Rev. Lett. **102**, 133903 (2009)
198. Y. Kominis, A. Papadopoulos, K. Hizanidis, Surface solitons in waveguide arrays: analytical solutions. Opt. Expr. **15**, 10041–10051 (2007)
199. Y. Kominis, A.K. Ram, K. Hizanidis, Quasilinear theory of electron transport by radio frequency waves and non-axisymmetric perturbations in toroidal plasmas. Phys. Plasmas **15**, 122501 (2008)
200. Y. Kominis, T. Bountis, K. Hizanidis, Breathers in a nonautonomous Toda lattice with pulsating coupling. Phys. Rev. E **81**, 066601 (2010)

201. Y. Kominis, A.K. Ram, K. Hizanidis, Kinetic theory for distribution functions of wave-particle interactions in plasmas. Phys. Rev. Lett. **104**, 235001 (2010)

202. G. Kopidakis, S. Komineas, S. Flach, S. Aubry, Absence of wave packet diffusion in disordered nonlinear systems. Phys. Rev. Lett. **100**, 084103 (2008)

203. Y.A. Kosevich, Nonlinear sinusoidal waves and their superposition in anharmonic lattices. Phys. Rev. Lett. **71**, 2058–2061 (1993)

204. T. Kotoulas, G. Voyatzis, Comparative study of the 2:3 and 3:4 resonant motion with Neptune: an application of symplectic mappings and low frequency analysis. Celest. Mech. Dyn. Astron. **88**, 343–363 (2004)

205. I. Kovacic, M.J. Brennan (eds.), *The Duffing Equation: Nonlinear Oscillators and Their Behaviour* (Wiley, Hoboken, 2011)

206. B. Kramer, A. MacKinnon, Localization: theory and experiment. Rep. Prog. Phys. **56**, 1469–1564 (1993)

207. D.O. Krimer, S. Flach, Statistics of wave interactions in nonlinear disordered systems. Phys. Rev. E **82**, 046221 (2010)

208. Y. Lahini, A. Avidan, F. Pozzi, M. Sorel, R. Morandotti, D.N. Christodoulides, Y. Silberberg, Anderson localization and nonlinearity in one-dimensional disordered photonic lattices. Phys. Rev. Lett. **100**, 013906 (2008)

209. L.D. Landau, E.M. Lifshitz, *Mechanics*, Third edn, Volume 1 of Course of Theoretical Physics (Butterworth-Heinemann, Amsterdam, 1976)

210. T.V. Laptyeva, J.D. Bodyfelt, D.O. Krimer, Ch. Skokos, S. Flach, The crossover from strong to weak chaos for nonlinear waves in disordered systems. Europhys. Lett. **91**, 30001 (2010)

211. J. Laskar, The chaotic motion of the Solar System: a numerical estimate of the size of the chaotic zones. Icarus **88**, 266–291 (1990)

212. J. Laskar, Frequency analysis of multi-dimensional systems. Global dynamics and diffusion. Phys. D **67**, 257–281 (1993)

213. J. Laskar, Introduction to frequency map analysis, in *Hamiltonian Systems of Three or More Degrees of Freedom*, ed. by C. Simó. NATO Advanced Study Institute, vol. 533 (Kluwer, Dordrecht, 1999), pp. 134–150

214. J. Laskar, C. Froeschlé, A. Celletti, The measure of chaos by the numerical analysis of the fundamental frequencies. Application to the standard map. Phys. D **56**, 253–269 (1992)

215. M. Lax, W.H. Louisell, W.B. McKnight, From Maxwell to paraxial wave optics. Phys. Rev. A **11**, 1365–1370 (1975)

216. F. Lederer, G.I. Stegeman, D.N. Christodoulides, G. Assanto, M. Segev, Y. Silberberg, Discrete solitons in optics. Phys. Rep. **463**, 1–126 (2008)

217. E. Lega, C. Froeschlé, Comparison of convergence towards invariant distributions for rotation angles, twist angles and local Lyapunov characteristic numbers. Planet. Space Sci. **46**, 1525–1534 (1998)

218. M. Leo, R.A. Leo, Stability properties of the $N/4$ ($\pi/2$-mode) one-mode nonlinear solution of the Fermi-Pasta-Ulam-β system. Phys. Rev. E **76**, 016216 (2007)

219. S. Lepri, R. Livi, A. Politi, Thermal conduction in classical low-dimensional lattices. Phys. Rep. **377**(1), 1–80 (2003)

220. S. Lepri, R. Livi, A. Politi, Studies of thermal conductivity in Fermi Pasta Ulam-like lattices. Chaos **15**, 015118 (2005)

221. B. Li, J. Wang, Anomalous heat conduction and anomalous diffusion in one-dimensional systems. Phys. Rev. Lett. **91**, 044301 (2003)

222. B. Li, L. Wang, B. Hu, Finite thermal conductivity in 1D models having zero Lyapunov exponents. Phys. Rev. Lett. **88**, 223901 (2002)

223. B. Li, G. Casati, J. Wang, Heat conductivity in linear mixing systems. Phys. Rev. E **67**, 021204 (2003)

224. B. Li, G. Casati, J. Wang, T. Prosen, Fourier Law in the alternate-mass hard-core potential chain. Phys. Rev. Lett. **92**, 254301 (2004)

225. B. Li, J. Wang, G. Casati, Thermal diode: rectification of heat flux. Phys. Rev. Lett. **93**, 184301 (2004)

226. A.J. Lichtenberg, M.A. Lieberman, *Regular and Chaotic Dynamics*, Second edn. (Springer, New York, 1992)
227. A. Lichtenberg, R. Livi, M. Pettini, S. Ruffo, Dynamics of oscillator chains. Lect. Notes Phys. **728**, 21–121 (2008)
228. R. Livi, M. Pettini, S. Ruffo, A. Vulpiani, Further results on the equipartition threshold in large nonlinear Hamiltonian systems. Phys. Rev. A **31**, 2740–2742 (1985)
229. R. Livi, A. Politi, S. Ruffo, Distribution of characteristic exponents in the thermodynamic limit. J. Phys. A-Math. Gen. **19**, 2033–2040 (1986)
230. W.C. Lo, L. Wang, B. Li, Thermal transistor: heat flux switching and modulating. J. Phys. Soc. Jpn, **77**(5), 054402 (2008)
231. E. Lohinger, C. Froeschlé, R. Dvorak, Generalized Lyapunov exponents indicators in Hamiltonian dynamics: an application to a double star system. Celest. Mech. Dyn. Astron. **56**, 315–322 (1993)
232. A.M. Lyapunov, *The General Problem of the Stability of Motion* (Taylor and Francis, London, 1992) (English translation from the French: A. Liapounoff, Problème général de la stabilité du mouvement. Annal. Fac. Sci. Toulouse **9**, 203–474 (1907). The French text was reprinted in Annals Math. Studies Vol.17 Princeton Univ. Press (1947). The original was published in Russian by the Mathematical Society of Kharkov in 1892)
233. M. Macek, P. Stránský, P. Cejnar, S. Heinze, J. Jolie, J. Dobeš, Classical and quantum properties of the semiregular arc inside the Casten triangle. Phys. Rev. C **75**, 064318 (2007)
234. M. Macek, J. Dobeš, P. Stránský, P. Cejnar, Regularity-induced separation of intrinsic and collective dynamics. Phys. Rev. Lett. **105**, 072503 (2010)
235. R.S. MacKay, S. Aubry, Proof of existence of breathers for time-reversible or Hamiltonian networks of weakly coupled oscillators. Nonlinearity **7**, 1623–1843 (1994)
236. R.S. Mackay, J.D. Meiss, *Hamiltonian Dynamical Systems* (Adam Hilger, Bristol, 1986)
237. M.C. Mackey, M. Tyran-Kaminska, Deterministic Brownian motion: the effects of perturbing a dynamical system by a chaotic semi-dynamical system. Phys. Rep. **422**, 167–222 (2006)
238. R.S. Mackay, J.D. Meiss, I.C. Percival, Transport in Hamiltonian systems. Phys. D **13**, 55–81 (1984)
239. N.P. Maffione, L.A. Darriba, P.M. Cincotta, C.M. Giordano, A comparison of different indicators of chaos based on the deviation vectors: application to symplectic mappings. Celest. Mech. Dyn. Astron. **111**, 285–307 (2011)
240. W. Magnus, S. Winkler, *Hill's Equation* (Wiley, New York, 1969) and 2nd edn. (Dover, New York, 2004)
241. P. Maniadis, T. Bountis, Quasiperiodic and chaotic breathers in a parametrically driven system without linear dispersion. Phys. Rev. E **73**, 046211 (2006)
242. T. Manos, E. Athanassoula, Regular and chaotic orbits in barred galaxies – I. Applying the SALI/GALI method to explore their distribution in several models. Mon. Not. R. Astron. Soc. **415**, 629–642 (2011)
243. T. Manos, S. Ruffo, Scaling with system size of the Lyapunov exponents for the Hamiltonian Mean Field model. Transp. Theor. Stat. **40**, 360–381 (2011)
244. T. Manos, Ch. Skokos, T. Bountis, Application of the Generalized Alignment Index (GALI) method to the dynamics of multi-dimensional symplectic maps, in *Chaos, Complexity and Transport: Theory and Applications. Proceedings of the CCT07*, ed. by C. Chandre, X. Leoncini, G. Zaslavsky (World Scientific, Singapore, 2008), pp. 356–364
245. T. Manos, Ch. Skokos, E. Athanassoula, T. Bountis, Studying the global dynamics of conservative dynamical systems using the SALI chaos detection method. Nonlinear Phenom. Complex Syst. **11**, 171–176 (2008)
246. T. Manos, Ch. Skokos, T. Bountis, Global dynamics of coupled standard maps, in *Chaos in Astronomy. Astrophysics and Space Science Proceedings*, ed. by G. Contopoulos, P.A. Patsis (Springer, Berlin/Heidelberg, 2009), pp. 367–371
247. T. Manos, Ch. Skokos, Ch. Antonopoulos, Probing the local dynamics of periodic orbits by the generalized alignment index (GALI) method. Int. J. Bifurc. Chaos (2012, In Press) E-print arXiv:1103.0700

248. J.L. Marín, S. Aubry, Breathers in nonlinear lattices: numerical calculation from the anticontinuous limit. Nonlinearity **9**, 1501–1528 (1996)
249. J.D. Meiss, E. Ott, Markov tree model of transport in area-preserving maps. Phys. D **20**, 387–402 (1986)
250. D.R. Merkin, *Introduction to the Theory of Stability*. Series: Texts in Applied Mathematics, vol. 24 (Springer, New York, 1997)
251. G. Miritello, A. Pluchino, A. Rapisarda, Central limit behavior in the Kuramoto model at the "edge of chaos". Phys. A **388**, 4818–4826 (2009)
252. M. Molina, Transport of localized and extended excitations in a nonlinear Anderson model. Phys. Rev. B **58**, 12547–12550 (1998)
253. M. Mulansky, A. Pikovsky, Spreading in disordered lattices with different nonlinearities. Europhys. Lett. **90**, 10015 (2010)
254. M. Mulansky, K. Ahnert, A. Pikovsky, D.L. Shepelyansky, Dynamical thermalization of disordered nonlinear lattices. Phys. Rev. E **80**, 056212 (2009)
255. M. Mulansky, K. Ahnert, A. Pikovsky, Scaling of energy spreading in strongly nonlinear disordered lattices. Phys. Rev. E **83**, 026205 (2011)
256. N.N. Nekhoroshev, An exponential estimate of the time of stability of nearly-integrable Hamiltoninan systems. Russ. Math. Surv. **32**(6), 1–65 (1977)
257. Z. Nitecki, *Differentiable Dynamics* (M.I.T., Cambridge, MA, 1971)
258. J.A. Núñez, P.M. Cincotta, F.C. Wachlin, Information entropy. An indicator of chaos. Celest. Mech. Dyn. Astron. **64**, 43–53 (1996)
259. V.I. Oseledec, A multiplicative ergodic theorem. Ljapunov characteristic numbers for dynamical systems. Trans. Mosc. Math. Soc. **19**, 197–231 (1968)
260. E.A. Ostrovskaya, Y.S. Kivshar, Matter-wave gap vortices in optical lattices. Phys. Rev. Lett. **93**, 160405 (2004)
261. A.A. Ovchinnikov, Localized long-lived vibrational states in molecular crystals. Sov. Phys. JETP-USSR **30**, 147 (1970)
262. P. Panagopoulos, T.C. Bountis, Ch. Skokos, Existence and stability of localized oscillations in one-dimensional lattices with soft spring and hard spring potentials. J. Vib. Acoust. **126**, 520–527 (2004)
263. P. Papagiannis, Y. Kominis, K. Hizanidis, Power- and momentum-dependent soliton dynamics in lattices with longitudinal modulation. Phys. Rev. A **84**, 013820 (2011)
264. R.E. Peierls, Quantum theory of solids, in *Theoretical Physics in the Twentieth Century*, ed. by M. Fierz, V.F. Weisskopf (Wiley, New York, 1961) 140–160
265. L. Perko, *Differential Equations and Dynamical Systems* (Springer, New York, 1995)
266. J.B. Pesin, Families of invariant manifolds corresponding to nonzero characteristic exponents. Math. USSR Izv. **10**, 1261–1305 (1976)
267. Ya.B. Pesin, Lyapunov characteristic indexes and ergodic properties of smooth dynamic systems with invariant measure. Dokl. Acad. Nauk. SSSR **226**, 774–777 (1976)
268. Ya.B. Pesin, Characteristic Lyapunov exponents and smooth ergodic theory. Russ. Math. Surv. **32**(4), 55–114 (1977)
269. Y.G. Petalas, C.G. Antonopoulos, T.C. Bountis, M.N. Vrahatis, Evolutionary methods for the approximation of the stability domain and frequency optimization of conservative maps. Int. J. Bifurc. Chaos **18**, 2249–2264 (2008)
270. M. Peyrard, The design of a thermal rectifier. Europhys. Lett. **76**, 49 (2006)
271. A. Pikovsky, D. Shepelyansky, Destruction of Anderson localization by a weak nonlinearity. Phys. Rev. Lett. **100**, 094101 (2008)
272. P. Poggi, S. Ruffo, Exact solutions in the FPU oscillator chain. Phys. D **103**, 251–272 (1997)
273. H. Poincaré, *Sur les Propriétés des Functions Définies par les Équations aux Différences Partielles* (Gauthier-Villars, Paris, 1879)
274. H. Poincaré *Les Méthodes Nouvelles de la Mécanique Céleste*, vol. 1 (Gauthier Villars, Paris, 1892) (English translation by D.L. Goroff, *New Methods in Celestial Mechanics* (American Institute of Physics, 1993))

275. A. Ponno, D. Bambusi, Korteweg-de Vries equation and energy sharing in Fermi-Pasta-Ulam. Chaos **15**, 015107 (2005)
276. A. Ponno, E. Christodoulidi, Ch. Skokos, S. Flach, The two-stage dynamics in the Fermi-Pasta-Ulam problem: from regular to diffusive behavior. Chaos, 21, 043127 (2011)
277. W.H. Press, S.A. Teukolsky, W.T. Vetterling, B.P. Flanney, *Numerical Recipes in Fortran 77. The Art of Scientific Computing*, Second edn. (Cambridge University Press, Cambridge/New York, 2001)
278. K. Pyragas, Continuous control of chaos by self-controlling feedback. Phys. Lett. A **170**, 421–428 (1992)
279. A. Ramani, B. Grammaticos, T. Bountis, The Painlevé property and singularity analysis of integrable and non-integrable systems. Phys. Rep. **180**, 159–245 (1989)
280. A.B. Rechester, R.B. White, Calculation of turbulent diffusion for the Chirikov-Taylor model. Phys. Rev. Lett. **44**, 1586–1589 (1980)
281. A. Rényi, On measures of information and entropy, in *Proceedings of the 4th Berkeley Symposium on Mathematics, Statistics and Probability 1960*, University of California Press, Berkeley/Los Angeles, 1961, pp. 547–561
282. J.A. Rice, *Mathematical Statistics and Data Analysis*, Second edn. (Duxbury Press, Belmont, 1995)
283. B. Rink, Symmetric invariant manifolds in the Fermi-Pasta-Ulam lattice. Phys. D **175**, 31–42 (2003)
284. G. Roati, C. D'Errico, L. Fallani, M. Fattori, C. Fort, M. Zaccanti, G. Modugno, M. Modugno, M. Inguscio, Anderson localization of a non-interacting Bose-Einstein condensate. Nature **453**, 895–899 (2008)
285. A. Rodríguez, V. Schwämmle, C. Tsallis, Strictly and asymptotically scale invariant probabilistic models of N correlated binary random variables having q-Gaussians as $N \to \infty$ limiting distributions. J. Stat. Mech.-Theory Exp. **2008**, P09006 (2008)
286. R.M. Rosenberg, The normal modes of nonlinear n-degree-of-freedom systems. J. Appl. Mech. **29**, 7–14 (1962)
287. V.M. Rothos, T. Bountis, Mel'nikov analysis of phase space transport in a N-degree-of-freedom Hamiltonian system. Nonlinear Anal. Theor. **30**, 1365–1374 (1997)
288. V.M. Rothos, T. Bountis, Mel'nikov's vector and singularity analysis of periodically perturbed 2 d.o.f. Hamiltonian systems, in *Hamiltonian Systems of Three or More Degrees of Freedom*, ed. by C. Simó. NATO Advanced Study Institute, vol. 533 (Kluwer, Dordrecht, 1999), pp. 544–548
289. D. Ruelle, Ergodic theory of differentiable dynamical systems. Publ. Math. IHÉS **50**, 27–58 (1979)
290. D. Ruelle, Measures describing a turbulent flow. Ann. NY Acad.Sci. **357**, 1–9 (1980)
291. D. Ruelle, Large volume limit of the distribution of characteristic exponents in turbulence. Commun. Math. Phys. **87**, 287–302 (1982)
292. G. Ruiz, C. Tsallis, Nonextensivity at the edge of chaos of a new universality class of one-dimensional unimodal dissipative maps. Eur. Phys. J. B **67**, 577–584 (2009)
293. G. Ruiz, T. Bountis, C. Tsallis, Time-evolving statistics of chaotic orbits of conservative maps in the context of the central limit theorem. Int. J. Bifurc. Chaos. (2012, In Press) arXiv:1106.6226
294. V.P. Sakhnenko, G.M. Chechin, Symmetrical selection rules in nonlinear dynamics of atomic systems. Sov. Phys. Dokl. **38**, 219–221 (1993)
295. V.P. Sakhnenko, G.M. Chechin, Bushes of modes and normal modes for nonlinear dynamical systems with discrete symmetry. Sov. Phys. Dokl. **39**, 625–628 (1994)
296. Zs. Sándor, B. Érdi, C. Efthymiopoulos, The phase space structure around L_4 in the restricted three-body problem. Celest. Mech. Dyn. Astron. **78**, 113–123 (2000)
297. Zs. Sándor, B. Érdi, A. Széll, B. Funk, The relative Lyapunov indicator: an efficient method of chaos detection. Celest. Mech. Dyn. Astron. **90**, 127–138 (2004)
298. K.W. Sandusky, J.B. Page, Interrelation between the stability of extended normal modes and the existence of intrinsic localized modes in nonlinear lattices with realistic potentials. Phys. Rev. B **50**, 866–887 (1994)

299. T. Schwartz, G. Bartal, S. Fishman, M. Segev, Transport and Anderson localization in disordered two-dimensional photonic lattices. Nature **446**, 52–55 (2007)
300. H. Segur, M.D. Kruskal, Nonexistence of small-amplitude breather solutions in ϕ^4 theory. Phys. Rev. Lett. **58**, 747–750 (1987)
301. V.D. Shapiro, R.Z. Sagdeev, Nonlinear wave-particle interaction and conditions for the applicability of quasilinear theory. Phys. Rep. **283**, 49–71 (1997)
302. H. Shiba, N. Ito, Anomalous heat conduction in three-dimensional nonlinear lattices. J. Phys. Soc. Jpn. **77**, 05400 (2008)
303. S. Shinohara, Low-dimensional solutions in the quartic Fermi-Pasta-Ulam system. J. Phys. Soc. Jpn. **71**, 1802–1804 (2002)
304. S. Shinohara, Low-dimensional subsystems in anharmonic lattices. Prog. Theor. Phys. Suppl. **150**, 423–434 (2003)
305. I.V. Sideris, Measure of orbital stickiness and chaos strength. Phys. Rev. E **73**, 066217 (2006)
306. A.J. Sievers, S. Takeno, Intrinsic localized modes in anharmonic crystals. Phys. Rev. Lett. **61**, 970–973 (1988)
307. Y.G. Sinai, Dynamical systems with elastic reflections. Russ. Math. Surv. **25**(2), 137–189 (1970)
308. Ya.G. Sinai, Gibbs measures in ergodic theory. Russ. Math. Surv. **27**(4), 21–69 (1972)
309. Ch. Skokos, Alignment indices: a new, simple method for determining the ordered or chaotic nature of orbits. J. Phys. A-Math. Gen. **34**, 10029–10043 (2001)
310. Ch. Skokos, The Lyapunov characteristic exponents and their computation. Lect. Notes Phys. **790**, 63–135 (2010)
311. Ch. Skokos, S. Flach, Spreading of wave packets in disordered systems with tunable nonlinearity. Phys. Rev. E **82**, 016208 (2010)
312. Ch. Skokos, E. Gerlach, Numerical integration of variational equations. Phys. Rev. E **82**, 036704 (2010)
313. Ch. Skokos, Ch. Antonopoulos, T.C. Bountis, M.N. Vrahatis, How does the smaller alignment index (SALI) distinguish order from chaos? Prog. Theor. Phys. Suppl. **150**, 439–443 (2003)
314. Ch. Skokos, Ch. Antonopoulos, T.C. Bountis, M.N. Vrahatis, Detecting order and chaos in Hamiltonian systems by the SALI method. J. Phys. A-Math. Gen. **37**, 6269–6284 (2004)
315. Ch. Skokos, T.C. Bountis, Ch. Antonopoulos, Geometrical properties of local dynamics in Hamiltonian systems: the generalized alignment index (GALI) method. Phys. D **231**, 30–54 (2007)
316. Ch. Skokos, T. Bountis, Ch. Antonopoulos, Detecting chaos, determining the dimensions of tori and predicting slow diffusion in Fermi-Pasta-Ulam lattices by the generalized alignment index method. Eur. Phys. J.-Spec. Top. **165**, 5–14 (2008)
317. Ch. Skokos, D.O. Krimer, S. Komineas, S. Flach, Delocalization of wave packets in disordered nonlinear chains. Phys. Rev. E **79**, 056211 (2009)
318. A. Smerzi, A. Trombettoni, Nonlinear tight-binding approximation for Bose-Einstein condensates in a lattice. Phys. Rev. A **68**, 023613 (2003)
319. P. Soulis, T. Bountis, R. Dvorak, Stability of motion in the Sitnikov 3-body problem. Celest. Mech. Dyn. Astron. **99**, 129–148 (2007)
320. P.S. Soulis, K.E. Papadakis, T. Bountis, Periodic orbits and bifurcations in the Sitnikov four-body problem. Celest. Mech. Dyn. Astron. **100**, 251–266 (2008)
321. M. Spivak, *Comprehensive Introduction to Differential Geometry*, vol. 1 (Perish Inc., Houston, 1999)
322. P. Stránský, P. Hruška, P. Cejnar, Quantum chaos in the nuclear collective model: classical-quantum correspondence. Phys. Rev. E **79**, 046202 (2009)
323. M. Strözer, P. Gross, C.M. Aegerter, G. Maret, Observation of the critical regime near Anderson localization of light. Phys. Rev. Lett. **96**, 063904 (2006)
324. Á. Süli, Motion indicators in the 2D standard map. PADEU **17**, 47–62 (2006)
325. A. Széll, B. Érdi, Z. Sándor, B. Steves, Chaotic and stable behaviour in the Caledonian symmetric four-body problem. Mon. Not. R. Astron. Soc. **347**, 380–388 (2004)
326. M. Terraneo, M. Peyrard, G. Casati, Controlling the energy flow in nonlinear lattices: a model for a thermal rectifier. Phys. Rev. Lett. **88**, 094302 (2002)

327. U. Tirnakli, C. Beck, C. Tsallis, Central limit behavior of deterministic dynamical systems. Phys. Rev. E **75**, 040106 (2007)
328. U. Tirnakli, C. Tsallis, C. Beck, Closer look at time averages of the logistic map at the edge of chaos. Phys. Rev. E **79**, 056209 (2009)
329. M. Toda, *Theory of Nonlinear Lattices*, (2nd edn.) (Springer, Berlin, 1989)
330. S. Trillo, W. Torruellas (eds.), *Spatial Solitons* (Springer, Berlin, 2001)
331. A. Trombettoni, A. Smerzi, Discrete solitons and breathers with dilute Bose-Einstein condensates. Phys. Rev. Lett. **86**, 2353–2356 (2001)
332. C. Tsallis, *Introduction to Nonextensive Statistical Mechanics: Approaching a Complex World* (Springer, New York, 2009)
333. C. Tsallis, U. Tirnakli, Nonadditive entropy and nonextensive statistical mechanics – Some central concepts and recent applications. J. Phys. Conf. Ser. **201**, 012001 (2010)
334. G.P. Tsironis, An algebraic approach to discrete breather construction. J. Phys. A-Math. Theor. **35**, 951–957 (2002)
335. S. Umarov, C. Tsallis, S. Steinberg, On a q-central limit theorem consistent with nonextensive statistical mechanics . Milan J. Math. **76**, 307–328 (2008)
336. S. Umarov, C. Tsallis, M. Gell-Mann, S. Steinberg, Generalization of symmetric α-stable Lévy distributions for $q > 1$. J. Math. Phys. **51**, 033502 (2010)
337. A.A. Vedenov, E.P. Velikhov, R.Z. Sagdeev, Nonlinear oscillations of rarified plasma. Nucl. Fusion **1**, 82–100 (1961)
338. H. Veksler, Y. Krivolapov, S. Fishman, Spreading for the generalized nonlinear Schrödinger equation with disorder. Phys. Rev. E **80**, 037201 (2009)
339. H. Veksler, Y. Krivolapov, S. Fishman, Double-humped states in the nonlinear Schrödinger equation with a random potential. Phys. Rev. E **81**, 017201 (2010)
340. N. Voglis, G. Contopoulos, Invariant spectra of orbits in dynamical systems. J. Phys. A-Math. Gen. **27**, 4899–4909 (1994)
341. N. Voglis, G. Contopoulos, C. Efthymiopoulos, Method for distinguishing between ordered and chaotic orbits in four-dimensional maps. Phys. Rev. E **57**, 372–377 (1998)
342. G. Voyatzis, S. Ichtiaroglou, On the spectral analysis of trajectories in near-integrable Hamiltonian systems. J. Phys. A-Math. Gen. **25**, 5931–5943 (1992)
343. J.-S. Wang, B. Li, Intriguing heat conduction of a chain with transverse motions. Phys. Rev. Lett. **92**, 074302 (2004)
344. E.T. Whittaker, G.N. Watson, *A Course in Modern Analysis*, 4th edn. (Cambridge University Press, Cambridge, 1927)/(Cambridge Mathematical Library, Cambridge, 2002)
345. D.S. Wiersma, P. Bartolini, A. Lagendijk, R. Righini, Localization of light in a disordered medium. Nature **390**, 671–673 (1997)
346. S. Wiggins, *Applied Nonlinear Dynamical Systems and Chaos* (Springer, New York, 1990)
347. S. Wiggins, *Chaotic Transport in Dynamical Systems* (Springer, New York, 1992)
348. N. Yang, G. Zhang, B. Li, Carbon nanocone: a promising thermal rectifier. Appl. Phys. Lett. **93**, 243111 (2008)
349. H. Yoshida, Construction of higher order symplectic integrators. Phys. Lett. A **150**, 262–268 (1990)
350. H. Yoshida, Recent progress in the theory and application of symplectic integrators. Celest. Mech. Dyn. Astr. **56**, 27–43 (1993)
351. K. Yoshimura, Modulational instability of zone boundary mode in nonlinear lattices: rigorous results. Phys. Rev. E **70**, 016611 (2004)
352. N.J. Zabusky, M.D. Kruskal, Interaction of "solitons" in a collisionless plasma and the recurrence of initial states. Phys. Rev. Lett. **15**, 240–243 (1965)
353. G.M. Zaslavsky, Chaos, fractional kinetics, and anomalous transport. Phys. Rep. **371**, 461–580 (2002)
354. Y. Zou, D. Pazó, M.C. Romano, M. Thiel, J. Kurths, Distinguishing quasiperiodic dynamics from chaos in short-time series. Phys. Rev. E **76**, 016210 (2007)
355. Y. Zou, M. Thiel, M.C. Romano, J. Kurths, Characterization of stickiness by means of recurrence. Chaos **17**, 043101 (2007)

Index

Action-angle variables, 21, 107
Additivity, 10
Anderson localization, 165, 180, 181
Area-preserving map, 212, 213
Asymptotic stability, 3
Autonomous dynamical system, 2

Bifurcation, 43
Bifurcation theory, 6
Bose Einstein condensation (BEC), 47, 61, 62
Bushes of NNMs, 72, 82, 84
Bushes of orbits, 70, 80

Canonical coordinates, 7
Canonical transformation, 20, 22, 25
Cantor set, 33
Chaos, 2, 8, 30
Chaotic motion, 99
Chaotic orbit, 24, 34, 53, 57, 95–97, 104, 112, 114
Compactness index, 184
Complex instability, 43
Conditional stability, 3
Conservative dynamical system, 3, 58
Constant of motion, 7
Control methods, 176, 179
Cyclic group, 71

Delocalization, 9
Deviation vector, 53, 57–59, 92, 94, 103, 104, 106–108
Diffusion, 9, 139, 157, 158, 160, 229
Dihedral group, 71, 72

Discrete breather, 9, 135, 165, 166, 174
Discreteness, 135, 166
Discrete nonlinear Schrödinger (DNLS) equation, 169
Discrete symmetries, 12, 67
Disorder, 9, 180, 183, 187
Disordered discrete nonlinear Schrödinger (DDNLS) equation, 183
Dissipation, 36, 58
Duffing oscillator, 30, 31, 38, 39, 61
Dynamical system, 2

Edge of chaos, 9, 205, 213
Elliptic fixed point, 16, 23, 34
Energy, 7
Energy surface, 8, 19
Equipartition, 9, 67
Escape energy, 30
Extended primitive cells, 73
Extensive quantity, 10, 56, 194
Exterior algebra, 124
Exterior product, 124

Feedback control, 176
Fermi Pasta Ulam (FPU) system, 9, 46, 61, 65, 77, 117, 133, 136, 139, 197
    paradox, 134, 136, 138
    recurrence, 9, 52, 66, 67, 222
Fixed boundary conditions, 48
Fixed point, 2, 3, 15, 34, 42
Floquet theory, 45
FPU. *See* Fermi Pasta Ulam (FPU) system
Frequency, 8
Fundamental matrix, 42, 45

T. Bountis and H. Skokos, *Complex Hamiltonian Dynamics*,
Springer Series in Synergetics, DOI 10.1007/978-3-642-27305-6,
© Springer-Verlag Berlin Heidelberg 2012